# Lecture Notes in Computer Sc.

T0238264

*Commenced Publication in 1973*
Founding and Former Series Editors:
Gerhard Goos, Juris Hartmanis, and Jan van Leeuwen

## Editorial Board

José C. Príncipe
Risto Miikkulainen (Eds.)

# Advances in Self-Organizing Maps

7th International Workshop, WSOM 2009
St. Augustine, Florida, June 8-10, 2009
Proceedings

 Springer

Volume Editors

José C. Príncipe
Computational NeuroEngineering Laboratory
University of Florida, Gainesville, FL, USA
E-mail: principe@cnel.ufl.edu

Risto Miikkulainen
Department of Computer Sciences
The University of Texas at Austin, TX, USA
E-mail: risto@cs.utexas.edu

Library of Congress Control Number: Applied for

CR Subject Classification (1998): F.1, I.2, D.2, J.3

LNCS Sublibrary: SL 1 – Theoretical Computer Science and General Issues

ISSN      0302-9743
ISBN-10   3-642-02396-7 Springer Berlin Heidelberg New York
ISBN-13   978-3-642-02396-5 Springer Berlin Heidelberg New York

springer.com

© Springer-Verlag Berlin Heidelberg 2009
Printed in Germany

Typesetting: Camera-ready by author, data conversion by Scientific Publishing Services, Chennai, India
Printed on acid-free paper      SPIN: 12697267      06/3180      5 4 3 2 1 0

# Preface

These proceedings contain refereed papers presented at the 7th WSOM held at the Casa Monica Hotel, St. Augustine, Florida, June 8–10, 2009. We designed the workshop to serve as a regular forum for researchers in academia and industry who are interested in the exciting field of self-organizing maps (SOM). The program includes excellent examples of the use of SOM in many areas of social sciences, economics, computational biology, engineering, time series analysis, data visualization and computer science as well a vibrant set of theoretical papers that keep pushing the envelope of the original SOM.

Our deep appreciation is extended to Teuvo Kohonen and Ping Li for the plenary talks and Amaury Lendasse for the organization of the special sessions. Our sincere thanks go to the members of the Technical Committee and other reviewers for their excellent and timely reviews, and above all to the authors whose contributions made this workshop possible. Special thanks go to Julie Veal for her dedication and hard work in coordinating the many details necessary to put together the program and local arrangements.

Jose C. Principe
Risto Miikkulainen

# Organization

## Technical Committee Members

Guilherme Barreto (Federal University of Ceara, Brazil)

James Bednar (University of Edinburgh, UK)

Yoonsuck Choe (Texas A&M University, USA)

Jose Alfredo F. Costa (Federal University, UFRN, Brazil)

Pablo Estevez (University of Chile, Santiago)

Adrian Flanagan (Nokia Research Center, Finland)

Tetsuo Furukawa (Kyushu Institute of Technology, Japan)

Colin Fyfe (University of Paisley, UK)

Barbara Hammer (Clausthal University of Technology, Germany)

Masumi Ishikawa (Kyushu Inst. of Technology, Japan)

Samuel Kaski (Helsinki University of Technology)

Thomas Martinetz (University of Lübeck, Germany)

Rudolf Mayer (Vienna University of Technology, Austria)

Daniel Polani (University of Hertfordshire, UK)

Bernardete Ribeiro (University of Coimbra, Portugal)

Olli Simula (Helsinki University of Technology, Finland)

Carme Torras (University of Catalonia, Spain)

Kadim Tasdemir (Yasar University, Turkey)

Alfred Ultsch (University of Marburg, Germany)

Marc van Hulle (KU Leuven, Belgium)

Michel Verleysen (Université catholique de Louvain, Belgium)

Thomas Villmann (University of Applied Sciences Mittweida, Germany)

Axel Wismueller (University of Rochester, USA)

# Table of Contents

# Batch-Learning Self-Organizing Map for Predicting Functions of Poorly-Characterized Proteins Massively Accumulated

Takashi Abe[1], Shigehiko Kanaya[2], and Toshimichi Ikemura[1]

[1] Nagahama Institute of Bio-Science and Technology, Tamura-cho 1266,
Nagahama-shi, Shiga-ken 526-0829, Japan
{takaabe,t_ikemura}@nagahama-i-bio.ac.jp
[2] Nara Institute of Science and Technology, Ikoma, Japan
skanaya@gtc.aist-nara.ac.jp

**Abstract.** As the result of the decoding of large numbers of genome sequences, numerous proteins whose functions cannot be identified by the homology search of amino acid sequences have accumulated and remain of no use to science and industry. Establishment of novel prediction methods for protein function is urgently needed. We previously developed Batch-Learning SOM (BL-SOM) for genome informatics; here, we developed BL-SOM to predict functions of proteins on the basis of similarity in oligopeptide composition of proteins. Oligopeptides are component parts of a protein and involved in formation of its functional motifs and structural parts. Concerning oligopeptide frequencies in 110,000 proteins classified into 2853 function-known COGs (clusters of orthologous groups), BL-SOM could faithfully reproduce the COG classifications, and therefore, proteins whose functions have been unidentified with homology searches could be related to function-known proteins. BL-SOM was applied to predict protein functions of large numbers of proteins obtained from metagenome analyses.

**Keywords:** batch-learning SOM, oligopeptide frequency, protein function, bioinformatics, high-performance supercomputer.

## 1  Introduction

Unculturable environmental microorganisms should contain a wide range of novel genes of scientific and industrial usefulness. Recently, a sequencing method for mixed genome samples directly extracted from environmental microorganism mixtures has become popular: metagenome analysis. A large portion of the environmental sequences thus obtained is registered in the International DNA Sequence Databanks (DDBJ/EMBL/GenBank) with almost no functional and phylogenetic annotation, and therefore, in the least useful manner. Homology searches for nucleotide and amino-acid sequences, such as BLAST, have become widely accepted as a basic bioinformatics tools not only for phylogenetic characterization of gene/protein sequences, but also for prediction of their biological functions when genomes and genomic segments are

J.C. Príncipe and R. Miikkulainen (Eds.): WSOM 2009, LNCS 5629, pp. 1–9, 2009.

decoded. Whereas the usefulness of the sequence homology search is apparent, it has became clear that homology searches can predict the protein function of only 50% or fewer of protein genes, when a novel genome or mixed genomes from environmental samples are decoded. In order to complement the sequence homology search, methods based on different principles are urgently required for predicting protein function.

Self-Organizing Map (SOM) is an unsupervised neural network algorithm developed by Kohonen and his colleagues [1-3], which provides an efficient and easy interpretation of the clustering of high-dimensional complex data using visualization on a two-dimensional plane. About 15 years ago, Ferran et al. [4] performed the pioneering and extensive SOM analysis of dipeptide composition in approximately 2000 human proteins stored in the SwissProt Database and reported clustering of the proteins according to both biological function and higher-order structure. Although this unsupervised learning method can be considered useful for predicting protein functions, the study was conducted long before decoding of genome sequences, and proteins of unknown function were rarely recognized at that time. Furthermore, because a long computation time was required for the SOM analysis of the dipeptide composition (400 dimensional vectorial data) even using high-performance computers at that time and because the final map was dependent on both the order of data input and the initial conditions, the conventional SOM method has rarely been used for prediction of protein function.

Previously, we developed a modified type SOM (batch-learning SOM: BL-SOM) for codon frequencies in gene sequences [5,6] and oligonucleotide frequencies in genome sequences [7-9] that depends on neither the order of data input nor the initial conditions. BL-SOM recognizes species-specific characteristics of codon or oligonucleotide frequencies in individual genomes, permitting clustering of genes or genomic fragments according to species without the need for species information during the BL-SOM learning. Various high-performance supercomputers are now available for biological studies, and the BL-SOM developed by our group is suitable for actualizing high-performance parallel-computing with high-performance supercomputers such as the Earth Simulator "ES" [10-12]. We previously used the BL-SOM for tetranucleotide frequencies for phylogenetic classification of genomic fragment sequences derived from mixed genomes of environmental microorganisms [13-16]. A large-scale phylogenetic classification was possible in a systematic way because genomic sequences were clustered (self-organized) according to species without species information or sequence alignment [7-9].

In the present report, we describe use of the BL-SOM method for prediction of protein function on the basis of similarity in composition of oligopeptides (di-, tri- and tetrapeptides in this study) of proteins. Oligopeptides are elementary components of a protein and are involved in the formation of functional motifs and structural organization of proteins. BL-SOM for oligopeptides may extract characteristics of oligopeptide composition, which actualize protein structure and function, and therefore, separate proteins according to their functions.

Sequences of approximately 8 million proteins are registered in the public databases, and about 500,000 proteins have been classified into approximately 5000 COGs (clusters of orthologous groups of proteins), which are the functional categories identified with bidirectional best-hit relationships between the completely sequenced genomes using the homology search of amino acid sequences [17]. The proteins belonging to a single COG exhibit significant homology of amino acid sequences over the whole range of the proteins and most likely have the same function. Therefore, COG is undoubtedly a useful categorization of proteins according to function while the biological functions of a half of COGs have not yet been identified conclusively. In the present study, we initially focused on oligopeptide compositions in the 110,000 proteins classified into 2853 function-known COGs and prepared BL-SOMs under various conditions to search for conditions that would faithfully reproduce the COG classification. Then, we applied the BLSOM method to predict the functions of a large number of proteins obtained from metagenome analyses.

## 2 Methods

SOM implements nonlinear projection of multi-dimensional data onto a two-dimensional array of weight vectors, and this effectively preserves the topology of the high-dimensional data space [1-3]. We modified previously the conventional SOM for genome informatics on the basis of batch-learning SOM (BL-SOM) to make the learning process and resulting map independent of the order of data input [5-8]. The initial weight vectors were defined by PCA instead of random values, as described previously [6]. Weight vectors ($w_{ij}$) were arranged in the 2D lattice denoted by i (= 0, 1, . . . , I - 1) and j (= 0, 1, . . . , J - 1). I was set as 300 in Fig 1, and J was defined by the nearest integer greater than (s2/s1) × I; s1 and s2 were the standard deviations of the first and second principal components, respectively. Weight vectors ($w_{ij}$) were set and updated as described previously [5-8]. The BLSOM was suitable for actualizing high-performance parallel-computing with high-performance supercomputers. Using 136 CPUs of "the Earth Simulator", calculations in this study could be performed primarily within two days.

Amino acid sequences were obtained from the NCBI COG database (http://www.ncbi.nlm.nih.gov/COG/). Proteins shorter than 200 amino acids in length were

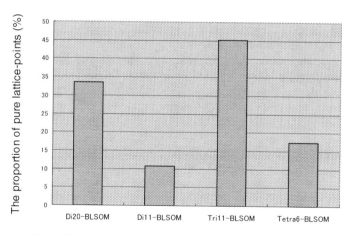

**Fig. 1.** The proportion of pure lattice-points for each condition

not included in the present study. We provided a window of 200 amino acids that is moved with a 50-amino acid step for proteins longer than 200 amino acids. To reduce the computation time, BL-SOM was constructed with tripeptide frequencies of the degenerate eleven groups of residues; {V, L, I}, {T, S}, {N, Q}, {E, D}, {K, R, H}, {Y, F, W}, {M}, {P}, {C}, {A} and {G}. BL-SOM was also constructed with tetrapeptide frequencies of degenerate six groups of residues; {V, L, I, M}, {T, S, P, G, A}, {E,D,N,Q}, {K,R,H}, {Y,F,W} and {C}.

# 3   Results and Discussion

## 3.1   BL-SOMs Constructed with Proteins Belonging to COGs

For the test dataset to examine whether proteins are clustered (i.e., self-organized) according to function by BL-SOM, we chose proteins that had been classified into function known COGs by NCBI [17]. Using BL-SOM, dipeptide composition ($20^2$ = 400 dimensional vectorial data) was investigated in 110,000 proteins belonging to the 2853 function-known COGs. In addition to this BL-SOM for the dipeptide composition of 20 amino acids (abbreviated as Di20-BLSOM), the BL-SOM for the dipeptide or tripeptide composition were constructed after categorizing amino acids into 11 groups based on the similarity of their physicochemical properties, $11^2$ (=121) or $11^3$ (=1331) dimensional data (abbreviated as Di11- or Tri11-BLSOM, respectively). BL-SOM was also constructed for the tetrapeptide composition after categorization into 6 groups, $6^4$ (=1296) dimensional data (abbreviated as Tetra6-BLSOM). These four different BL-SOM conditions were examined to establish which gave the best accuracy and to what degree similar results were obtained among the four conditions. It should be noted that BL-SOMs for much higher dimensional data, such as those for the tripeptide composition of 20 amino acids (8000-dimensional data) and for the tetrapeptide composition after grouping 11 categories (14641-dimensional data), was difficult in the present study because of the limitations of ES resources available to our group.

   In order to introduce a method that is less dependent on the sequence length of proteins, we provided a window of 200-amino acids that is moved with a 50-amino acid step for proteins longer than 200 amino acids, and the BL-SOM was constructed for these overlapped 200-amino acid sequences (approximately 500,000 sequences in total). Introduction of a window with a shifting step enabled us to analyze both multifunctional multidomain proteins primarily originating from the fusion of distinct proteins during evolution and smaller proteins, collectively. The 200-amino acid window was tentatively chosen in the present study because sizes of fictional domains of a wide range of proteins range from 100 to 300 amino acids.

   One important issue of the present method is at what level each lattice-point on a BL-SOM contains 200-amino acid fragments derived from a single COG. The number of the function-known COG categories is 2835, and the size of the BL-SOM was chosen so as to provide approximately eight fragments per lattice-point on average. If fragments were randomly chosen, the probability that all fragments associated with one lattice-point were derived from a single COG by chance should be extremely low, e.g. $(1/2853)^8 = 2.3 \times 10^{-28}$, while ensuring that this value depends on the total number of fragments derived from proteins belonging to the respective COG. We designate here the lattice-point that contained fragments derived only from a single COG category as a "pure lattice-point".

We compared the occurrence level of pure lattice-points among four different BL-SOMs. Although no COG information was given during BL-SOM learning, a high percentage of pure lattice-points (i.e., correct self-organization of sequences according to the COG category) was obtained (Fig. 1), despite the fact that the occurrence probability of a pure lattice-point as an accidental event is extremely low. The highest occurrence level of pure lattice-points was observed on the Tri11-BLSOM; approximately 45% of lattice-points on the Tri11-BLSOM contained sequences derived from only a single COG (Fig. 1). To graphically show the difference among these BL-SOMs, pure lattice-points were colored in red (Fig. 2A-C). The finding that the COG clustering (self-organization) with high accuracy was achieved indicates BL-SOM to be a powerful tool for function prediction of function-unknown proteins.

In Fig. 3, the number of sequences at each pure lattice-point on the Tri11-BLSOM was shown with the height of the vertical bar with a color representing each of the

(A) Di20-BLSOM        (B) Tri11-BLSOM        (C) Tetra6-BLSOM

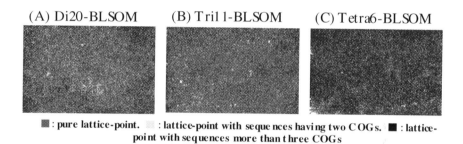

■: pure lattice-point.    : lattice-point with sequences having two COGs.   ■ : lattice-point with sequences more than three COGs

**Fig. 2.** The distribution of pure lattice-points colored in red

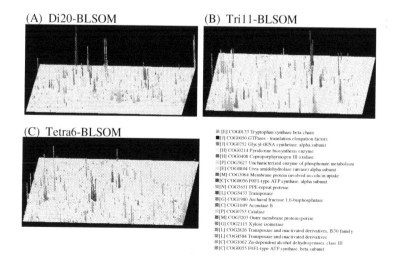

(A) Di20-BLSOM        (B) Tri11-BLSOM

(C) Tetra6-BLSOM

※ [E] COG0133 Tryptophan synthase beta chain
■ [J] COG0050 GTPases - translation elongation factors
※ [J] COG0752 Glycyl-tRNA synthetase, alpha subunit
[H] COG0214 Pyridoxine biosynthesis enzyme
■ [H] COG0408 Coproporphyrinogen III oxidase
※ [E] COG3627 Uncharacterized enzyme of phosphonate metabolism
※ [E] COG0804 Urea amidohydrolase (urease) alpha subunit
■ [M] COG3064 Membrane protein involved in colicin uptake
■ [C] COG0056 F0F1-type ATP synthase, alpha subunit
※ [N] COG5651 PPE-repeat proteins
■ [L] COG5433 Transposase
※ [G] COG1980 Archaeal fructose 1.6-bisphosphatase
■ [C] COG1049 Aconitase B
※ [P] COG0753 Catalase
■ [M] COG3203 Outer membrane protein (porin)
※ [G] COG2115 Xylose isomerase
■ [L] COG2826 Transposase and inactivated derivatives, IS30 family
■ [L] COG4584 Transposase and inactivated derivatives
※ [C] COG1062 Zn-dependent alcohol dehydrogenases, class III
※ [C] COG0055 F0F1-type ATP synthase, beta subunit

**Fig. 3.** Clustering of protein sequences according to COG (20 samples)

20 COG examples. Not only for these 20 examples, but also for a large portion of COG categories, sequences belonging to a single COG were localized in the neighboring points, resulting in a high peak composed of neighboring, high bars. In Fig. 3, a few high peaks with the same color located far apart from each other are also observed. Detailed inspection showed that these detached high peaks were mostly due to the different 200-amino acid segments (e.g., anterior and posterior portions) derived from one protein, which have distinct oligopeptide compositions and possibly represented distinct structural and functional domains of the respective protein. This type of major but distinct peaks appears to be informative for the prediction of functions of multifunctional multidomain proteins.

## 3.2  Function Prediction of Proteins Obtained from Metagenome Analyses

Most environmental microorganisms cannot be cultured under laboratory conditions. Genomes of the unculturable microorganisms have remained mostly uncharacterized but are believed to contain a wide range of novel protein genes of scientific and industrial usefulness. Metagenomic approaches that decode the sequences of the mixed genomes of uncultured environmental microbes [18-20] have been developed recently for finding a wide variety of novel and industrially useful genes. Venter et al. [21] applied large-scale metagenome sequencing to mixed genomes collected from the Sargasso Sea near Bermuda and deposited a large number of sequence fragments in the International DNA Databanks.

The most important contribution of the present alignment-free and unsupervised clustering method, BLSOM, should be the prediction of the functions of an increasingly large number of function-unknown proteins derived from the less characterized genomes, such as those studied in the metagenomic approaches. To test the feasibility of BL-SOM for function prediction of environmental sequences, we searched in advance the Sargasso proteins that showed significant global homology with the NCBI COG proteins by using the conventional sequence homology search. Based on a criterion that in 80% or more of the region, 80% or more identity of the amino acid sequence was observed, 3924 Sargasso proteins (> 200 amino acids) could be related to NCBI COG categories (designated Sargasso COG sequences). Then, we mapped the 200-amino acid fragments derived from these Sargasso COG proteins onto Di20- and Tri11-BLSOMs, which were previously constructed with NCBI COG sequences in Figs. 1 and 2. For each lattice point on which Sargasso COG fragments were mapped, the most abundant NCBI COG sequences were identified, and the mapped Sargasso segments were tentatively assumed to belong to this most abundant NCBI COG category. After summing up these tentative COGs for each Sargasso protein, individual Sargasso proteins were finally classified into one NCBI COG category, if more than 60% of the 200-amino acid fragments derived from one Sargasso protein gave the same COG category. By mapping on Tri11-, Di20- or Tet6-BLSOM, 87.5, 86.8 or 79.0% of the 3924 Sargasso COG proteins showed the COG category identical to that had been identified by the sequence homology search in advance. The highest identity level was found on Tri11-BLSOM. In Fig. 4, the number of Sargasso fragments thus classified into COGs on Di20-, Tri11- and Tetra6-BLSOM was shown by the height of the vertical bar. In the next step, when the false prediction for the Sargasso COG proteins was checked in detail, the pairs of real and

(A)  Di20-BLSOM

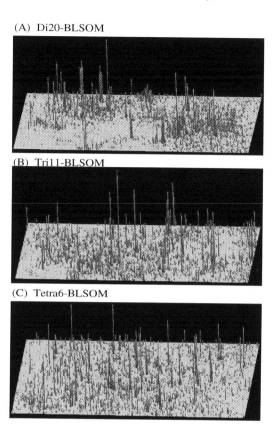

(B)  Tri11-BLSOM

(C)  Tetra6-BLSOM

**Fig. 4.** Mapping of the Sargasso COG fragments on Di20- (A), Tri11- (B) and Tetra6- (C) BLSOM. The height of the vertical bar shows the number of fragments.

falsely-assigned COGs corresponded to those that have the functions closely related with each other, such as those with paralogous relationships. According to the definition of COG (clusters of orthologous groups of proteins), paralogous gene proteins should belong to different COGs in spite of the similarity of functions. COG categorization appears be too strict to be used for function predictions of a wide variety of proteins.

In the final analysis, we mapped the residual Sargasso proteins, which could not be classified into NCBI COGs using the sequence homology search, onto Di20-, Tri11- and Tetra6-BLSOMs. Approximately 15% of the Sargasso proteins (i.e., approximately 90,000 proteins) were associated with an NCBI COG category. For Sargasso proteins for which the consistency of the predicted function is obtained by separate analyses of di-, tri- and tetrapeptide frequencies, the reliability of the prediction should be very high. We plan to publicize the results of the assignments obtained concordantly with three BLSOM conditions (Tri11-, Di20- and Tetra6-BLSOMs).

To identify functions of a large number of function-unknown proteins accumulated in databases systematically, we have to construct a large scale-BL-SOM in advance that analyzes all function-known proteins available in databases utilizing a high-performance supercomputer such as ES [22,23]. This approach should serve as a new and powerful strategy to predict functions of a large number of novel proteins collectively, systematically and efficiently. The BLSOM data obtained by high-performance supercomputers are unique in fields of genomics and proteomics and provide a new guideline for research groups, including those in industry, for the study of function identification of novel genes through experiment.

**Acknowledgements.** This work was supported by Grant-in-Aid for Scientific Research (C) and for Young Scientists (B) from the Ministry of Education, Culture, Sports, Science and Technology of Japan. The present computation was done with the Earth Simulator of Japan Agency for Marine-Earth Science and Technology.

# References

1. Kohonen, T.: Self-organized formation of topologically correct feature maps. Biol. Cybern. 43, 59–69 (1982)
2. Kohonen, T.: The self-organizing map. Proc. IEEE 78, 1464–1480 (1990)
3. Kohonen, T., Oja, E., Simula, O., Visa, A., Kangas, J.: Engineering applications of the self-organizing map. Proc. IEEE 84, 1358–1384 (1996)
4. Ferran, E.A., Pflugfelder, B., Ferrara, P.: Self-organized neural maps of human protein sequences. Protein Sci. 3, 507–521 (1994)
5. Kanaya, S., Kudo, Y., Abe, T., Okazaki, T., Carlos, D.C., Ikemura, T.: Gene classification by self-organization mapping of codon usage in bacteria with completely sequenced genome. Genome Inform. 9, 369–371 (1998)
6. Kanaya, S., Kinouchi, M., Abe, T., Kudo, Y., Yamada, Y., Nishi, T., Mori, H., Ikemura, T.: Analysis of codon usage diversity of bacterial genes with a self-organizing map (SOM): characterization of horizontally transferred genes with emphasis on the E. coli O157 genome. Gene. 276, 89–99 (2001)
7. Abe, T., Kanaya, S., Kinouchi, M., Ichiba, Y., Kozuki, T., Ikemura, T.: A novel bioinformatic strategy for unveiling hidden genome signatures of eukaryotes: Self-organizing map of oligonucleotide frequency. Genome Inform. 13, 12–20 (2002)
8. Abe, T., Kanaya, S., Kinouchi, M., Ichiba, Y., Kozuki, T., Ikemura, T.: Informatics for unveiling hidden genome signatures. Genome Res. 13, 693–702 (2003)
9. Abe, T., Kozuki, T., Kosaka, Y., Fukushima, S., Nakagawa, S., Ikemura, T.: Self-organizing map reveals sequence characteristics of 90 prokaryotic and eukaryotic genomes on a single map. In: WSOM 2003, pp. 95–100 (2003)
10. Abe, T., Sugawara, H., Kinouchi, M., Kanaya, S., Matsuura, Y., Tokutaka, H., Ikemura, T.: A large-scale Self-Organizing Map (SOM) constructed with the Earth Simulator unveils sequence characteristics of a wide range of eukaryotic genomes. In: WSOM 2005, pp. 187–194 (2005)
11. Abe, T., Sugawara, H., Kinouchi, M., Kanaya, S., Ikemura, T.: A large-scale Self-Organizing Map (SOM) unveils sequence characteristics of a wide range of eukaryote genomes. Gene. 365, 27–34 (2006)
12. Abe, T., Sugawara, H., Kanaya, S., Ikemura, T.: Sequences from almost all prokaryotic, eukaryotic, and viral genomes available could be classified according to genomes on a large-scale Self-Organizing Map constructed with the Earth Simulator. J. Earth Simulator 6, 17–23 (2006)

13. Abe, T., Sugawara, H., Kinouchi, M., Kanaya, S., Ikemura, T.: Novel phylogenetic studies of genomic sequence fragments derived from uncultured microbe mixtures in environmental and clinical samples. DNA Res. 12, 281–290 (2005)
14. Hayashi, H., Abe, T., Sakamoto, M., et al.: Direct cloning of genes encoding novel xylanases from human gut. Can. J. Microbiol. 51, 251–259 (2005)
15. Uchiyama, T., Abe, T., Ikemura, T., Watanabe, K.: Substrate-induced gene-expression screening of environmental metagenome libraries for isolation of catabolic genes. Nature Biotech. 23, 88–93 (2005)
16. Abe, T., Sugawara, H., Kanaya, S., Ikemura, T.: A novel bioinformatics tool for phylogenetic classification of genomic sequence fragments derived from mixed genomes of environmental uncultured microbes. Polar Bioscience 20, 103–112 (2006)
17. Tatsusov, R.L., Koonin, E.V., Lipman, D.J.: A genomic perspective on protein families. Science 278, 631–637 (1997)
18. Amann, R.I., Ludwig, W., Schleifer, K.H.: Phylogenetic identification and in situ detection of individual microbial cells without cultivation. Microbiol. Rev. 59, 143–169 (1995)
19. Hugenholtz, P., Pace, N.R.: Identifying microbial diversity in the natural environment: a molecular phylogenetic approach. Trends Biotechnol. 14, 190–197 (1996)
20. Rondon, M.R., August, P.R., Bettermann, A.D., et al.: Cloning the soil metagenome: a strategy for accessing the genetic and functional diversity of uncultured microorganisms. Appl. Environ. Microbiol. 66, 2541–2547 (2000)
21. Venter, J.C., et al.: Environmental genome shotgun sequencing of the Sargasso Sea. Science 304, 66–74 (2004)
22. Abe, T., Ikemura, T.: A large-scale batch-learning Self-Organizing Maps for function prediction of poorly characterized proteins progressively accumulating in sequence databases. Annual Report of the Earth Simulator, April 2006 - March 2007, pp. 247–251 (2007)
23. Abe, T., Ikemura, T.: A large-scale genomics and proteomics analyses conducted by the Earth Simulator. Annual Report of the Earth Simulator, April 2007 - March 2008, pp. 245–249 (2008)

# Incremental Unsupervised Time Series Analysis Using Merge Growing Neural Gas

Andreas Andreakis, Nicolai v. Hoyningen-Huene, and Michael Beetz

Technische Universität München, Intelligent Autonomous Systems Group,
Boltzmannstrasse 3, 85747 Garching, Germany
{andreaki,hoyninge,beetz}@cs.tum.edu

**Abstract.** We propose Merge Growing Neural Gas (MGNG) as a novel unsupervised growing neural network for time series analysis. MGNG combines the state-of-the-art recursive temporal context of Merge Neural Gas (MNG) with the incremental Growing Neural Gas (GNG) and enables thereby the analysis of unbounded and possibly infinite time series in an online manner. There is no need to define the number of neurons a priori and only constant parameters are used. In order to focus on frequent sequence patterns an entropy maximization strategy is utilized which controls the creation of new neurons. Experimental results demonstrate reduced time complexity compared to MNG while retaining similar accuracy in time series representation.

**Keywords:** time series analysis, unsupervised, self-organizing, incremental, recursive temporal context.

## 1 Introduction

Time series represent the vast majority of data in everyday life and their handling is a key feature of living organisms. Automatic processing of sequential data has received a broad interest in action recognition, DNA analysis, natural language processing and CRM systems, to name just a few.

The analysis of time series aims at two main goals, namely the identification of the nature of the underlying phenomenon and the forecasting of future values. Both goals depend upon a good representation of sequential patterns observed from a given time series. One approach, that has proven to be successful, is to build a temporal quantization of sequences in order to form a compact representation in an unsupervised way.

We propose Merge Growing Neural Gas (MGNG) as a recursive growing self-organizing neural network for time series analysis. MGNG extends the state-of-the-art Merge Neural Gas (MNG) [1] to an incremental network by utilizing Growing Neural Gas (GNG) [2]. The theoretically founded recursive temporal context of MNG represents the input history as an exponentially decayed sum and is inherited for MGNG.

MGNG allows for online quantization of time series in previously unknown and possibly infinite data streams by the use of constant parameters only. There's

J.C. Príncipe and R. Miikkulainen (Eds.): WSOM 2009, LNCS 5629, pp. 10–18, 2009.

no need to know the number of neurons in advance due to the growing of the network. The proposed algorithm exhibits faster runtime performance than MNG while keeping similar accuracy in the time series representation.

The remainder of the paper is organized as follows: After a review of related work in unsupervised time series analysis in Section 2, the MGNG algorithm is explained in Section 3 in detail. We evaluate the performance of the proposed approach in three experiments comparing our results with MNG and other models in Section 4. Conclusions and future prospects round off the work in Section Section 5.

## 2   Related Work

Based on well known unsupervised models like the Self Organizing Map (SOM) [3] and Neural Gas (NG) [4] several extensions have been proposed for sequential input data. Common approaches use hierarchies [5], non-Euclidean sequence metrics [6], time-window techniques [4], mapping to spatial correlations [7] and there exist a wider field of recursive sequence models.

Recursive sequence models extend unsupervised neural networks by recursive dynamics such as leaky integrators [8]. Hammer et al. give an overview over recursive models [9] and present a unifying notation [10]. Temporal Kohonen Map (TKM) [11], Recurrent SOM (RSOM) [12], Recursive SOM (RecSOM) [13], SOM for structured data (SOM-SD) [14], Merge SOM (MSOM) and Merge Neural Gas (MNG) [1] represent popular recursive models which have been applied in several applications [15,16,17,18,19]. The specific models differ mainly in their internal representation of time series, which influences the capacity of the model, the flexibility with respect to network topology and the processing speed. MNG has shown superior performance to the other recursive models for acceptable time complexity.

All extensions towards quantization of temporal data have in common that the optimal number of neurons has to be known in advance. However, too many neurons waste resources and may lead to overfitting and too less neurons cannot represent the input space well enough and might therefor lead to high quantization errors.

Growing Neural Gas (GNG) was introduced by Fritzke [2] as an incremental unsupervised neural network for non-temporal data. Starting with two neurons GNG grows in regular time intervals up to a maximum size. Connections between neurons are created by topology preserving Competitive Hebbian Learning [20]. Only the best matching neuron and its direct topological neighbors are updated for each input signal leading to lower time complexity than SOM or NG where the whole network is updated. All used learning parameters are constant and enable the handling of infinite input streams. Also the variant GNG-U has been introduced for non-stationary data distributions [21].

Kyan and Guan proposed Self-Organized Hierarchical Variance Map (SO-HVM) [22] as another incremental network that unfortunately lacks the capability of online processing due to declining parameters but shows better accuracy with a higher complexity than GNG.

To the best of our knowledge there exists no growing model which utilizes recursive dynamics and their associated advantages yet.

# 3   Merge Growing Neural Gas

Merge Growing Neural Gas transfers the features of GNG into the domain of time series analysis by utilizing the state-of-the-art temporal dynamics of MNG.

MGNG is a self-organizing neural network consisting of a set of neurons $\mathcal{K}$ which are connected by edges $\mathcal{E}$. Each neuron $n \in \mathcal{K}$ comprises of a weight vector $\mathbf{w}_n$ representing the current time step and a context vector $\mathbf{c}_n$ representing all past time steps of a sequence, both having the dimensionality of the input space.

An input sequence $\mathbf{x}_1, \ldots, \mathbf{x}_t$ is assigned to the best matching neuron (also called winner or best matching unit) by finding the neuron $n$ with lowest distance $d_n$ in time step $t$:

$$d_n(t) = (1 - \alpha) \cdot \|\mathbf{x}_t - \mathbf{w}_n\|^2 + \alpha \cdot \|\mathbf{C}_t - \mathbf{c}_n\|^2 \tag{1}$$

The parameter $\alpha \in [0, 1]$ weights the importance of the current input signal over the past. $\mathbf{C}_t$ is called the global temporal context and is computed as a linear combination (merge) of the weight and context vector from the winner $r$ of time step $t - 1$:

$$\mathbf{C}_t := (1 - \beta) \cdot \mathbf{w}_r + \beta \cdot \mathbf{c}_r \tag{2}$$

The parameter $\beta \in [0, 1]$ controls the influence of the far over the recent past and $\mathbf{C}_1 := 0$.

When the network is trained, $\mathbf{C}_t$ converges to the optimal global temporal context vector $\mathbf{C}_t^{opt}$ that can be written as [1]:

$$\mathbf{C}_t^{opt} := \sum_{j=1}^{t-1} (1 - \beta) \cdot \beta^{t-1-j} \cdot \mathbf{x}_j \tag{3}$$

The training algorithm for MGNG is depicted in Figure 1. Hebbian learning takes place by adapting the winner neuron towards the recent input signal $\mathbf{x}_t$ and the past $\mathbf{C}_t$ based on the learning rate $\epsilon_b$. Also, the winner's direct neighbors are adapted using the learning rate $\epsilon_n$ (see line 13). The connection between the best and second best matching unit is created or refreshed following a competitive Hebbian learning approach (see line 8 and 9). All other connections are weakened and too infrequent connections are deleted depending on $\gamma$ (see line 10 and 11). The network grows at regular time intervals $\lambda$ up to a maximum size $\theta$ by the insertion of new nodes based on entropy maximization (see lines 15a-e).

## 3.1   Entropy Maximization

In time series analysis we are usually interested in a representation of frequent sequence patterns. This is achieved by an entropy maximization strategy for node insertion because the entropy of a network is highest if the activation of

1. time variable $t := 1$
2. initialize neuron set $\mathcal{K}$ with 2 neurons having counter $e := 0$ and random weight and context vectors
3. initialize connection set $\mathcal{E} \subseteq \mathcal{K} \times \mathcal{K} := \emptyset$
4. initialize global temporal context $\mathbf{C}_1 := \mathbf{0}$
5. read / draw input signal $\mathbf{x}_t$
6. find winner $r := \arg\min_{n \in \mathcal{K}} d_n(t)$
   and second winner $s := \arg\min_{n \in \mathcal{K} \setminus \{r\}} d_n(t)$
   where
   $$d_n(t) = (1 - \alpha) \cdot \|\mathbf{x}_t - \mathbf{w}_n\|^2 + \alpha \cdot \|\mathbf{C}_t - \mathbf{c}_n\|^2$$
7. increment counter of $r$: $e_r := e_r + 1$
8. connect $r$ with $s$: $\mathcal{E} := \mathcal{E} \cup \{(r, s)\}$
9. $age_{(r,s)} := 0$
10. increment the age of all edges connected with $r$
    $$age_{(r,n)} := age_{(r,n)} + 1 \quad (\forall n \in \mathcal{N}_r \setminus \{s\})$$
11. remove old connections $\mathcal{E} := \mathcal{E} \setminus \{(a, b) | age_{(a,b)} > \gamma\}$
12. delete all nodes with no connections
13. update neuron $r$ and its direct topological neighbors $\mathcal{N}_r$:
    $$\mathbf{w}_r := \mathbf{w}_r + \epsilon_b \cdot (\mathbf{x}_t - \mathbf{w}_r) \text{ and } \mathbf{c}_r := \mathbf{c}_r + \epsilon_b \cdot (\mathbf{C}_t - \mathbf{c}_r)$$
    $$\forall n \in \mathcal{N}_r : \mathbf{w}_n := \mathbf{w}_n + \epsilon_n \cdot (\mathbf{x}_t - \mathbf{w}_i) \text{ and } \mathbf{c}_n := \mathbf{c}_n + \epsilon_n \cdot (\mathbf{C}_t - \mathbf{c}_i)$$
14. calculate the global temporal context for the next time step
    $$\mathbf{C}_{t+1} := (1 - \beta) \cdot \mathbf{w}_r + \beta \cdot \mathbf{c}_r$$
15. create new node if $t \bmod \lambda = 0$ and $|\mathcal{K}| < \theta$
    a. find neuron $q$ with the greatest counter: $q := \arg\max_{n \in \mathcal{K}} e_n$
    b. find neighbor $f$ of $q$ with $f := \arg\max_{n \in \mathcal{N}_q} e_n$
    c. initialize new node $l$
       $$\mathcal{K} := \mathcal{K} \cup \{l\}$$
       $$\mathbf{w}_l := \tfrac{1}{2}(\mathbf{w}_q + \mathbf{w}_f)$$
       $$\mathbf{c}_l := \tfrac{1}{2}(\mathbf{c}_q + \mathbf{c}_f)$$
       $$e_l := \delta \cdot (e_f + e_q)$$
    d. adapt connections: $\mathcal{E} := (\mathcal{E} \setminus \{(q, f)\}) \cup \{(q, n), (n, f)\}$
    e. decrease counter of $q$ and $f$ by the factor $\delta$
       $$e_q := (1 - \delta) \cdot e_q$$
       $$e_f := (1 - \delta) \cdot e_f$$
16. decrease counter of all neurons by the factor $\eta$:
    $$e_n := \eta \cdot e_n \quad (\forall n \in \mathcal{K})$$
17. $t := t + 1$
18. if more input signals available goto step 5 else terminate

**Fig. 1.** Pseudocode for training of Merge Growing Neural Gas (MGNG)

all neurons is balanced. At high entropy more neurons are used for frequent sequences reducing the representation capacity for rare ones. This helps to focus on quantization of important information beside the usually combinatorial explosion of time series.

Following Fritzke's [23] proposed strategy we insert a new node in regions with high activation frequency leading to an increase of the entropy of the network. Frequency is tracked by a counter of every neuron that is incremented every time the neuron is selected as the winner (see line 7). New nodes are inserted between the most active neuron $q$ and its most frequent topological neighbor $f$ reducing the likelihood of both nodes $q$ and $f$ to be selected as the winner and therefor increasing the overall entropy of the network. The new node $l$ is initialized as the mean of the two selected nodes and inserted in-between them. The counters of $q$ and $f$ are reduced to reflect the expected decrease of activation while the new neuron takes over this activation. The parameter $\delta$ controls the amount of the activation shift (see lines 15a-e). All counters are subject to exponential decay by the parameter $\eta$ in order to give recent changes a higher relevance (see line 16). To further increase the entropy, nodes with no connections are deleted because the last selection as the first or second best matching unit was too long ago (see line 12).

In the entropy maximization strategy of MNG [1] the parameter $\alpha$ of the distance function (see eq. (1)) is gradually increased and decreased based on the entropy of the network describing a zigzag curve with diminishing amplitude. This results in a training based alternately on weight and context vectors and unfortunately causes a temporary destabilization of the network for extreme values of $\alpha$ close to one, because winner selection is based on the past only omitting the recent time step of a sequence. Our entropy maximization strategy avoids these problems and is suitable for an online learning setting with constant parameters only.

## 4    Experimental Results

Three experiments were conducted to evaluate MGNG's performance in temporal quantization, density estimation and representation capacity.

### 4.1    Mackey Glass

The Mackey Glass time series is a 1-dimensional, continuous and chaotic function defined by the differential equation $\frac{dx}{d\tau} = -0.1x(\tau) + \frac{0.2x(\tau-17)}{1+x(\tau-17)^{10}}$ and is commonly used to evaluate the temporal quantization of recursive models [13,1,24].

The experiment was conducted on the models SOM, NG, GNG, MNG and MGNG which were trained with a time series of $150,000$ elements. After training, the temporal quantization error [13] for up to 30 past time steps into the past was calculated by re-inserting the time series and saving the winner unit for each time step without modification of the network.

MGNG was configured using the following parameters: $\alpha = 0.5$, $\beta = 0.75$, $\theta = 100$, $\lambda = 600$, $\gamma = 88$, $\epsilon_b = 0.05$, $\epsilon_n = 0.0006$, $\delta = 0.5$, $\eta = 0.9995$. We

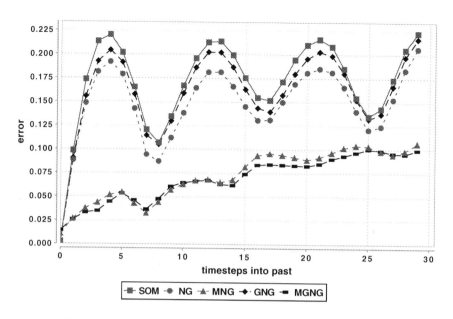

**Fig. 2.** Experimental results for the Mackey Glass time series

used the same $\alpha$ and $\beta$ for MNG and inherited the other parameters from [1]. Analogous parameters have been used for the non-temporal models SOM, NG and GNG, where a 10 to 10 rectangular lattice was used for the SOM.

Figure 2 visualizes the temporal quantization error of the models for up to 30 time steps into the past.

MNG and MGNG have learned the temporal context and show a similar quantization performance. However MGNG required just 8,018s for the computation in contrast to 20,199s needed by MNG. The runtime advantage of MGNG is based on the underlying GNG model, which in comparison to NG requires less adaptions for each input signal.

Results for the recursive models RSOM, RecSOM and SOM-SD can be obtained from [8]. As expected they are superior to non-temporal models SOM, NG and GNG, but their performance lies below MNG's and MGNG's.

## 4.2 Binary Automata

The Binary automaton experiment has been proposed by Voegtlin [13] in order to evaluate the representation capacity of temporal models. The experiment uses a Markov automaton with states 0 and 1 and the probabilities $P(0) = \frac{4}{7}$, $P(1) = \frac{3}{7}$, $P(0|1) = 0.4$, $P(1|1) = 0.6$, $P(1|0) = 0.3$, $P(0|0) = 0.7$. A sequence with $10^6$ elements was generated and trained in a network with 100 neurons. After training the winner units for the 100 most probable sequences are determined and for multiple winners only the longest sequence is associated. An optimal result would be achieved if each of the 100 sequences has an unique winner. The experiment was carried out with MNG and MGNG using $\alpha = 0.5$, $\beta = 0.45$ and the other parameters are inherited from the previous experiment.

MGNG shows a slight improvement against MNG in the representation capacity and a clear advantage in computation time. A total number of 64 longest sequences could be discriminated by MGNG requiring 69,143s and 62 sequences were discriminated by MNG in 131,177s.

However, both models cannot discriminate all 100 sequences, because recursive models that represent the temporal context as a weighted sum cannot discriminate between sequences with repeated '0' signals (such as: 0, 00, 0000...0). Choosing other values like '1' and '2' would improve the results.

### 4.3   Noisy Automata

The noisy automaton experiment originates from Hammer et al. [24] and its goal is to evaluate the density estimating capabilities of a temporal model by reconstructing the transition probabilities of a second order Markov model. Two-dimensional input signals are generated from three normal distributions with the means $a = (0,0)$, $b = (1,0)$ and $c = (0,1)$ and a common standard deviations $\sigma$. Figure 3 visualizes the Markov automaton where the transition probabilities can be configured with the parameter $x$.

In the first part, the experiment was carried out with different transition probabilities $x \in \{0, 0.1, 0.2, 0.3, 0.4, 0.5, 0.6, 0.7, 0.8\}$ and with a constant $\sigma = 0.1$ using a total count of $10^6$ input signals. In the second part $x = 0.4$ was set constant while different $\sigma \in \{0, 0.1, 0.2, 0.3, 0.4, 0.5\}$ were used. MNG was configured with $\alpha = 0$ and $\beta = 0.5$ MGNG with $\alpha = 0.5$ and the same $\beta$. The other parameters are equal to the previous experiments.

The transition probabilities are reconstructed based on the backtracking method described in [24]. As can be seen in Tables 1(a) and 1(b) the transition probabilities for both MNG and MGNG could be reconstructed accurately in the first part of the experiment. In the second part, MNG shows better results for $\sigma > 0.3$. However unlike MNG, MGNG is able to identify clusters without any

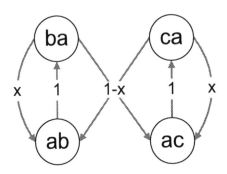

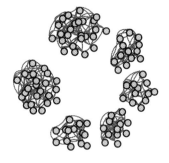

**Fig. 3.** Transition probabilities of the noisy automaton

**Fig. 4.** Kamada-Kawai based MGNG Topology for $x = 0.5$

**Table 1.** Experimental results for the noisy automaton experiment

(a) MNG varying $x$ and $\sigma$ separately.

| $x$ | 0.0 | 0.1 | 0.2 | 0.3 | 0.4 | 0.5 | 0.6 | 0.7 | 0.8 | 0.4 | | | |
|---|---|---|---|---|---|---|---|---|---|---|---|---|---|
| $\sigma$ | | | | | 0.1 | | | | | 0.2 | 0.3 | 0.4 | 0.5 |
| $P(a\|ba)$ | 0.0 | 0.0 | 0.0 | 0.0 | 0.0 | 0.0 | 0.0 | 0.0 | 0.0 | 0.0 | 0.0 | 0.203 | 0.227 |
| $P(b\|ba)$ | 0.0 | 0.11 | 0.211 | 0.31 | 0.399 | 0.513 | 0.595 | 0.687 | 0.791 | 0.471 | 0.312 | 0.515 | 0.248 |
| $P(c\|ba)$ | 1.0 | 0.889 | 0.788 | 0.689 | 0.6 | 0.48 | 0.404 | 0.312 | 0.208 | 0.528 | 0.687 | 0.28 | 0.523 |
| $P(a\|ca)$ | 0.0 | 0.0 | 0.0 | 0.0 | 0.0 | 0.0 | 0.0 | 0.0 | 0.0 | 0.0 | 0.122 | 0.0 | 0.0 |
| $P(b\|ca)$ | 1.0 | 0.884 | 0.787 | 0.689 | 0.599 | 0.482 | 0.404 | 0.341 | 0.205 | 0.55 | 0.387 | 0.554 | 0.735 |
| $P(c\|ca)$ | 0.0 | 0.115 | 0.212 | 0.31 | 0.4 | 0.517 | 0.595 | 0.658 | 0.794 | 0.449 | 0.49 | 0.445 | 0.264 |

(b) MGNG varying $x$ and $\sigma$ separately.

| $x$ | 0.0 | 0.1 | 0.2 | 0.3 | 0.4 | 0.5 | 0.6 | 0.7 | 0.8 | 0.4 | | | |
|---|---|---|---|---|---|---|---|---|---|---|---|---|---|
| $\sigma$ | | | | | 0.1 | | | | | 0.2 | 0.3 | 0.4 | 0.5 |
| $P(a\|ba)$ | 0.0 | 0.0 | 0.0 | 0.0 | 0.0 | 0.0 | 0.0 | 0.0 | 0.0 | 0.0 | 0.0 | 0.216 | 0.409 |
| $P(b\|ba)$ | 0.0 | 0.098 | 0.201 | 0.302 | 0.398 | 0.498 | 0.603 | 0.699 | 0.796 | 0.501 | 0.591 | 0.576 | 0.389 |
| $P(c\|ba)$ | 1.0 | 0.901 | 0.798 | 0.697 | 0.601 | 0.501 | 0.396 | 0.3 | 0.203 | 0.498 | 0.408 | 0.207 | 0.201 |
| $P(a\|ca)$ | 0.0 | 0.0 | 0.0 | 0.0 | 0.0 | 0.0 | 0.0 | 0.0 | 0.0 | 0.0 | 0.174 | 0.21 | 0.438 |
| $P(b\|ca)$ | 1.0 | 0.9 | 0.798 | 0.7 | 0.6 | 0.498 | 0.398 | 0.3 | 0.2 | 0.587 | 0.557 | 0.34 | 0.145 |
| $P(c\|ca)$ | 0.0 | 0.099 | 0.201 | 0.299 | 0.399 | 0.501 | 0.601 | 0.699 | 0.799 | 0.412 | 0.267 | 0.448 | 0.416 |

further post-processing. Every subgraph in MGNG can be interpreted as a cluster and Figure 4 visualizes the six identified clusters using the Kamada-Kawai graph drawing algorithm [25], representing all possible sequence triples.

## 5   Conclusions

We introduced MGNG as a novel unsupervised neural network for time series analysis. By combining the advantages of growing networks and recursive temporal dynamics our model exhibits state-of-the-art accuracy in quantization, density estimation and representation of sequences. In contrast to all other recursive networks no a-priori knowledge is required and only constant parameters are used. Our approach is simple to implement and shows better runtime performance than MNG. In future research we plan to investigate the extension of MGNG with different node creation and deletion strategies based on the entropy and maximum local variance to further improve the accuracy.

**Acknowledgments.** The work described in this paper was partially funded by the German Research Foundation DFG.

## References

1. Strickert, M., Hammer, B.: Merge SOM for temporal data. Neurocomputing 64, 39–71 (2005)
2. Fritzke, B.: A Growing Neural Gas network learns topologies. In: Tesauro, G., Touretzky, D.S., Leen, T.K. (eds.) NIPS, pp. 625–632. MIT Press, Cambridge (1995)

3. Kohonen, T.: Self-Organizing Maps, 3rd edn. Springer, Berlin (2001)
4. Martinetz, T., Martinetz, T., Berkovich, S., Schulten, K.: "Neural-gas" Network for Vector Quantization and its Application to Time-Series Prediction. IEEE-Transactions on Neural Networks 4(4), 558–569 (1993)
5. Carpinteiro, O.A.S.: A Hierarchical Self-Organizing Map Model for Sequence Recognition. In: Proc. of ICANN, vol. 2, pp. 815–820. Springer, London (1998)
6. Hammer, B., Villmann, T.: Classification using non standard metrics. In: Proc. of ESANN, pp. 303–316 (2005)
7. Euliano, N.R., Principe, J.C.: A Spatio-Temporal Memory Based on SOMs with Activity Diffusion. In: Oja (ed.) Kohonen Maps, pp. 253–266. Elsevier, Amsterdam (1999)
8. Hammer, B., Micheli, A., Sperduti, A., Strickert, M.: Recursive self-organizing network models. Neural Networks 17(8-9), 1061–1085 (2004)
9. Hammer, B., Micheli, A., Neubauer, N., Sperduti, A., Strickert, M.: Self Organizing Maps for Time Series. In: Proc. of WSOM, Paris, France, pp. 115–122 (2005)
10. Hammer, B., Micheli, A., Sperduti, A.: A general framework for unsupervised processing of structured data. In: Proc. of ESANN, vol. 10, pp. 389–394 (2002)
11. Chappell, G.J., Taylor, J.G.: The Temporal Kohonen map. Neural Networks 6(3), 441–445 (1993)
12. Koskela, T., Varsta, M., Heikkonen, J., Kaski, K.: Temporal sequence processing using Recurrent SOM. In: Proc. of KES, pp. 290–297. IEEE, Los Alamitos (1998)
13. Voegtlin, T.: Recursive Self-Organizing Maps. Neural Networks 15(8-9) (2002)
14. Hagenbuchner, M., Sperduti, A., Tsoi, A.C.: A Self-Organizing Map for adaptive processing of structured data. Neural Networks 14(3), 491–505 (2003)
15. Lambrinos, D., Scheier, C., Pfeifer, R.: Unsupervised Classification of Sensory-Motor states in a Real World Artifact using a Temporal Kohonen Map. In: Proc. of ICANN, vol. 2, EC2, pp. 467–472 (1995)
16. Farkas, I., Crocker, M.: Recurrent networks and natural language: exploiting self-organization. In: Proc. of CogSci (2006)
17. Farka, I., Crocker, M.W.: Systematicity in sentence processing with a recursive Self-Organizing Neural Network. In: Proc. of ESANN (2007)
18. Trentini, F., Hagenbuchner, M., Sperduti, A., Scarselli, F., Tsoi, A.: A Self-Organising Map approach for clustering of XML documents. In: WCCI (2006)
19. Estevez, P.A., Zilleruelo-Ramos, R., Hernandez, R., Causa, L., Held, C.M.: Sleep Spindle Detection by Using Merge Neural Gas. In: WSOM (2007)
20. Martinetz, T.: Competitive Hebbian Learning Rule Forms Perfectly Topology Preserving Maps. In: Proc. of ICANN, pp. 427–434. Springer, Heidelberg (1993)
21. Fritzke, B.: A self-organizing network that can follow non-stationary distributions. In: Gerstner, W., Hasler, M., Germond, A., Nicoud, J.-D. (eds.) ICANN 1997. LNCS, vol. 1327, pp. 613–618. Springer, Heidelberg (1997)
22. Kyan, M., Guan, L.: Local variance driven self-organization for unsupervised clustering. In: Proc. of ICPR, vol. 3, pp. 421–424 (2006)
23. Fritzke, B.: Vektorbasierte Neuronale Netze. PhD thesis, Uni Erlangen (1998)
24. Strickert, M., Hammer, B., Blohm, S.: Unsupervised recursive sequence processing. Neurocomputing 63, 69–97 (2005)
25. Kamada, T., Kawai, S.: An algorithm for drawing general undirected graphs. Inf. Process. Lett. 31(1), 7–15 (1989)

# Clustering Hierarchical Data Using Self-Organizing Map: A Graph-Theoretical Approach

Argyris Argyrou

HANKEN School of Economics
Arkadiankatu 22, 00101 - Helsinki, Finland
argyris.argyrou@hanken.fi

**Abstract.** The application of Self-Organizing Map (SOM) to hierarchical data remains an open issue, because such data lack inherent quantitative information. Past studies have suggested binary encoding and Generalizing SOM as techniques that transform hierarchical data into numerical attributes. Based on graph theory, this paper puts forward a novel approach that processes hierarchical data into a numerical representation for SOM-based clustering. The paper validates the proposed graph-theoretical approach via complexity theory and experiments on real-life data. The results suggest that the graph-theoretical approach has lower algorithmic complexity than Generalizing SOM, and can yield SOM having significantly higher cluster validity than binary encoding does. Thus, the graph-theoretical approach can form a data-preprocessing step that extends SOM to the domain of hierarchical data.

**Keywords:** Clustering, hierarchical data, SOM, graph theory.

## 1   Introduction

The Self-Organizing Map (SOM) [1] represents a type of artificial neural network that is based on unsupervised learning; it has been applied extensively in the areas of dimensionality reduction, data visualization, and clustering [2]. The original formulation of SOM uses the Euclidean distance as a similarity metric [3, p.4], and hence its domain of application is restricted to metric spaces [4]. SOM has been extended to non-metric spaces by using generalized means and medians as the distance measures and the batch variant of SOM [4]; for example, speech recognition [5], and clustering of protein sequences [6]. An online algorithm for SOM of symbol strings was provided by [7]. However, neither a metric distance nor a string metric (e.g. Levenshtein distance) can yield meaningful results in the domain of hierarchical data, and thus the application of SOM in this domain remains an open issue. For example, consider clustering the data: {cat, rat, mouse}. A string metric would find that {cat} and {rat} are more closely related to each other than {rat} and {mouse} are, while a metric distance would produce meaningless results.

J.C. Príncipe and R. Miikkulainen (Eds.): WSOM 2009, LNCS 5629, pp. 19–27, 2009.
© Springer-Verlag Berlin Heidelberg 2009

To address this issue, prior studies have suggested two main techniques that transform hierarchical attributes into numerical attributes. First, the most prevalent technique encodes a categorical attribute in binary terms {1,0}, where 1 and 0 denote the presence and absence of an attribute respectively. The binary encoding is then treated as a numerical attribute in the range {1,0}. Second, Hsu [8] introduced Generalizing SOM (GSOM), whereby a domain expert describes a set of categorical data by means of a concept hierarchy, and then extends it to a distance hierarchy in order to represent and calculate distances between the categorical data. However, both techniques suffer from theoretical and practical limitations.

Motivated by this open issue, the paper puts forward a graph-theoretical approach that processes hierarchical data into a numerical representation, and thus renders them amenable for clustering using SOM. To elaborate, based on graph theory, the paper encodes a set of hierarchical data in the form of a rooted and ordered tree. The root vertex represents the complete set of the hierarchical data, and each vertex represents a sub-set of its "parent" vertex. An edge between a pair of vertices is assigned a weight, which can be any positive real number, representing the distance between the two vertices. Thus, the distance between a pair of vertices, $v_i$ and $v_j$, is the sum of the weighted-edges that exist in the path from $v_i$ to $v_j$. The paper uses a level-order traversal algorithm to calculate the distances between each vertex and all other vertices. This process yields a symmetric distance matrix $D = (d_{ij})_{nn}$, where $n$ is the number of vertices, and $d_{ij}$ the distance between $v_i$ and $v_j$.

In the present case, the paper encodes the animals that are contained in the zoo-dataset [9] in the form of a rooted and ordered tree, and calculates the distances between all pairs of animals by using a level-order traversal of the tree, as shown in Fig. 1. The symmetric distance matrix $D = (d_{ij})_{nn}$ thus derived forms the numerical representation of the zoo-dataset, where $n = 98$ reflecting the number of animals, and $d_{ij}$ denotes the distance between a pair of animals. The distance metric $d_{ij}$ satisfies the conditions of a metric space, as follows [10, p.65]: (i) $d_{ij} \geq 0$, (ii) $d_{ij} = 0$ if and only if $i = j$, (iii) $d_{ij} = d_{ji}$, and (iv) $d_{iz} \leq d_{ij} + d_{jz}$. Each row in $D$ represents an animal, and becomes an input vector $- x_j \in \mathbb{R}^{98}$, $j = 1, 2, \ldots 98$ – to SOM.[1]

The paper trains two SOMs, batch and sequence, for each of the two representations of the zoo-dataset, original binary encoding and paper's graph-theoretical approach. For each of the four combinations, the paper selects one hundred samples by using bootstrap; and for each of the 400 bootstrapped samples, it trains a SOM with a Gaussian neighborhood and an 8 x 5 hexagonal lattice. The paper evaluates the quality of each SOM in terms of: (i) the entropy of clustering, (ii) quantization error, (iii) topographic error, and (iv) the Davies-Bouldin index. Based on these quality measures, the paper uses the Wilcoxon rank-sum test at the one-tailed 5% significance level to assess whether the graph-theoretical

---

[1] The distance matrix $D$ is symmetric, and hence the number of observations (i.e. animals) is equal to the number of dimensions (i.e. 98), and selecting either rows or columns as input vectors to SOM would yield the same result.

approach can yield significantly better SOM than binary encoding does. Further, the paper compares the algorithmic complexity of the graph-theoretical approach with that of GSOM.

The results suggest that the graph-theoretical approach enjoys a lower algorithmic complexity than Generalizing SOM does, and can yield SOM having significantly higher cluster validity than binary encoding does.

The paper's novelty and contribution lie in the formulation of the graph-theoretical approach, and its application as a data-preprocessing step that can extend SOM to the domain of hierarchical data.

The paper proceeds as follows. Section 2 describes briefly the SOM algorithm, binary encoding, and Generalizing SOM. Section 3 formulates the graph-theoretical approach. Section 4 outlines the design of experiments, and section 5 presents and discusses the results. Section 6 presents the conclusions.

## 2   Background and Related Work

### 2.1   The SOM Algorithm

In the context of this study, the SOM algorithm performs a non-linear projection of the probability density function of the 98-dimensional input space to an 8 x 5 2-dimensional hexagonal lattice. A neuron $i$, $i = 1, 2, \ldots 40$, is represented by XY coordinates on the lattice, and by a codevector, $m_i \in \mathbb{R}^{98}$, in the input space. The formation of a SOM involves three processes [11, p.447]: (i) competition, (ii) co-operation, and (iii) adaptation. First, each input vector, $x \in \mathbb{R}^{98}$, is compared with all codevectors, $m_i \in \mathbb{R}^{98}$, and the best match in terms of the smallest Euclidean distance, $\| x - m_i \|$, is mapped onto neuron $i$, which is termed the best-matching unit (BMU):

$$BMU = \operatorname*{argmin}_{i} \{\| x - m_i \|\} \ . \tag{1}$$

In the co-operation process, the BMU locates the center of the neighborhood kernel $h_{ci}$:

$$h_{ci} = a\,(t) \cdot \exp\left[-\frac{\| r_c - r_i \|^2}{2\sigma^2\,(t)}\right] \ . \tag{2}$$

where $r_c$, $r_i \in \mathbb{R}^2$ are the radius of BMU and node $i$ respectively, $t$ denotes discrete time, $a\,(t)$ is a learning rate, and $\sigma\,(t)$ defines the width of the kernel; $a(t)$ and $\sigma(t)$ are monotonically decreasing functions of time [3, p.5].

In the adaptive process, the sequence-training SOM updates the BMU codevector as follows:

$$m_i(t + 1) = m_i(t) + h_{ci}(t)\,[x(t) - m_i(t)] \ . \tag{3}$$

The batch-training SOM estimates the BMU according to (1), but updates the BMU codevector as [12, p.9]:

$$m_i(t + 1) = \frac{\sum_{j=1}^{n} h_{ci}(t)x_j}{\sum_{j=1}^{n} h_{ci}(t)} \ . \tag{4}$$

To carry out the experiments, the paper uses both sequence-training (3) and batch-training (4) SOM.

## 2.2  Binary Encoding and Generalizing SOM

Binary encoding converts a categorical variable into a numerical representation consisting of values in the range $\{1,0\}$, where 1 and 0 denote the presence and absence of an attribute respectively. The binary encoding of each categorical datum is then treated as a numerical attribute for SOM-based clustering.

To overcome the limitations associated with binary encoding, Hsu [8] introduced Generalizing SOM (GSOM). Briefly, a domain expert extends a concept hierarchy, which describes a data domain, to a distance hierarchy by associating a weight for each link on the former. The weight represents the distance between the root and a node of a distance hierarchy. For example, a point $X$ in distance hierarchy $dh(X)$ is described by $X = (N_X, d_X)$, where $N_X$ is a leaf node and $d_X$ is the distance from the root to point $X$. The distance between points $X$ and $Y$ is defined as follows:

$$|X - Y| = d_X + d_Y - 2d_{LCP(X,Y)} . \qquad (5)$$

where $d_{LCP(X,Y)}$ is the distance between the root and the least common point of $X$ and $Y$.

## 3  The Graph-Theoretical Approach

### 3.1  Preliminaries

A comprehensive review of graph theory lies beyond the scope of this paper; a textbook account on this subject can be found in [10]. For the purposes of this paper, it suffices to define a tree as a special type of graph, $G = (V, E, w)$, that satisfies at least two of the following three necessary and sufficient properties: (i) $G$ is acyclic, (ii) $G$ is connected, and (iii) $|E| = |V| - 1$; any two of these properties imply the third [10, p.8]. Let $T = (V, E, w)$ be a tree that is: (i) rooted, with $v_0$ the root vertex, and (ii) ordered, which means that there is a

**Table 1.** Notations and definitions

| | |
|---|---|
| $G = (V, E, w)$ | A graph |
| $V = \{v_1, v_2, \dots v_n\}$ | Set of vertices |
| $E = \{e_1, e_2, \dots e_m\}$ | Set of edges |
| $w : E \to \mathbb{R}^+$ | Function assigning a positive real number to an edge |
| $|V|$ | Degree of graph, cardinality of $V$ |
| $|E|$ | Order of graph, cardinality of $E$ |
| $e = \{v_i, v_j\}$ | Edge connecting vertices $v_i$ and $v_j$ |
| $d_{ij} = w(e)$ | Distance between $v_i$ and $v_j$ |
| $D = (d_{ij})_{nn}$ | Distance matrix |

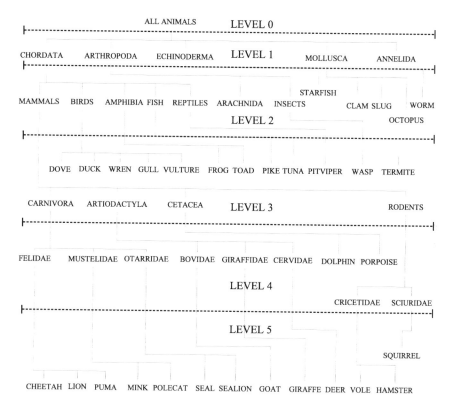

**Fig. 1.** Extract from the graph-theoretical representation of the zoo-dataset

linear ordering of its vertices such that for each edge $e = \{v_i, v_j\}$ then $v_i < v_j$. It can be easily deduced that in tree $T$: (i) all vertices excluding $v_0$ have at most one "parent" vertex, (ii) at least one vertex has no "child" vertices, and (iii) there is a unique path between any two vertices. A tree can be traversed in a level-order way; such a traversal starts from the root vertex, $v_0$, and proceeds from left-to-right to visit each vertex at distance $d$ from $v_0$ before it visits any vertex at distance $d + 1$, as shown in Fig. 1.

## 3.2 Description

The graph-theoretical approach is motivated by the observation that hierarchical variables have a set of states that can be ranked in a meaningful order. For example, consider the variable "size" having five states: {very big, big, medium, small, very small}. It is obvious that {very big} matches {big} more closely than it matches {very small}. However, this piece of information is lost if binary encoding is used, because such an encoding produces a dichotomous output: a state either matches another state or does not.

The graph-theoretical approach operates in three phases. First, it encodes a set of hierarchical data in the form of a rooted and ordered tree. The root vertex represents the complete set of hierarchical data, and all other vertices are ordered in such a way that each vertex represents a sub-set of its "parent" vertex. The edges indicate the covering relation between the vertices. For example, consider a finite order set $P$; $x, y \in P$; $T = (V, E, w)$; and $v_x, v_y \in V$ correspond to $x$ and $y$ respectively. If $x$ is covered by $y$ (i.e. $x \prec y$), then $v_x$ is a "child" vertex of $v_y$. Each edge is assigned a weight, which can be any positive real number (i.e. $w : E \rightarrow \mathbb{R}^+$).

Second, the graph-theoretical approach traverses the tree in a level-order manner in order to calculate the distances between the root vertex and all other vertices. The distance between the root vertex $v_o$ and a vertex $v_i$ is the sum of the weighted-edges that exist in the unique path between $v_o$ and $v_i$. This calculation has an algorithmic complexity of $O(|V|)$. To calculate the distances for all pairs of vertices, the graph-theoretical approach designates each vertex as the root vertex and repeats the level-order traversal. Thus, the all-pairs distances can be obtained in $O(|V|^2)$. This process yields a symmetric distance matrix $D = (d_{ij})_{nn}$, where $d_{ij}$ denotes the distance between vertex $v_i$ and vertex $v_j$, $d_{ij} > 0$ for all $i \neq j$, $d_{ij} = 0$ if and only if $i = j$, $d_{ij} = d_{ji}$, and $d_{iz} \leq d_{ij} + d_{jz}$.

Finally, the distance matrix $D$ constitutes the numerical representation of the set of hierarchical data and each of its rows becomes an input vector to SOM.

## 4   Data and Experiments

The design of experiments consists of six steps. First, the zoo-dataset [9] contains 101 animals that are described by one numerical attribute and 15 binary attributes, and classified into seven groups. The paper eliminates the instances "girl" and "vampire" for obvious but unrelated reasons, and one instance of "frog", because it appears twice.

Second, to apply the graph-theoretical approach to the zoo-dataset, the paper uses none of the original attributes. Instead, it uses a "natural" taxonomy that classifies animals based on their "phylum", "class", and "family". This taxonomy can be expressed as a tree (Fig. 1), where the root vertex stands for the complete set of animals. For the experiments, the weight for each edge is set to 1 (i.e. $w : E \rightarrow 1$), though it can be any positive real number and different for each edge. The paper calculates the distances of all pairs of vertices by using a level-order traversal of the tree, and thus derives a distance matrix that makes up the numerical representation of the zoo-dataset.

Third, for each representation of the zoo-dataset, original binary encoding and the paper's graph-theoretical approach, the paper uses bootstrap to draw one hundred random samples with replacement. Fourth, for each bootstrapped sample, the paper trains two SOMs, batch and sequence, with a Guassian neighborhood and an 8 x 5 hexagonal lattice. Fifth, the paper evaluates each SOM in terms of four quality measures: (i) the entropy of clustering, (ii) quantization error, (iii) topographic error, and (iv) the Davies-Bouldin index. Sixth, based on the quality measures, the paper uses the Wilcoxon rank-sum test at the one-tailed 5% significance level to assess whether the graph-theoretical approach can yield significantly better SOMs than binary encoding does.

**Table 2.** Wilcoxon rank-sum test

| SOM-Training | H(Z) | QE | TE | DBI |
|---|---|---|---|---|
| Batch | A<B | A<B | N.S | A<B |
| Sequence | A<B | A<B | N.S | A<B |

Further, the paper compares the algorithmic complexity of the proposed graph-theoretical approach with that of Generalizing SOM [8]. An experimental comparison was not possible, because GSOM was not available.[2]

### 4.1 Quality Measures

The quantization error, QE, and topographic error, TE, have been extensively reviewed in the literature pertinent to SOM. Thus, this section concentrates on two cluster validity indices: (i) the Davies-Bouldin index, and (ii) the entropy of clustering.

The Davies-Bouldin index [13], DBI, is defined as:

$$DBI = \frac{1}{C} \sum_{i=1}^{C} \max_{i \neq j} \left\{ \frac{\Delta(C_i) + \Delta(C_j)}{\delta(C_i, C_j)} \right\} .$$  (6)

where $C$ is the number of clusters produced by SOM, $\delta(C_i, C_j)$, and $\Delta(C_i)$ and $\Delta(C_j)$ the intercluster and intracluster distances respectively.

Following [14], the entropy of clustering Z, H(Z), can be defined as:

$$H(Z) = - \sum_{j=1}^{C} \frac{m_j}{m} \sum_{i=1}^{K} \frac{m_{ij}}{m_j} log_2 \frac{m_{ij}}{m_j} .$$  (7)

where C is the number of clusters produced by SOM, $K = 7$, the number of groups of animals in the zoo-dataset, $m_{ij}$ is the number of animals in group $i$ that are clustered by SOM in cluster $j$, $m_j$ is the size of cluster $j$, and $m$ is the size of all clusters.

## 5   Results and Discussion

The results (Table 2) suggest that the graph-theoretical approach yields SOMs having statistically significant lower entropy of clustering, quantization error, and Davies-Bouldin index than binary encoding does. In contrast, the difference in topographic error is not significant. Further, the results are invariant to the two SOM-training algorithms, batch and sequence.

Referring to Table 2, A and B stand for the graph-theoretical approach and binary encoding respectively, $A < B$ denotes that the difference between the

---

[2] Personal correspondence with the author of Generalizing SOM.

two approaches is statistically significant at the one-tailed 5% significance level, whereas N.S implies that a significant difference does not exist.

To compare the algorithmic complexity of the graph-theoretical approach with that of GSOM [8], the paper assumes that GSOM is applied to the zoo-dataset, and that GSOM uses this paper's tree (Fig. 1) as its distance hierarchy. As discussed in Sect. 2.2, GSOM entails the following three tasks: (i) calculate distances from the root to all nodes, a level-order traversal of the tree has $O(|V|)$ complexity; (ii) find the all-pairs least common point (LCP), the current fastest algorithm has $O\left(|V|^{2.575}\right)$ complexity [15]; and (iii) calculate distances from the root to all LCPs, this takes $O(l)$, where $l$ is the number of LCPs.

Therefore, the algorithmic complexity of GSOM is $O\left(|V|^{2.575}\right)$, and hence higher than the quadratic complexity, $O\left(|V|^{2}\right)$, of the graph-theoretical approach.

## 5.1  Critique

The proposed graph-theoretical approach is not impervious to criticism. Like binary encoding, it increases the dimensionality of the input space in direct proportion to the number of states a hierarchical variable has. In turn, the dimensionality of the search space increases exponentially with the dimensionality of the input space, a phenomenon aptly named "the curse of dimensionality" [16, p.160]. Further, it assumes that the hierarchical data are static, and hence a deterministic approach is sufficient. To deal with this limitation, future research may explore a probabilistic variant of the graph-theoretical approach.

## 6  Conclusions

The paper's novelty and contribution lie in the development and application of a data-preprocessing step that is based on graph theory and can extend SOM to the domain of hierarchical data. The results suggest that the proposed graph-theoretical approach has lower algorithmic complexity than Generalizing SOM, and can yield SOM having significantly higher cluster validity than binary encoding does. Further, the graph-theoretical approach is not confined only to SOM, but instead it can be used by any algorithm (e.g. k-means) to process hierarchical data into a numerical representation. Future research may consider a probabilistic variant of the graph-theoretical approach as well as its application in the area of hierarchical clustering. Notwithstanding its limitations, the paper presents the first attempt that uses graph theory to process hierarchical data into a numerical representation for SOM-based clustering.

**Acknowledgments and Dedication.** I am much indebted to Mr. Tom Linström, Prof. Anders Tallberg, Dr. Andriy Andreev, and Dr. Sofronis K. Clerides for their insightful comments and suggestions. This paper is dedicated to my first mentor, the late Mr. Marios Christou.

# References

1. Kohonen, T.: Self-Organizing Maps, 2nd edn. Springer Series in Information Sciences, vol. 30. Springer, Heidelberg (1997)
2. Vesanto, J.: Data Exploration Process Based on the Self-Organizing Map. Doctoral dissertation, Helsinki University of Technology, Espoo, Finland (May 2002)
3. Kohonen, T., Hynninen, J., Kangas, J., Laaksonen, J.: Som-pak: The self-organizing map program package. Technical Report A31, Helsinki University of Technology, Laboratory of Computer and Information Science, Espoo, Finland (1996)
4. Kohonen, T., Somervuo, P.: Self-organizing maps of symbol strings. Neurocomputing 21(1-3), 19–30 (1998)
5. Kohonen, T., Somervuo, P.: Self-organizing maps of symbol strings with application to speech recognition. In: Proceedings of the First International Workshop on Self-Organizing Maps (WSOM 1997), pp. 2–7 (1997)
6. Kohonen, T., Somervuo, P.: How to make large self-organizing maps for nonvectorial data. Neural Networks 15(8-9), 945–952 (2002)
7. Somervuo, P.J.: Online algorithm for the self-organizing map of symbol strings. Neural Networks 17(8-9), 1231–1239 (2004)
8. Hsu, C.C.: Generalizing self-organizing map for categorical data. IEEE Transactions on Neural Networks 17(2), 294–304 (2006)
9. Asuncion, A., Newman, D.: UCI Machine Learning Repository. School of Information and Computer Sciences, University of California, Irvine (2007), http://archive.ics.uci.edu/ml/datasets/Zoo
10. Jungnickel, D.: Graphs, Networks and Algorithms. Algorithms and Computation in Mathematics, vol. 5. Springer, Berlin (English edition, 2002)
11. Haykin, S.: Neural Networks. A Comprehensive Foundation, 2nd edn. Prentice Hall International, Upper Saddle River (1999)
12. Vesanto, J., Himberg, J., Alhoniemi, E., Parhankangas, J.: Som toolbox for matlab 5. Technical Report A57, SOM Toolbox Team, Helsinki University of Technology, Espoo, Finland (2000)
13. Davies, D., Bouldin, D.: A cluster separation measure. IEEE Transactions on Pattern Analysis and Machine Intelligence 1(2), 224–227 (1979)
14. Shannon, C.E.: A mathematical theory of communication. The Bell System Technical Journal 27, 379–423, 623–656 (1948)
15. Czumaj, A., Kowaluk, M., Lingas, A.: Faster algorithms for finding lowest common ancestors in directed acyclic graphs. Theoretical Computer Science 380, 37–46 (2007)
16. Maimon, O., Rokash, L. (eds.): The Data Mining and Knowledge Discovery Handbook, 1st edn. Springer, New York (2005)

# Time Series Clustering for Anomaly Detection Using Competitive Neural Networks

Guilherme A. Barreto[1] and Leonardo Aguayo[2]

[1] Department of Teleinformatics Engineering, Federal University of Ceará
Av. Mister Hull, S/N, Center of Technology, Campus of Pici
CP 6005, CEP 60455-970, Fortaleza, Ceará, Brazil
`guilherme@deti.ufc.br`
[2] Nokia Institute of Technology (INdT)
SCS Bl. 1, Camargo Corrêa Bld., 6th floor, CEP 70397-900, Brasília, Brazil
`leonardo.aguayo@indt.org.br`

**Abstract.** In this paper we evaluate competitive learning algorithms in the task of identifying anomalous patterns in time series data. The methodology consists in computing decision thresholds from the distribution of quantization errors produced by normal training data. These thresholds are then used for classifying incoming data samples as normal/abnormal. For this purpose, we carry out performance comparisons among five competitive neural networks (SOM, Kangas' Model, TKM, RSOM and Fuzzy ART) on simulated and real-world time series data.

## 1   Introduction

In recent years, it has been observed an increasing number of applications of the Self-Organizing Map (SOM) to anomaly detection tasks [1,2,3,4], most of them dealing with static data only. However, several real-world applications provide data in a time-ordered fashion, usually in the form of successive measurements of several variables of interest, giving rise to time series data. In industry, for example, many process monitoring procedures involve the measurement of various sensor readings continuously in time to track the state of the system [5,6].

Anomaly detection in time series is particularly challenging due to the usual presence of noise, inserted by the measurement device, as well as of deterministic features - such as trend and seasonality - that can mask the character of novelty that may be present in data. Inherent non-stationary processes, such as regime-switching time series, also impose additional limitations on time series modeling. Furthermore, time-critical applications, such as fault detection and surveillance, require on-line anomaly detection.

Despite the recent interest in unsupervised learning for time series analysis [7], few clustering-based algorithms for anomaly detection are currently available in the literature. This assertion is even stronger if we consider the use of SOM algorithm as a clustering tool for anomaly detection systems. Most of the SOM-based approaches usually converts the time series into a non-temporal representation (e.g. spectral features computed through Fourier transform) and use it as an

J.C. Príncipe and R. Miikkulainen (Eds.): WSOM 2009, LNCS 5629, pp. 28–36, 2009.

input to the standard SOM architecture [8]. Another common approach is to use fixed-length tapped delay lines at the input of the SOM, again converting the time series into a spatial representation.

Since the early 1990's several temporal variants of the SOM algorithm have been proposed to deal with time series data. However, to the best of our knowledge, such temporal SOMs have never been used for anomaly/novelty detection purposes. Thus, this paper aims at answering two questions: (1) Do temporal competitive neural networks perform better than static ones in detecting anomalies in time series data? (2) Do the types of memory used by the temporal SOMs influence their performances in this task? For this purpose, we present temporal variants of the standard SOM, such as the Kangas' model [9], TKM-Temporal Kohonen Map [10] and RSOM-Recursive SOM [11]. Then, the Fuzzy ART network [12], which is a static competitive model with an inherent mechanism of novelty detection, is described. All these algorithms were trained on-line and computer simulations carried out in order to compare their performances.

The remainder of the paper is divided as follows. In Section 2 we describe the self-organizing algorithms used in this work to perform anomaly/novelty detection in time series. In this section, we also present in detail the decision-support methodology used to run the simulations. In Section 4 the numerical results and comments on the performance of all the simulated algorithms are reported. Finally, the conclusions are presented in Section 5.

## 2     Time Series Clustering for Anomaly Detection

In this section we describe competitive learning algorithms adapted to perform anomaly detection in time series. It is assumed that the algorithms are trained on-line as the data samples are being collected. Thus, at time step $t$, input vectors are built as fixed-length window as

$$\mathbf{x}^{+}(t) = [x(t)\ \ x(t-1)\ \ \cdots\ \ x(t-p+1)]^{T}, \tag{1}$$

where $p \geq 1$ is the memory-depth parameter. Weight updating is allowed for a fixed number of steps, $T_{max}$. The first four algorithms to be described are based on the SOM algorithm, while the third one belongs to the family of ART (Adaptive Resonance Theory) architectures. Once the networks are trained, decision thresholds are computed based on the quantization errors for the SOM-based methods. ART-based models have an intrinsic novelty-detection mechanism, which can also be used for anomaly detection purposes.

### 2.1     Static Competitive Neural Networks

**Standard SOM:** SOM training is carried out using the set of vectors $\{\mathbf{x}^{+}(t)\}_{t=p}^{T_{max}}$ sequentially presented as inputs to the network. At time step $t$, the winning neuron, $i^{*}(t)$, is given by

$$i^{*}(t) = \arg\min_{\forall i} \|\mathbf{x}^{+}(t) - \mathbf{w}_{i}(t)\|, \quad i = 1, \ldots, Q, \tag{2}$$

where $\mathbf{w}_i \in \mathbb{R}^p$ is the prototype vector of the $i$-th neuron and $Q$ is the number of neurons. The weight vectors are updated by the following learning rule:

$$\mathbf{w}_i(t+1) = \mathbf{w}_i(t) + \eta(t)h(i^*, i; t)[\mathbf{x}^+(t) - \mathbf{w}_i(t)], \tag{3}$$

where $h(i^*, i; t)$ is a gaussian function which control the degree of change imposed to the weight vectors of those neurons in the neighborhood of the winning neuron:

$$h(i^*, i; t) = \exp\left(-\frac{\|\mathbf{r}_i(t) - \mathbf{r}_{i^*}(t)\|^2}{2\gamma^2(t)}\right), \tag{4}$$

where $\mathbf{r}_i \in \mathbb{R}^2$ and $\mathbf{r}_{i^*} \in \mathbb{R}^2$ denote the coordinates of neurons $i$ and $i^*$ in the output array, while $\gamma(t)$ defines the radius of the neighborhood function at iteration $t$. The learning rate $(0 < \eta(t) < 1)$ and the neighborhood width $(\gamma(t) > 0)$ must decay in time to guarantee convergence of the network. In this paper we use $\eta(t) = \eta_0 (\eta_T/\eta_0)^{-(t/T_{max})}$, where $\eta_0$ and $\eta_T$ are, respectively, the initial and final values of $\eta$, assuming $T_{max}$ training steps. We use the same annealing method for the parameter $\gamma(t)$.

**The Fuzzy ART Algorithm:** We also evaluate the performance of the Fuzzy ART algorithm [12] on anomaly detection in time series, due to its simplicity of implementation and low computational cost. The input vector $\mathbf{x}^+(t)$ is presented to a competitive layer of $Q$ neurons. The winning neuron $i^*$ is selected if its *choice function* $T_{i^*}$ is the highest one among all neurons:

$$i^*(t) = \arg\max_{\forall i} \{T_i(t)\}, \tag{5}$$

where the choice function $T_i$ is computed as follows:

$$T_i(t) = \frac{|\mathbf{x}^+(t) \wedge \mathbf{w}_i(t)|}{\varepsilon + |\mathbf{w}_i(t)|}, \tag{6}$$

where $0 < \varepsilon \ll 1$ is a very small constant, and $|\mathbf{u}|$ denotes the $L_1$-norm of the vector $\mathbf{u}$. The symbol $\wedge$ denotes the component-wise minimum operator. The next step involves a test for *resonance*. If $|\mathbf{x}^+(t) \wedge \mathbf{w}_{i^*}(t)| \geq \rho|\mathbf{x}^+(t)|$, then the weights of the winning neuron $i^*(t)$ are updated as follows:

$$\mathbf{w}_{i^*}(t+1) = \eta\left(\mathbf{x}^+(t) \wedge \mathbf{w}_{i^*}(t)\right) + (1 - \eta)\mathbf{w}_{i^*}(t) \tag{7}$$

where the parameters $0 < \rho < 1$ and $0 < \eta < 1$ are the vigilance parameter and the learning rate, respectively.

If the resonance test for the current winning neuron $i^*(t)$ fails, then another neuron is selected as the winner, usually the one with the second highest value for $T_i(t)$. If this neuron also fails, then the one with the third highest value for $T_i(t)$ is selected, and so on until one of the selected winning neurons $i^*(t)$ matches the resonance test. If none of the existing prototype vectors *resonates* with the current input vector, then the input vector is declared *novel* and turned into a new prototype vector. The parameter $\rho$ controls the sensitivity of the Fuzzy ART algorithm to new input vectors. If $\rho \to 1$, more prototypes are created in the competitive layer. If $\rho \to 0$, the number of prototypes decreases.

## 2.2   Temporal Competitive Neural Networks

**Kangas' Model:** Kangas' model [9] is one of the simplest temporal SOM algorithms available. The underlying idea of this model is to perform a temporal smoothing on the input vector $\mathbf{x}^+(t)$:

$$\overline{\mathbf{x}}(t) = (1 - \lambda)\overline{\mathbf{x}}(t - 1) + \lambda\mathbf{x}^+(t), \tag{8}$$

where $0 < \lambda < 1$ is a memory decay parameter. The filtered vector $\overline{\mathbf{x}}(t)$ is then presented to the standard SOM algorithm, which follows its usual training procedure as described above.

**Temporal Kohonen Map (TKM):** The TKM model introduces a short-term memory mechanism for the activation $a_i(t)$ of each neuron in the map:

$$a_i(t) = \lambda a_i(t - 1) - \frac{1}{2}\|\mathbf{x}(t) - \mathbf{w}_i(t)\|^2, \tag{9}$$

with $0 < \lambda < 1$ being equivalent to the decay parameter defined for Kangas' model. The winner $i^*(t)$ is the one with the highest activation:

$$i^*(t) = \arg\max_{\forall i}\{a_i(t)\}. \tag{10}$$

The weight vectors are through Eq. (3). We set $a_i(0) = 0$, $i = 1, 2, \ldots, Q$.

**Recurrent SOM (RSOM):** In this variant, a temporal smoothing mechanism acts over the difference vector $\mathbf{d}(t) = \mathbf{x}^+(t) - \mathbf{w}_i(t)$:

$$\mathbf{y}_i(t) = (1 - \lambda)\mathbf{y}_i(t - 1) + \lambda\mathbf{d}(t). \tag{11}$$

The winning neuron is then redefined as

$$i^*(t) = \arg\min_{\forall i}\{\mathbf{y}_i(t)\}, \tag{12}$$

and the learning rule in Eq. (3) is rewritten as

$$\mathbf{w}_i(t + 1) = \mathbf{w}_i(t) + \eta(t)h(i^*, i; t)\mathbf{y}_i(t), \tag{13}$$

where the memory is now taken into account when updating the weights of the winning neuron. We set $\mathbf{y}_i(0) = \mathbf{0}$, $\forall i$.

## 3   Detection Methodology

Unlike the Fuzzy ART algorithm, the SOM-based methods previously described do not have an intrinsic mechanism to detect anomalous data. However, it has become common practice to use the quantization error

$$e_q(i^*, t) = \|\mathbf{x}^+(t) - \mathbf{w}_{i^*}(t)\|, \tag{14}$$

as a measure of the degree of proximity of $\mathbf{x}^+(t)$ to a statistical representation of normal behavior encoded in the weight vectors of the SOM variants. Once the network has been trained, we present the training data vectors once again to this network. From the resulting quantization errors $\{e_q(i^*, t)\}_{t=1}^{T_{max}}$, computed for all training vectors, we compute decision thresholds for the anomaly detection tests. For a successfully trained network the sample distribution of these quantization errors should reflect the 'normal' (i.e. not anomalous) behavior of the input variable whose time series model is being constructed.

In order to compute decision thresholds we apply the method proposed in [2]. For a given significance level $\alpha$ (e.g. $\alpha$=0.05 or 0.01), we are interested in an interval within which we can certainly find a percentage of $100(1-\alpha)\%$ of normal values of the quantization error. We then proceed with the computation of the lower and upper boundaries of the detection interval as follows:

- **Lower Limit** $(\tau^-)$: This is the $100\frac{\alpha}{2}$th percentile[1] of the distribution of quantization errors associated with the training data vectors.
- **Upper Limit** $(\tau^+)$: This is the $100(1 - \frac{\alpha}{2})$th percentile of the distribution of quantization errors associated with the training vectors.

Once the decision interval $[\tau^-, \tau^+]$ has been computed, any anomalous behavior of the time series can be detected on-line by means of the simple rule:

$$\begin{aligned} \text{IF} \quad & e_q(i^*, t) \in [\tau^-, \tau^+] \\ \text{THEN} \quad & \mathbf{x}^+(t) \text{ is NORMAL} \\ \text{ELSE} \quad & \mathbf{x}^+(t) \text{ is ABNORMAL} \end{aligned} \qquad (15)$$

## 4   Simulations

The aforementioned competitive neural neworks are evaluated as anomaly detectors using simulated and real-world time series data. The simulated time series is built from four different dynamic systems, three of them being realizations of chaotic series. The first one is composed by the $x$ component of Lorenz equations

$$\dot{x} = \sigma_L(y - x), \ \dot{y} = x(\alpha_L - z) - y, \ \dot{z} = xy - \epsilon_L z, \qquad (16)$$

which exhibits chaotic dynamics for $\sigma_L = 10$, $\alpha_L = 28$ and $\epsilon_L = 8/3$. The second and third realizations are generated from the Mackey-Glass dynamical system, for two distinct values of the delay $\tau$:

$$\dot{x} = Rx(t) + P\frac{x(t - \tau)}{(1 + x(t - \tau)^{10})}, \qquad (17)$$

with $P = 0.2$, $R = -0.1$ and $\tau = 17$ (MG17) or $\tau = 35$ (MG35). The fourth realization is generated from a second order autoregressive process (AR2):

$$x(t + 1) = 1.9x(t - 1) - 0.99x(t - 2) + a(t), \qquad (18)$$

with $a(t)$ is an additive gaussian white noise process with zero mean and variance $\sigma_a^2 = 10^{-3}$. Figure 1 depicts 300 samples of each signal.

---

[1] The percentile of a distribution of values is a number $N_\alpha$ such that a percentage $100(1 - \alpha)$ of the sample values are less than or equal to $N_\alpha$.

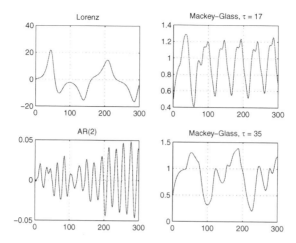

**Fig. 1.** Samples of simulated time series used in the simulations

The first experiment is designed to perform on-line detection of anomalous signal samples, after training the networks with a sequence representing normal (i.e. expected) behavior of the system being monitored. Each realization is presented one after the other to emulate a longer time series with different dynamic regimes. The role of normal behavior is assigned to the Lorenz series, leaving the Mackey-Glass and the AR(2) realizations as representing anomalous (i.e. unknown) data. The three different testing sequences are presented sequentially in this order: MG17, MG35 and AR2.

As a typical result, Figure 2 shows the quantization errors produced by the Kangas' model, for the first $T_{max} = 1000$ samples of the training set generated by the Lorenz equations. As expected, the model produced lower quantization errors for known data (first 1000 samples), while for the remaining 3000 samples the resulting quantization errors are considerably higher. For this experiment we set $Q = 20$, $p = 10$, $\eta_0 = 0.5$, $\eta_T = 10^{-3}$, $T_{max} = 1000$, $\gamma_0 = 10$, $\gamma_T = 0.1$, $\lambda = 0.8$, $\varepsilon = 10^{-2}$ and $\rho = 0.8$.

The second experiment involves a comprehensive evaluation of the performances of all competitive models on the same time series data through the analysis of the *Receiver Operating Characteristic* (ROC) curves of the classifiers. Let TP and FP denote the true positives and false positive ratios of a classifier, respectively[2]. The coordinates (FP, TP) is a point in ROC space, and can be used to visually identify good and bad binary classifiers. A perfect binary classifier ideally achieve the (0,1) coordinate at ROC space. Now, if we change the percentile $N_\alpha$, the decision interval $[\tau^-, \tau^+]$ in Eq. (15) is modified, and a set of points in ROC space can be derived, allowing to verify the performances of the classifiers under different degrees of tolerance for the quantization error.

---

[2] A true positive is the classification of an incoming vector $\mathbf{x}^+(t)$ as abnormal, when it is truly an anomalous one, and a false positive is the classification of an incoming vector $\mathbf{x}^+(t)$ as abnormal when it is a normal one.

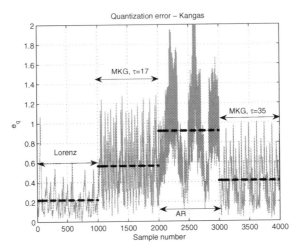

**Fig. 2.** Quantization errors for the Kangas' model at each regime of the time series

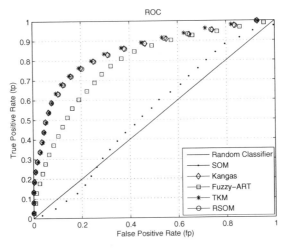

**Fig. 3.** ROC curves for the evaluated competitive neural networks

Figure 3 shows a typical ROC curves for the comparative performance among the networks, obtained with the same parameters of the first experiment.

Analysing the ROC curves, we arrive at the following conclusions: ($i$) the original SOM algorithm performed poorly in detecting anomalous behavior in simulated time series data, with a performance equivalent to a random classifier; ($ii$) all the temporal variants of the SOM achieved very good performances; ($iii$) there were no significant differences among the performances of the the temporal variants of the SOM; and ($iv$) unlike the standard SOM, the Fuzzy-ART network (which is a static model!) performed quite well on the detection task. From the exposed, we can conclude that the temporal variants of the SOM are indeed more

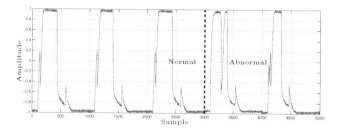

**Fig. 4.** Energizing cycles of a space shuttle solenoid

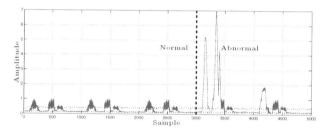

**Fig. 5.** Quantization error for the NASA valve data set

suitable to deal with time series data than static competitive neural networks. Furthermore, since the temporal SOMs had equivalent performances, preference is given to the Kangas's model due to its implementational simplicity.

The last experiment involves testing the Kangas' model in a real-world time series data from the NASA shuttle program. The NASA valve data set [13] consists of solenoid current measurements recorded on Marrotta series MPV-41 valves as they are remotely opened and closed in a laboratory. These small valves are used to actuate larger, hydraulic valves that control the flow of fuel to the space shuttle engines. Sensor readings were recorded using either a shunt resistor or a Hall effect sensor under varying conditions of voltage, temperature, or blockage or forced movement of the poppet to simulate fault conditions.

Each opening/closing cycle of the valve consists of 1000 samples at a rate of 1 ms per sample. The whole time series comprises 5 cycles, with the first three cycles being considered normal ones while the last two are anomalous cycles (see Figure 4). Using the same training parameters of previous experiments, the Kangas' model is trained on-line with the first $T_{max} = 1000$ samples and tested with the remaining 4000 samples. The corresponding quantization error signal is shown at Figure 5, including the thresholds $[\tau^-, \tau^+]$ corresponding to the 5% and 95% percentiles of $\{e_q(i^*, t)\}_{t=1}^{1000}$. As expected, the method performed satisfactorily, since it correctly detected all abnormal valve states.

## 5    Conclusions

This paper described some results on the application of competitive neural networks for detecting novelties in time series data. We compared the performances of five unsupevised competitive learning architectures (SOM, Kangas' Model, TKM, RSOM and Fuzzy ART) on simulated and real-world time series data. The obtained results indicate that temporal variants of the SOM are more suitable to deal with time series data than static competitive neural networks.

**Acknowledgments.** The authors thank CNPq (grant 474843/2008-4) for its financial support.

# References

1. Sarasamma, S.T., Zhu, Q.A.: Min-max hyperellipsoidal clustering for anomaly detection in network security. IEEE Transactions on Systems, Man and Cybernetics B-36(4), 887–901 (2006)
2. Barreto, G.A., Mota, J.C.M., Souza, L.G.M., Frota, R.A., Aguayo, L.: Condition monitoring of 3G cellular networks through competitive neural models. IEEE Transactions on Neural Networks 16(5), 1064–1075 (2005)
3. Lee, H.J., Cho, S.: SOM-based novelty detection using novel data. In: Gallagher, M., Hogan, J.P., Maire, F. (eds.) IDEAL 2005. LNCS, vol. 3578, pp. 359–366. Springer, Heidelberg (2005)
4. Singh, S., Markou, M.: An approach to novelty detection applied to the classification of image regions. IEEE Transactions on Knowledge and Data Engineering 16(4), 1041–1047 (2004)
5. Zorriassatine, F., Al-Habaibeh, A., Parkin, R., Jackson, M., Coy, J.: Novelty detection for practical pattern recognition in condition monitoring of multivariate processes: a case study. International Journal of Advanced Manufacturing Technology 25(9-10), 954–963 (2005)
6. Jamsa-Jounela, S.L., Vermasvuori, M., Enden, P., Haavisto, S.: A process monitoring system based on the kohonen self-organizing maps. Control Engineering Practice 11(1), 83–92 (2003)
7. Hammer, B., Micheli, A., Sperduti, A., Strickert, M.: Recursive self-organizing network models. Neural Networks 17, 1061–1086 (2004)
8. Wong, M., Jack, L., Nandi, A.: Modified self-organising map for automated novelty detection applied to vibration signal monitoring. Mechanical Systems and Signal Processing 20(3), 593–610 (2006)
9. Kangas, J.A., Kohonen, T.K., Laaksonen, J.: Variants of self-organizing maps. IEEE Transactions on Neural Networks 1(1), 93–99 (1990)
10. Chappell, G.J., Taylor, J.G.: The temporal Kohonen map. Neural Networks 6(3), 441–445 (1993)
11. Koskela, T., Varsta, M., Heikkonen, J., Kaski, K.: Time series prediction using recurrent SOM with local linear models. International Journal of Knowledge-based Intelligent Engineering Systems 2(1), 60–68 (1998)
12. Carpenter, G.A., Grossberg, S., Rosen, D.B.: Fuzzy ART: Fast stable learning and categorization of analog patterns by an adaptive resonance system. Neural Networks 4(6), 759–771 (1991)
13. Ferrell, B., Santuro, S.: NASA shuttle valve data (2005),
   http://www.cs.fit.edu/~pkc/nasa/data/

# Fault Prediction in Aircraft Engines Using Self-Organizing Maps

Marie Cottrell[1], Patrice Gaubert[1], Cédric Eloy[2], Damien François[2],
Geoffroy Hallaux[2], Jérôme Lacaille[3], and Michel Verleysen[2]

[1] SAMOS-MATISSE, UMR CNRS CES, Université Paris 1 Panthéon Sorbonne, France
[2] Machine Learning Group, Université catholique de Louvain, Belgium
[3] SNECMA YYE, Villaroche, France
Marie.cottrell@univ-paris1.fr

**Abstract.** Aircraft engines are designed to be used during several tens of years. Their maintenance is a challenging and costly task, for obvious security reasons. The goal is to ensure a proper operation of the engines, in all conditions, with a zero probability of failure, while taking into account aging. The fact that the same engine is sometimes used on several aircrafts has to be taken into account too.

The maintenance can be improved if an efficient procedure for the prediction of failures is implemented. The primary source of information on the health of the engines comes from measurement during flights. Several variables such as the core speed, the oil pressure and quantity, the fan speed, etc. are measured, together with environmental variables such as the outside temperature, altitude, aircraft speed, etc.

In this paper, we describe the design of a procedure aiming at visualizing successive data measured on aircraft engines. The data are multi-dimensional measurements on the engines, which are projected on a self-organizing map in order to allow us to follow the trajectories of these data over time. The trajectories consist in a succession of points on the map, each of them corresponding to the two-dimensional projection of the multi-dimensional vector of engine measurements. Analyzing the trajectories aims at visualizing any deviation from a normal behavior, making it possible to anticipate an operation failure.

However rough engine measurements are inappropriate for such an analysis; they are indeed influenced by external conditions, and may in addition vary between engines. In this work, we first process the data by a General Linear Model (GLM), to eliminate the effect of engines and of measured environmental conditions. The residuals are then used as inputs to a Self-Organizing Map for the easy visualization of trajectories.

**Keywords:** aircraft engine maintenance, fault detection, general linear models, self-organizing maps.

## 1 Introduction

Security issues in the aircrafts are a major concern for obvious reasons. Among the many aspects of security issues, ensuring a proper operation of engines over their lifetime is an important task.

J.C. Príncipe and R. Miikkulainen (Eds.): WSOM 2009, LNCS 5629, pp. 37–44, 2009.
© Springer-Verlag Berlin Heidelberg 2009

Aircraft engines are built with a high level of security norms. They undergo regularly a full maintenance with disassembling, replacement of parts, etc. In addition, between two such maintenances, many parameters are measured on the engines during the flights. These parameters are recorded, and used both at short and long terms for immediate action and alarm generation respectively.

In this work, we are interested in the long-term monitoring of aircraft engines. Measurements on the engines during flights are used to detect any deviation from a "normal" behavior, making it possible to anticipate possible faults. This fault anticipation is aimed to facilitate the maintenance of aircraft engines.

Self-Organizing Maps are here used to provide experts a supplementary tool to visualize easily the evolution of the data measured on the engines. The evolution is characterized by a trajectory on the two-dimensional Self-Organizing Map. Abnormal aging and fault appearance will result in deviation of this trajectory, with respect to normal conditions. The output of this data mining study is therefore a visual tool that can be used by experts, in addition to their traditional tools based on quantitative inspection of some measured variables. Self-Organizing Maps are useful tools for fault detection and prediction in plants and machines (see [1], [2], [3], [4], [5], for example).

Analyzing the rough variables measured on the engines during flights is however not appropriate. Indeed these measurements may vary from one engine to another, and may also vary according to "environmental" conditions (such as the altitude, the outside temperature, the speed of the aircraft, etc.). In this work, we first remove the effects of environmental (measured) variables, and the engine effects, from the rough measurements. The residuals of the regression are then used for further analysis by Self-Organizing maps.

The following of this paper is organized as follows. In Section 2, the data are described and notations are defined. Section 3 presents the methodology: Section 3.1 describes how the effects of engines and of environmental variables are removed by a General Linear Model, and Section 3.2 shows the visual analysis of the GLM residuals by Self-Organizing Maps. Section 4 describes the experimental results, before some conclusions in Section 5.

## 2  Data

Measurements are collected on a set of $I$ engines. On each engine $i$ ($1 \leq i \leq I$), $n_i$ sets of measurements are performed successively. Usually one set is measured during each flight; there is thus no guarantee that the time intervals between two sets of measures are approximately equal. Each set of observations is denoted by $Z_{ij}$, with $1 \leq i \leq I$ and $1 \leq j \leq n_i$.

Each set $Z_{ij}$ contains both variables related to the behavior of the engine, and variables that are related to the environment. Let us denote the $p$ engine variables by $Y_{ij}^1, \ldots, Y_{ij}^p$ and the $q$ environmental variables by $X_{ij}^1, \ldots, X_{ij}^q$. Each set of measurements is thus a vector $Z_{ij}$, where

$$Z_{ij} = (Y_{ij}, X_{ij}) = (Y_{ij}^1, \ldots, Y_{ij}^p, X_{ij}^1, \ldots, X_{ij}^q) . \tag{1}$$

In this study, the variables at disposal are those listed in Table 1.

**Table 1.** Engine and environmental variables

| Engine variables | | Environmental variables | |
|---|---|---|---|
| $Y_{ij}^{1}$ | core speed | $X_{ij}^{1}$ | Mach |
| $Y_{ij}^{2}$ | oil pressure | $X_{ij}^{2}$ | Engine bleed valve 1 |
| $Y_{ij}^{3}$ | HPC discharge stat. pres. | $X_{ij}^{3}$ | Engine bleed valve 2 |
| $Y_{ij}^{4}$ | HPC discharge temp. | $X_{ij}^{4}$ | Engine bleed valve 3 |
| $Y_{ij}^{5}$ | Exhaust gas temp. | $X_{ij}^{5}$ | Engine bleed valve 4 |
| $Y_{ij}^{6}$ | Oil temperature | $X_{ij}^{6}$ | Isolation valve left |
| $Y_{ij}^{7}$ | Fuel flow | $X_{ij}^{7}$ | Altitude |
| | | $X_{ij}^{8}$ | HPT active clearance |
| | | $X_{ij}^{9}$ | LPT active clearance |
| | | $X_{ij}^{10}$ | Total air temperature |
| | | $X_{ij}^{11}$ | Nacelle temperature |
| | | $X_{ij}^{12}$ | ECS Pack 1 flow |
| | | $X_{ij}^{13}$ | ECS Pack 2 flow |

The goal of this study is to visualize the $Y_{ij}$ vectors. The visualization of the successive measurements $j$ for a specific engine $i$ corresponds to a trajectory.

## 3  Methodology

Rough $Y_{ij}$ measurements of the engine variables cannot be used as such for the analysis. Indeed the $Y_{ij}$ strongly depend on

- engine effects, i.e. the fact that the engines may differ, and on
- environmental effects, i.e. the dependence of the engine variables $Y_{ij}$ on the environmental conditions $X_{ij}$.

Both dependences lead to differences in observed variables that have nothing to do with aging or fault anticipation. It is therefore important to remove these effects before further analysis.

In this work, we use a GLM (General Linear Model) [6] to remove these effects, since the independent variables are of two types : categorical (engine effect) and real-valued (environmental variables). The use of GLM implies two hypotheses. First, it is assumed that the effect of the environment is effectively measured in the environmental variables $X_{ij}$; obviously, non-measured effects cannot be removed. Secondly, it is assumed that the relation between the engine variables $Y_{ij}$ and the environmental variables $X_{ij}$ is linear; this last assumption is probably not perfectly correct, but it will be shown in the experimental section that even under this hypothesis, the statistical significance of the $X_{ij}$ effects is high; this justifies a posteriori to remove at least the linear part (first-order approximation) of the relation.

The residuals of the regression of the $Y_{ij}$ variables over the $X_{ij}$.ones and the motor effects are then used for the analysis. A Self-Organizing Map is used to visualize the two-dimensional projection of the residuals corresponding to each vector $Y_{ij}$. Then, the different states $j$ ($1 \leq j \leq n_i$) of a single engine are linked together to form a trajectory.

The next two subsections detail how to perform the GLM regression on the engine variables, and how to use the Self-Organizing Maps on the GLM residuals.

## 3.1 Computation of the Residuals (So-Called Corrected Values)

The computation of the values obtained by removing the effects of the environment variables and of the engine is done by using a General Linear Model, where the explanatory variables are of two kinds: one variable is categorical (the engine number), the others are real-valued variables (the environment variables).

For each engine variable $m = 1, ..., p$, the GLM model can be written as:

$$Y_{ij}^m = \mu^m + \alpha_i^m + \lambda_1^m X_{ij}^1 + ... + \lambda_q^m X_{ij}^q + \varepsilon_{ij}^m, \tag{2}$$

where $i = 1, ..., I$, is the engine number, $j$ is the flight number, $\alpha_i^m$ is the engine effect on the $m$-th variable, $X_{ij}^1, ..., X_{ij}^q$ are the environmental variables, $\lambda_1^m, ..., \lambda_q^m$ are the regression coefficients for the $m$-th variable, and the error term $\varepsilon_{ij}^m$ is centered with variance $\sigma_m^2$. The parameters $\alpha_i^m, \lambda_1^m, ..., \lambda_q^m$, are estimated by the least squares method, and in order to avoid colinearity, we have to add the constraint $\sum_{i=1}^{I} n_i \alpha_i^m = 0$.

Note that it is possible to model the motor effect by a random term $A_i^m$ instead of the fixed effect $\alpha_i^m$ ; $A_i^m$ is also supposed to be centered with variance $\sigma_A^2$. Even if the model is slightly different, the residuals are the same.

Fisher statistics allows us to verify the significance of the models and to confirm the interest of the adjustment of engine variable for the environmental ones and the motor effect.

Let us denote by $R_{ij}^m$ , $m = 1, ..., p$ the residuals (2), equal to the estimated values $\hat{\varepsilon}_{ij}^m$ .

The residuals are the values adjusted for the motor effect and the environment variables.

## 3.2 Self-Organizing Maps on the Residuals

Next we consider a $n$ by $n$ Kohonen map [7] and train it with the $p$-dimensional residuals $R_{ij}^m$ ($m = 1, ..., p$). We use the SOM toolbox for Matlab [8] for the experiments. In that way, each flight $j$ of each engine $i$ is projected on a Kohonen class on the map. We can identify the different locations on the map by looking at the corresponding code-vectors and at their components, and then give a description of the clusters. For each engine $i$, we define the sequence of the class numbers corresponding to the successive flights $j = 1, ..., n_i$. This sequence is the trajectory of engine $i$. In this way we get a visual representation of the successive states of the engines on the Kohonen map. Then we can compare these trajectories by introducing a measure of distance between them.

# 4  Experiments

We consider real data which consist in the observation of $I = 91$ engines. Each engine is measured for a number of flights between 500 and 800. There are 7 engine variables and 13 environment variables, as illustrated in Table 1.

## 4.1  Justification of the Computation of Adjusted Variables

To justify the computation of the residuals (i.e. the values adjusted for engine effect and environment variables), we can for example show the result of a PCA on the raw data and use different colors for 5 different engines. We see (Fig.1, left) that each engine clearly defines a cluster in the projection on the first two principal components. Fig.1, right also shows that the histograms of the engine variables ($Y_{ij}4$ is illustrated) depend on the engine.

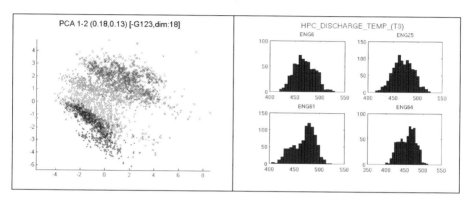

**Fig. 1.** Left: the first two principal components for five engines. The data are the 7-dimensional engine variables. Right; the values of variable $Y^4$ (HPC discharge temperature) for 4 engines.

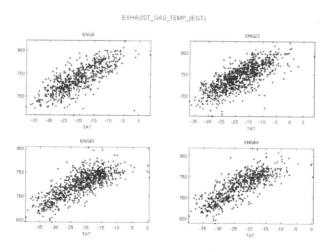

**Fig. 2.** Almost linear dependence between variable $Y^5$ (EGT) and variable $X^{10}$ (Total Air Temperature)

The correlation between variables can be illustrated too. As an example, Figure 2 shows variable $Y^5$ (EGT) as a function of variable $X^{10}$ (Total Air Temperature) in four engines. It is obvious that both variables are strongly dependent.

These few examples clearly show that it is necessary to remove the effects of the engine and of the environmental variables, by computing the residuals in model (2).

## 4.2  Self-Organizing Maps on Adjusted Variables and Trajectories

After the extraction of the residuals $R_{ij}$ as detailed in Section 3.1, the second step consists in training a 20 by 20 Kohonen map on these residuals. Figure 3 shows the map obtained, colored according to each of the 7 engine variables. It is clearly visible that the organization of the map is successful (all variables are smoothly varying on the map).

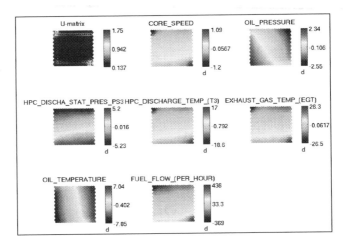

**Fig. 3.** 20x20 self-organizing map on the residuals. The first plot shows the U-matrix, the other ones display the distribution of the 7 engine variables $R^1 - R^7$ over the map.

We can see that variables $R^1$, $R^3$, $R^4$, $R^5$, $R^7$ on one hand, and $R^2$ and $R^6$ on another hand, form high-correlated groups of variables (his property can be verified by computing the correlation matrix).

The 400 classes are then grouped (hierarchical clustering) into 5 super-classes, as shown in Figure 4. Finally, Figure 5 shows the trajectories of the engines. As examples, the trajectories of engines 6, 25 and 88 are illustrated.

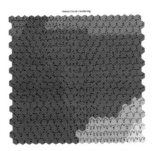

**Fig. 4.** Five super-classes are shown after hierarchical clustering of the 400 classes. The centroids are also shown inside each class.

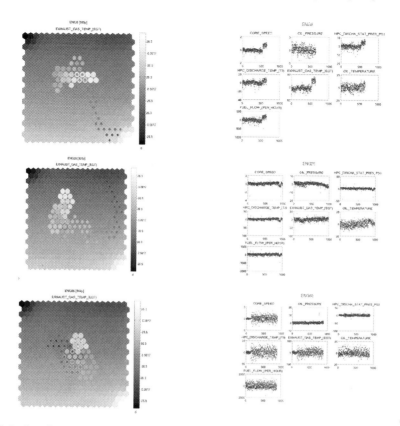

**Fig. 5.** Left, trajectories of engines 6, 25 and 88 on the Kohonen map; the dots color indicates the evolution along the trajectory (from red to blue, through yellow and green). The background shows the level of the EGT variable ($R^5$). Right: the residuals for the same engines.

We observe that the trajectories have different shapes. Looking at the graphs of the 7 adjusted engine variables (Figure 5 right), we conclude that the visual representations on the Kohonen map provide a synthetic representation for the temporal evolution of the engines.

The next step is then to characterize the different shapes of trajectories, to define a suitable distance measure between these trajectories, and to define typical behaviors related to typical faults.

## 5  Conclusions

The proposed method is a useful tool to summarize and represent the temporal evolution of an aircraft engine flight after flight. Further work will consist in defining classes for the trajectories and in associating each class to some specific behavior. Using the maintenance reports which contain the a posteriori measured data related to each engine, it will be possible to identify the classes with possible failures. So the visual examination of such trajectories will help anticipating faults in aircraft engines.

# References

1. Goser, K., Metzen, S., Tryba, V.: Designing of basic Integrated Circuits by Self-organizing Feature Maps, Neuro-Nîmes (1989)
2. Barreto, G.A., Mota, J.C.M., Souza, L.G.M., Frota, R.A., Aguayo, L.: Condition monitoring of 3G cellular networks through competitive neural models. IEEE Transactions on Neural Networks 16(5), 1064–1075 (2005)
3. Sarasamma, S.T., Zhu, Q.A.: Min-max hyperellipsoidal clustering for anomaly detection in network security. IEEE Transactions on Systems, Man and Cybernetics, Part B 36(4), 887–901 (2006)
4. Svensson, M., Byttner, S., Rögnvaldsson, T.: Self-organizing maps for automatic fault detection in a vehicle cooling system. In: Intelligent Systems, 2008. IS 2008. 4th International IEEE Conference, September 6-8, vol. 3, pp. 24-8–24-12 (2008), doi:10.1109/IS.2008.4670481
5. Alhoniemi, E., Simula, O., Vesanto, J.: Process monotoring and modeling using the self-organizing. Integrated Computer Aided Engineering 6(1), 3–14 (1999)
6. Draper, N.R., Smith, H.: Applied Regression Analysis. John Wiley & Sons, New York (1966)
7. Kohonen, T.: Self-Organizing Maps. Springer Series in Information Sciences, vol. 30. Springer, Heidelberg (1995)
8. Laboratory of Computer and Information Science, Helsinky University of Technology, SOM Toolbox for Matlab, http://www.cis.hut.fi/projects/somtoolbox

# Incremental Figure-Ground Segmentation Using Localized Adaptive Metrics in LVQ

Alexander Denecke[1,2], Heiko Wersing[2], Jochen J. Steil[1], and Edgar Körner[2]

[1] Bielefeld University - CoR-Lab, P.O. Box 10 01 31, D-33501 Bielefeld, Germany
adenecke@cor-lab.uni-bielefeld.de
http://www.cor-lab.de
[2] Honda Research Institute Europe, Carl-Legien-Str. 30,
D-63073 Offenbach/Main, Germany

**Abstract.** Vector quantization methods are confronted with a model selection problem, namely the number of prototypical feature representatives to model each class. In this paper we present an incremental learning scheme in the context of figure-ground segmentation. In presence of local adaptive metrics and supervised noisy information we use a parallel evaluation scheme combined with a local utility function to organize a learning vector quantization (LVQ) network with an adaptive number of prototypes and verify the capabilities on a real world figure-ground segmentation task.

## 1  Introduction

The appropriate choice of the number of model neurons is a principle problem in vector quantization networks. In particular incremental learning offers a solution to adjust the amount of resources needed versus classification performance to find a tradeoff between representation quality and the avoidance of over-fitting. Vector quantization methods provide a simple algorithmic yet powerful framework with applications, for example in image processing [1] or life-long learning [2,3]. We investigate such methods in the context of online figure-ground segmentation where homogenous image regions are represented by single feature representatives. Ideally the dimensionality of the network should represent the meaningful entities in the data. As this problem is ill-posed (subjective), several researchers have addressed this problem with heuristics in supervised or unsupervised settings. One main criterion used for unsupervised setups is the distance of the features to their representatives, namely the quantization error. The criterion in Growing Neural Gas [4] (and similar for the Growing Cell structures [2]) aims at a minimization of the quantization error and introduces new prototypes where the quantization error is large, guaranteeing that the introduction of new prototypes reduces this error. Supervised LVQ primarily aims at the minimization of the classification error which offers another source of information. For example Kirstein et al. [3] propose a heuristics to insert new prototypes at the decision boundary using the misclassified data points together with a distance criterion to determine the location for new prototypes.

J.C. Príncipe and R. Miikkulainen (Eds.): WSOM 2009, LNCS 5629, pp. 45–53, 2009.
© Springer-Verlag Berlin Heidelberg 2009

In this contribution we investigate Generalized Learning Vector Quantization (GLVQ [5]) with adaptive metrics and propose a framework for incremental and online figure-ground segmentation which faces two problems. Firstly the local adaptive metrics complicates distance-based criteria to place new prototypes, where we use the confidence of the classification instead. Secondly the method has to cope with noisy supervised information, that is, the labels to adapt the networks are not fully confident. In particular we address the second problem by using a parallel evaluation method on the basis of a local utility function, which does not rely on global error optimization. After a short problem statement, we describe the overall method to allow incremental learning in the presence of non-confident supervised information together with the criteria to introduce and remove prototypes from the network. Finally we evaluate the method on a real world segmentation task and compare to previous results.

## 2   Method

### 2.1   Scenario

Our proposed method addresses the problem of online figure-ground segmentation for object learning and recognition. The application scenario consists of a human presenter showing objects to a pan-tilt stereo camera system, which is controlled by an attention system [6]. Using the concept of peripersonal space, the depth estimation of the region in front of the system is analyzed with a blob-detection within a specified depth interval (50cm-80cm). The most salient object in front of the system is continuously tracked and centered in view by setting the gaze direction. Additionally a square region of interest (ROI) is defined based on a distance estimate of the tracked blob and normalized to a size of $\mathcal{I} \times \mathcal{I}$ pixels, where we use $\mathcal{I} = 144$. Both methods assure an approximate invariance to position and size for the incoming stream of images showing the object in front of cluttered background. A first cue for parts of the scene that belong to the object can be derived from depth estimation which we call *object hypothesis*. Because extracting 3D information from 2D images in general is an ill-posed problem, the resulting hypothesis is characterized by a partially inconsistent overlap with the outline/region of the object. That is, some regions of the background are indicated as foreground and vice versa. Since the learning and recognition can be improved as the quality of the figure-ground segmentation is enhanced, we follow the concept of hypothesis refinement to derive the significant object parts (according to the underlying image features) from this initial guess. In [1] we investigated methods for object segmentation that use prototypical feature representatives to model figure and ground. In particular, we used binarized depth hypotheses as a supervised label for the image features to train a classifier for figure and ground with GLVQ. In our previous setup an empirically predefined number of prototypes were used, while the model selection problem was left open.

## 2.2   Generalized Learning Vector Quantization

From the camera system the following data is available for each frame. A stack of $M = 5$ feature maps $\mathcal{F} := \{F_i^{x,y}|i = 1..M\}$ corresponding to the RGB color and position information of the pixels forms the dataset $\mathcal{D} := \{\boldsymbol{\xi}|\boldsymbol{\xi}^{x,y} = (F_1^{x,y}..F_M^{x,y})^T, 1 \leq x, y \leq \mathcal{I}\}$, where every pixel defines a feature vector. To take advantage from the temporal character of the data the features of $T = 2$ frames are combined to one dataset. Switching to a new frame effectively replaces 50% (or less if $T$ is increased) of the data from one to another adaptation step. Additionally to the data the hypothesis $\mathcal{H}$ is available indicating which pixels belong to foreground $\mathcal{H}^{x,y} = 1$ or background $\mathcal{H}^{x,y} = 0$ which is used as label $c(\boldsymbol{\xi}^{x,y}) := \mathcal{H}^{x,y}$ for the image features. Assume that $\mathcal{H}$ is partially wrong (i.e. only a small portion of the data is wrongly labeled), the goal is to derive a classifier $\mathcal{A}_{\mathcal{F},\mathcal{H}}^{x,y}(\boldsymbol{\xi}^{x,y})$ for the pixel features that generalizes to the relevant foreground (object) features.

The method of GLVQ is defined by a network of $N$ class-specific prototypical feature representatives $\mathcal{P} := \{\boldsymbol{w}_p \in \mathbb{R}^M | p = 1..N\}$. For figure-ground segmentation a two class setup is used where $c(\boldsymbol{w}_p) \in \{0, 1\}$ encodes the user assigned class-membership of every prototype. The goal of the learning dynamics is to find the representatives in feature space to represent the data by minimizing the classification error defined by the functional $E[\mathcal{D}, \mathcal{P}] = \sum_{\boldsymbol{\xi}^{x,y} \in \mathcal{D}} \sigma(\mu(d))$ with $\sigma(x) = \frac{1}{1+e^{-x}}$, $\mu(d) = \frac{d_J - d_K}{d_J + d_K}$. Here the variables $d_J = d(\boldsymbol{\xi}^{x,y}, \boldsymbol{w}_J)$ and $d_K = d(\boldsymbol{\xi}^{x,y}, \boldsymbol{w}_K)$ represent the distance between $\boldsymbol{\xi}^{x,y}$ and the most similar prototype $\boldsymbol{w}_J$ from the correct class with $\mathcal{H}^{x,y} = c(\boldsymbol{w}_J)$ and the distance to the most similar prototype $\boldsymbol{w}_K$ from an incorrect class. Since similarity-based clustering and classification crucially depends on the underlying metrics, recently several adaptive metrics were proposed [7]. In the most general case the similarity metrics is extended towards a Mahalanobis metrics $d(\boldsymbol{\xi}, \boldsymbol{w}_p) = (\boldsymbol{\xi} - \boldsymbol{w}_p)^T \Lambda_p (\boldsymbol{\xi} - \boldsymbol{w}_p)$, where the distance computation of the features to the representatives is extended towards a prototype specific $M \times M$ matrix $\Lambda_p$ of relevance factors (Localized Generalized Matrix LVQ, LGMLVQ). In general, using the kernelized distance computation introduces non-linear decision boundaries. As described in Crammer et al. [8] this allows for a reduced number of prototypes while achieving a comparable performance to standard LVQ with multiple prototypes.

The prototypes $\boldsymbol{w}_J$ and $\boldsymbol{w}_K$ as well as the corresponding relevance factors $\Lambda_J$ and $\Lambda_K$ are optimized by means of gradient descent according to $E$ on $10000 \cdot T$ randomly chosen pairs $(\boldsymbol{\xi}^{x,y}, \mathcal{H}^{x,y})$, which is described in more detail in [1]. Since $\Lambda_p$ has to be positive semi-definite to yield a valid metrics, i.e. $d(\boldsymbol{\xi}, \boldsymbol{w}_p) = (\boldsymbol{\xi} - \boldsymbol{w}_p)^T \Omega_p \Omega_p^T (\boldsymbol{\xi} - \boldsymbol{w}_p) = (\Omega_p^T(\boldsymbol{\xi} - \boldsymbol{w}_p))^2 \geq 0$, this is assured by adapting $\Omega_p$, where $\Lambda_p = \Omega_p \Omega_p^T$. Additionally, the diagonal elements are normalize by $\sum_{i=1}^M \Lambda_{i,i} = 1$. In general the prototypes are kept from one to the consecutive frame and adapted to the new data.

To segment an image on the basis of such a network, it is partitioned into $N$ segments (binary maps) $V_p \in \{0, 1\}$ by assigning all feature vectors $\boldsymbol{\xi}^{x,y}$ (i.e. pixels of a particular frame) independently to the prototype $\boldsymbol{w}_p$ with the smallest distance $d(\boldsymbol{\xi}^{x,y}, \boldsymbol{w}_p)$. Using a prototype-based representation, the final segmentation $\mathcal{A}$ (binary map) is combined by choosing the binary maps from the prototypes assigned to the foreground $\mathcal{A} = \sum_p^N c(\boldsymbol{w}_p)V_p$. For object learning and recognition now $\mathcal{A}$ is used instead of $\mathcal{H}$.

## 2.3  Incremental Framework

We showed [1] that this method is robust in the presence of the noisy $\mathcal{H}$ and the increased model complexity using the adaptive metrics yields an improved segmentation quality in absence of over-fitting effects when an appropriate network dimensionality is chosen. Therefore the number of prototypes is an important parameter which determines the performance of the network with respect to runtime and generalization capability. Our main goal is here to adapt the number of prototypes during online processing of the data to use as many prototypes as necessary for the segmentation.

**Fig. 1.** The general algorithm to adapt the size of the network follows three main parts. Standard adaptation of a network using the LGMLVQ update rules (yellow circles), consisting of step 1 to 3 in Sec. 2.3. The green circles (plus) indicate an additional step to add a new prototype. This step yields two networks which are evaluated in parallel on the consecutive frame. Finally the red circle (minus) indicate an additional contraction step, where one of the prototypes (if appropriate) is removed.

*Incremental Online Processing:* The proposed method consists of three parts, a standard adaptation step, one method to add new prototypes and a local criterion to remove prototypes from the network. To stabilize the incremental learning of the network in presence of the noisy supervised information, we use the temporal aspect of the data for a sequential processing together with a parallel evaluation scheme. That is, to avoid the adaptation to the hypothesis on a particular frame, adding and removing prototypes are applied in a consecutive manner where on a single frame only one prototype is added or removed. Additionally due to the risk of disturbing the network with such operations while online segmenting the image, we use a parallel evaluation scheme Fig. 1. The prototypes are added to a second network, which is an exact copy of the first one. After the evaluation a decision is applied whether the original network or the modified network is kept for the following frame.

*Controlling the Network Size:* Incremental learning in prototype-based networks needs a mechanism to control the growing process to determine an appropriate number of prototypes. A widely used possibility is a global quality assessment. This information can be used to select the best performing set of prototypes after the network grew until a predefined maximum number of prototypes was reached [9], or to stop if the change in a quality measure does not significantly vary by adding further prototypes. An online scenario as well as noisy supervised information, which corrupt global quality assessments, prohibits such methods. In our approach the network size is controlled by a local utility function without a criterion of global classification performance or measure for model complexity. In comparison to the work of Hamker [2] we avoid to use (non-normalized) distance-based error criteria for the insertion and removal of prototypes from the network which is attributed to the local metrics of the prototypes. We place new prototypes according to a confidence criterion on the decision boundary and rate this placement afterwards by the utility criterion.

*Network Expansion:* Since a confident global measurement of the representation quality is not available to determine when it is necessary to introduce new prototypes the network is expanded in specified time intervals. To decide where a new prototype can be added possible criteria are random insertion, a placement on false classified data or on the decision boundary. In prototype-based networks the decision boundary can be characterized by a similar distance of a feature to two prototypes from different classes. In particular the objective of GLVQ is to minimize an error functional which represents not only the classification error but also introduces an error term for unconfidently classified data points which bases on the difference (the margin) in the nominator of the function $\mu(d)$. Using the margin for learning for example was proposed in the context of active learning by Schleif et al. [10]. Here new data points for learning are acquired on the basis of the margin criterion. But this information was not used in the context of incremental learning before. Since the margin is implicitly optimized by the GLVQ error function, we decide to add new prototypes in these regions of low confidence, respectively directly on the decision boundary. Therefore for each expansion step a new prototype is positioned at the training vector with the minimum normalized margin $m(\boldsymbol{\xi}^i) = \frac{\|d_J^i - d_K^i\|}{d_J^i + d_K^i}$. The label of the new prototype is initialized according to the supervised information, while the relevance matrix is taken from the best matching correct prototype according to this label. Since the network size is not adapted on a single frame and the data is changing from one to another frame, adding prototypes does not affect the optimization of the margin but provides a better initialization of the network for the adaptation on the next frame.

*Network Contraction:* To rate the importance of every single prototype in the network a local utility criterion can be used. In the context of vector quantization Fritzke [11] proposes to rate single neurons according to the quantization error of a prototype by the following utility function $U(w_p) := E[\mathcal{D}, \mathcal{P} \setminus w_p] - E[\mathcal{D}, \mathcal{P}] = \sum_{\boldsymbol{\xi} \in \mathcal{D}} \| \boldsymbol{\xi} - \boldsymbol{w}_s \|^2 - \| \boldsymbol{\xi} - \boldsymbol{w}_p \|^2$ where $\boldsymbol{w}_s$ is the winning prototype from the set

$\mathcal{P} \setminus \{\boldsymbol{w}_p\}$. As the quantization error (which is also exploited by Hamker [2] for a local utility function) is based on a global consistent metrics this method is not appropriate for localized adaptive metrics. Therefore this inspires a utility $u(\boldsymbol{w}_p)$ function on the basis of the classification error. For a single training example $\boldsymbol{\xi}$ this function is:

$$u(\boldsymbol{w}_p, \boldsymbol{w}_s, \xi) = \begin{cases} 1 & c(\boldsymbol{w}_p) = c(\boldsymbol{\xi}), \ c(\boldsymbol{w}_s) \neq c(\boldsymbol{\xi}) \\ 0 & \text{else} \end{cases}$$

Finally the utility of the prototype on the whole dataset is normalized by the number of activations $n(\boldsymbol{w}_p) = |\{\xi | d(\boldsymbol{w}_p, \boldsymbol{\xi}) = \min_{q \in \mathcal{P}} d(\boldsymbol{w}_q, \boldsymbol{\xi})\}|$ of this proto-type: $U(\boldsymbol{w}_p) = \frac{1}{n(\boldsymbol{w}_p)} \sum_{\xi \in \mathcal{D}} u(\boldsymbol{w}_p, \boldsymbol{w}_s, \xi)$. If the value $U(\boldsymbol{w}_p)$ falls below a given threshold $t_u = 0.01$ in our experiments, the prototype is regarded as a removal candidate. After an expansion step, the new prototype is kept, if this one and all other current prototypes are useful (i.e. $U(\boldsymbol{w}_p) > t_u \forall p \in \mathcal{P}$)), which assures to avoid unnecessary instabilities of the network. Independent of the utility function to evaluate the success of an expansion step, we use this function for separate contraction steps of the whole network to determine possibly spare prototypes or misplaced prototypes. Spare prototypes can be replaced by another prototype without impairing the performance. Misplaced prototypes can be characterized by an assignment to the wrongly labeled subset of data by the initial hypothesis $\mathcal{H}$. Usually this causes in the application/segmentation step that more image portions of the background are assigned to the foreground. These badly placed prototypes can be identified to cause a large classification error even on correctly labeled data and therefore reduce the overall segmentation quality. Together with the recorded activation $n(\boldsymbol{w}_p)$ we use the utility criterion to remove such prototypes. That is, additional to the utility criterion a prototype is removed if $\frac{n(\boldsymbol{w}_p)}{|\mathcal{D}|} < t_n$, where $t_n = 0.005$.

*Algorithm*

1. Input and preprocessing:
   - feature maps and hypothesis from object ROI: $\mathcal{F}^{x,y} := \{F_i^{x,y} | i = 1..M\}$, $\mathcal{H}^{x,y} \in \{0, 1\}$
   - Preprocessing of feature maps $\mathcal{F}$ and hypothesis $\mathcal{H}$, see Sec. 3
   - Init codebook and metric (on first frame only) $\mathcal{P} = \{\boldsymbol{w}_p | p = 1, .., N\}$ where $N = 2$, $\forall \boldsymbol{w}_p \in \mathcal{P} : \boldsymbol{w}_p = \frac{1}{|L|} \sum_L \boldsymbol{\xi}$, $L := \{\boldsymbol{\xi} | c(\boldsymbol{\xi}) = c(\boldsymbol{w}_p)\}$
   - Replace the data of the oldest frame by the data of current feature maps $\mathcal{F}$ in the short term history $\mathcal{D}$
2. Adaptation (for $T$ update steps)
   - Find best matching prototypes $\boldsymbol{w}_J$ for the correct label, $\boldsymbol{w}_K$ for the incorrect label according to a randomly selected $\boldsymbol{\xi}^i \in \mathcal{D}$. e.g. $\boldsymbol{w}_J = \{\boldsymbol{w}_p \in \mathcal{P} | d(\boldsymbol{w}_p, \boldsymbol{\xi}^i) = \min_{q, c(\boldsymbol{w}_q) = \mathcal{H}^i} d(\boldsymbol{w}_q, \boldsymbol{\xi}^i)\}$
   - Update prototypes $\boldsymbol{w}_{J,K}$ by means of $\boldsymbol{w}_{J,K} \leftarrow \boldsymbol{w}_{J,K} + \alpha \cdot \Delta \boldsymbol{w}_{J,K}$ with learning rate $\alpha = 0.05$ and similar the relevance factors $\Lambda_{J,K}$ with $\alpha = 0.005$

3. Evaluation: for all pixels $i \in \mathcal{D}$
   - $\forall \boldsymbol{w}_p \in \mathcal{P}, V_p^i := \begin{cases} 1 & \text{if } d(\boldsymbol{\xi}^i, \boldsymbol{w}_p) < d(\boldsymbol{\xi}^i, \boldsymbol{w}_r), \forall r \neq p, \{r, p\} \in \mathcal{P}, \\ 0 & \text{else} \end{cases}$
   - Determine the binary foreground segmentation $\mathcal{A} = \sum_p^N c(\boldsymbol{w}_p) \cdot V_p$
   - Compute margin for every feature $m(\boldsymbol{\xi}^i) = \frac{d_J^i - d_K^i}{d_J^i + d_K^i}$
   - Compute utility $U(\boldsymbol{w}_p)$ and prototype activation $n(\boldsymbol{w}_p)$, Sec. 2.3
4. (Optional) Network Expansion
   - $\boldsymbol{w}_{new} = \boldsymbol{\xi}^i$ where $i = \arg\min_{\xi_i \in \mathcal{D}} m(\boldsymbol{\xi}^i), c(\boldsymbol{w}_{new}) = \mathcal{H}^i, \Lambda_{new} = \Lambda_J$
   - $\mathcal{P} = \{\mathcal{P}, w_{new}\}, N = N + 1$
5. (Optional) Network Contraction
   - select $\boldsymbol{w}_p$ with the smallest utility $p = \arg\min_{p \in \mathcal{P}} U(\boldsymbol{w}_p)$
   - remove $\boldsymbol{w}_p$ if $U(\boldsymbol{w}_p) < t_u$ or $n(\boldsymbol{w}_p) < t_n, \mathcal{P} = \mathcal{P} \setminus \{\boldsymbol{w}_p\}, N = N - 1$

# 3 Results

*Data:* Finally we evaluate the capabilities of this approach on challenging real world image data and investigate the effort of the derived object segmentations in the context of online object learning and recognition. Here we are using the data from [6] consisting of 50 natural, view centered objects with 300 training and 100 testing images. After the acquisition of the feature maps $\mathcal{F}$ and the hypothesis $\mathcal{H}$ a pre-processing $F_i^{x,y} \leftarrow T_F(F_i^{x,y})$ of the feature maps $F_i^{x,y}$ (a gamma correction and white balancing on the maps representing the RGB image data) is performed first. From the available depth and skin information the hypothesis $\mathcal{H}$ is computed where all skin-colored areas $\mathcal{S}, \mathcal{S}^{x,y} \in \{0,1\}$ are removed from the hypothesis $\mathcal{H} \leftarrow T_H(\mathcal{H})$, where $(T_H(\mathcal{H}) := \mathcal{H} \leftarrow \mathcal{H} - (\mathcal{H} \cap \mathcal{S}))$. This is necessary because the hand is strongly connected to every object/hypothesis and can be regarded as systematic noise violating our assumptions. To compare the results with previous work, the image regions defined by the foreground classification (i.e. the presented objects) are fed into a hierarchical feature processing stage [6]. For object learning and recognition a separate nearest neighbor classifier is applied to the derived high dimensional shape features. The separation of training and test data is used for the object classifier, while the incremental segmentation is adapted on a subset of the pixel data for every single frame.

*Network Dimensionality:* First the behavior of the algorithm is analyzed on an example of the training-dataset in Fig. 2. The change in object identity yields an adaptation of the number of prototypes in particular for the foreground, which shows significant differences for some of the objects dependent on their subjective visual complexity. To avoid an influence from the sequence of the presented objects, the order of the 50 objects was randomly rearranged for the eight repetitions of the experiment. In contrast to previous work a reduced complexity of the representation finally allows for a more efficient processing of single frames.

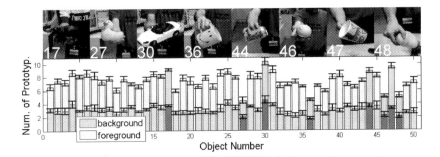

**Fig. 2.** Number of prototypes for an application to the training-dataset (50 objects with 300 views, each bar is the average of 8 repetitions and 300 views each). On average over all objects 4.35 prototypes are used for foreground and 3.07 are used for background. Additional the object specific std. dev. of the average number of prototypes for multiple repetitions is drawn, which shows that this number for a particular object is consistent over multiple repetitions of the experiment. On top, examples for eight objects with the highest and lowest number of prototypes are shown.

**Table 1.** Results of the incremental segmentation scheme compared to previous results (average of 8 repetitions, except the last column). Dependent on the derived number of prototypes (on average 3 for background and 4 for foreground) the proposed method achieves a comparable performance to a predefined prototype setup, whereby the variance of the results is significantly reduced.

| N (#bg/#fg) | 2(1/1) | 7(3/4) | 20(15/5) [1] | adaptive | hypothesis[1] |
|---|---|---|---|---|---|
| mean | 0.7442 | 0.8715 | 0.8828 | 0.8742 | 0.755 |
| std. dev. | 0.0132 | 0.0110 | 0.0252 | 0.0036 | n.a. |

*Classification Performance:* Compared to a predefined number of prototypes in previous results (see Table 1) three aspects are important: i) the general performance of the object classifier on the basis of the segmentation (an indirect quality assessment, verified in [1]), ii) the used resources to derive the results and iii) the variance in the results. Therefore we compare the incremental method to the results derived by predefined number of prototypes (chosen according to the average number of the incremental method). On the basis of the same resources, a comparable performance can be achieved. Remarkably the variance of the results is significantly decreased which indicates a higher robustness to the noisy supervised data by discarding misplaced prototypes. Together with a faster adaptation to the changing image data the incremental method also reduce the dependency on the initialization of the prototypes. Since the initialization for the fixed prototype setup was purely random this can explain the beneficial effect. Compared to an offline parameter search the incremental segmentation might not be able to reach the potentially maximum performance (for 20 prototypes, 15 background - 5 foreground on this dataset), but offers an application to data with unknown "optimal" number of prototypes.

# 4    Conclusion

In this paper we present an incremental learning scheme for GLVQ in the context of figure-ground segmentation. In presence of local adaptive metrics and supervised noisy information we use a parallel evaluation scheme combined with a local utility function to organize a learning vector quantization with an adaptive number of prototypes. On our real world benchmark dataset we show, that the incremental network is capable to achieve a comparable (to the results from [1]) performance in hypothesis refinement while maintaining a significantly smaller variance of the results, thus is more robust. Due to the parallel evaluation scheme the expansion of the network is free of additional computational load and does not impair the current performance of the network.

# References

1. Denecke, A., Wersing, H., Steil, J.J., Körner, E.: Online figure-ground segmentation with adaptive metrics in generalized LVQ. Neurocomputing 72(7-9), 1470–1482 (2009)
2. Hamker, F.H.: Life-long learning cell structures —continuously learning without catastrophic interference. Neural Networks 14(4-5), 551–573 (2001)
3. Kirstein, S., Wersing, H., Körner, E.: A biologically motivated visual memory architecture for online learning of objects. Neural Networks 1, 65–77 (2008)
4. Fritzke, B.: A growing neural gas network learns topologies. In: Tesauro, G., Touretzky, D.S., Leen, T.K. (eds.) NIPS, pp. 625–632. MIT Press, Cambridge (1994)
5. Sato, A., Yamada, K.: Generalized learning vector quantization. In: Advances in Neural Information Processing Systems, vol. 7, pp. 423–429 (1995)
6. Wersing, H., Kirstein, S., Götting, M., Brandl, H., Dunn, M., Mikhailova, I., Goerick, C., Steil, J.J., Ritter, H., Körner, E.: Online learning of objects in a biologically motivated visual architecture. Int. J. Neur. Syst. 17(4), 219–230 (2007)
7. Schneider, P., Biehl, M., Schleif, F.-M., Hammer, B.: Advanced metric adaptation in Generalized LVQ for classification of mass spectrometry data. In: Proceedings of 6th International Workshop on Self-Organizing Maps (WSOM) (2007)
8. Crammer, K., Gilad-Bachrach, R., Navot, A., Tishby, N.: Margin analysis of the LVQ algorithm. In: NIPS (2002)
9. Jirayusakul, A., Auwatanamongkol, S.: A supervised growing neural gas algorithm for cluster analysis. Int. J. Hybrid Intell. Syst. 4(4), 217–229 (2007)
10. Schleif, F.M., Hammer, B., Villmann, T.: Margin-based active learning for LVQ networks. Neurocomput 70(7-9), 1215–1224 (2007)
11. Fritzke, B.: The LBG-U method for vector quantization – an improvement over LBG inspired from neural networks. Neural Process. Lett. 5(1), 35–45 (1997)

# Application of Supervised Pareto Learning Self Organizing Maps and Its Incremental Learning

Hiroshi Dozono[1], Shigeomi Hara[1], Shinsuke Itou[1], and Masanori Nakakuni[2]

[1] Faculty of Science and Engineering, Saga University,
1-Honjyo Saga 840-8502, Japan
hiro@dna.ec.saga-u.ac.jp
[2] Information Technology Center, Fukuoka University,
8-19-1, Nanakuma, Jonan-ku, Fukuoka 814-0180, Japan
nak@fukuoka-u.ac.jp

**Abstract.** We have proposed Supervised Pareto Learning Self Organizing Maps(SP-SOM) based on the concept of Pareto optimality for the integration of multiple vectors and applied SP-SOM to the biometric authentication system which uses multiple behavior characteristics as feature vectors. In this paper, we examine performance of SP-SOM for the generic classification problem using iris data set. Furthermore, we propose the incremental learning algorithm for SP-SOM and examine effectiveness in a classification problem and adaptation ability to the change of the behavior biometric features by time.

## 1 Introduction

Recently, biometric authentication systems are often used to cover the weak point of password authentication. Password authentication can be easily hacked by illegal users who get the password phrase by peeping, guessing from personal information or using key logger software because password is a simple text phrase. Biometric authentication uses biometric character of the user himself, so the features used for the authentication can not be copied easily.

Biometric authentication is classified to 2 types, biometric authentication using biological characteristics and biometric authentication using behavior characteristics. As the biological characteristics, fingerprint, face, vein patterns and iris patterns are used. The authentication systems using biological characteristics perform high accuracy, but for obtaining biological characteristics, special hardware is required. Furthermore, some issues concerning the spoofing of the fingerprint are reported. This weak point of biological characteristics comes from the static nature of biological characteristics. Biological characteristics can not be easily copied, but if it is copied once, the authentication system can not detect spoofing users. As the behavior characteristics, key stroke timing[1], hand written sign or pattern[2], sound spectrogram of voice and so on are used. Some of the behavior characteristics can be obtained from the standard input devices equipped to computers. Furthermore, behavior characteristics are the dynamic features of the users' own, so can not be easily imitated by illegal users. But, the

J.C. Príncipe and R. Miikkulainen (Eds.): WSOM 2009, LNCS 5629, pp. 54–62, 2009.
© Springer-Verlag Berlin Heidelberg 2009

accuracy of the authentication is inferior to that of biological characteristics because of the drift of the behavior at each authentication time. For this problem, we proposed the authentication method using multiple behavior characteristics, such as, combination of keystroke timing and key typing sound and combination of keystroke timing and handwritten pattern[3]. For the integration of multiple vectors, we used Self Organizing Map(SOM)s. But using the conventional SOM, the weight values for each feature vector should be set according to the validity and magnitude. The accuracy of the authentication differs depending on the weight values. For this problem, we proposed Pareto learning Self Organizing Map(P-SOM)s. P-SOM organizes the input data composed of the multiple independent vectors based on the Pareto optimal concept. Additionally, we proposed Supervised Pareto learning SOM(SP-SOM)s to improve the accuracy of authentication with supervised learning of the user id as a feature vector[4].

Apart from the application to the authentication system, P-SOM and SP-SOM can be generally applicable to the integration of multi-modal vectors. Using conventional SOM, multi-modal input vectors are learned as a composed vectors. P-SOM and SP-SOM can learn the multi-modal input vectors directly without composing vectors and the multi-modal vectors are integrated on the map based on the concept of Pareto optimality. In this paper, we examine the classification ability of SP-SOM with the benchmark of classification problem of iris data set registered in the UCI Machine learning repository. The iris data set is composed of the multiple attributes of different features. To examine the effect from the composition of vectors, we make some experiments that change the combination of the attributes composed in each vector. After learning the reference data, incremental learning of the test data may improve the accuracy of classification during the test process of the classification problem. We introduce the incremental learning algorithm to SP-SOM and confirm the effectiveness of incremental learning. For the authentication system using behavior characteristics, the change of the behavior over time should be considered. For example, typing speed will be faster as the user accustomed to the computer, so the keystroke timings will be changed. To adapt to the changes, incremental learning of the input data at each authentication time is considered to be effective. We make authentication experiment with changing the input data artificially and confirmed the adaptation ability of SP-SOM with incremental learning.

## 2   Pareto Learning Self Organizing Maps

### 2.1   Conventional SOM

Conventional SOM can be used for integrating the multi-modal vectors. The multi-modal vectors $x_1, x_2, \ldots, x_n$ are simply composed as an vector as follows.

$$x = (x_1, x_2, \ldots, x_n) \tag{1}$$

In this case, all vectors are learned in same weight. Considering the feature of each vector, multi-modal vectors should be composed in a vector $x$ using the weight value for each vector as follows.

$$\mathbf{x} = (w_1\mathbf{x_1}, w_2\mathbf{x_2}, \ldots, w_n\mathbf{x_n}) \tag{2}$$

where $w_i$ is the weight value for vector $x_i$. Using this method, the error $e$ between the vector $\mathbf{m} = (\mathbf{m_1}, \mathbf{m_2}, \ldots, \mathbf{m_n})$ assigned to the i-th unit on the map and input vector $\mathbf{x}$ is given by

$$e = \sqrt{\sum_{j=1}^{n} e_j^2} \tag{3}$$

$$e_j = |w_j\mathbf{x_j} - \mathbf{m_j}| \tag{4}$$

where $e_j$ is error between the $\mathbf{x_j}$ and $\mathbf{m_j}$. Because the map is organized according to this error function, the resulting map is heavily depending on the weight values $w_i$.

## 2.2  Pareto Optimality

Pareto optimality is a concept in multi-modal optimization problem. Consider the problem finding the vector $\mathbf{x} \in Y$ which minimize the set of objective functions $f_i(\mathbf{x}), i = 1, \ldots, m$. To solve the problem finding the vector which minimize an objective function, weighted sum of the functions $f_i(\mathbf{x})$ can be used as the objective function. But, how to set the weight values of each function dominates the quality of the solution.

The Pareto set $P$ which is composed of the vectors $\mathbf{x}$ that are not inferior to others is defined as follows.

**Definition 1.** *The Pareto set $P$ is the set of the vectors in $Y$ that are not strictly dominated by any vector in $Y$. A vector $\mathbf{x}$ is said to be strictly dominated by $\mathbf{y}$, if $f_i(\mathbf{x}) \leq f_i(\mathbf{y})$ for any $i$ and $f_i(\mathbf{x}) < f_i(\mathbf{y})$ for some $i$.*

All of the members in the Pareto set are the candidates of optimal solution.

## 2.3  Pareto learning SOM

Pareto learning SOM(P-SOM) is the SOM which uses the concept of Pareto optimality for integrating multi-modal vectors. For Pareto learning SOM, the input vector is given as the composition of distinct vectors as follows.

$$\mathbf{x} = (\{\mathbf{x_1}\}, \{\mathbf{x_2}\}, \ldots, \{\mathbf{x_n}\}) \tag{5}$$

The algorithm of P-SOM is as follows.

**P-SOM Algorithm**

1. Initialization of the map
   Initialize the vector $\mathbf{m^{ij}}$ which are assigned to unit $U^{ij}$ on the map using the 1st and 2nd principal components as base vectors of 2-dimensional map.

2. Batch learning phase
   (1) Clear all learning buffer of units $U^{ij}$.
   (2) For each vector $x^i$, search for the pareto optimal set of the units $P = \{U_p^{ab}\}$. $U_p^{ab}$ is an element of pareto optimal set P, if for all units $U_{kl} \in P - U_p^{ab}$, existing h such that $e_h^{ab} \le e_h^{kl}$ where

$$e_h^{kl} = |\mathbf{x_h^i} - \mathbf{m_h^{kl}}| \tag{6}$$

   (3) Add $x^i$ to the learning buffer of all units $U_p^{ab} \in P$.
3. Batch update phase
   For each unit $U^{ij}$ update the associated vector $m^{ij}$ using the weighted average of the vectors recorded in the buffer of $U^{ij}$ and its neighboring units as follows.
   (1)For all vectors $x$ recorded in the buffer of $U^{ij}$ and its neighboring units in distance $d \le Sn$, calculate weighted sum $\mathbf{S}$ of the updates and the sum of weight values W.

$$\mathbf{S} = \mathbf{S} + \eta fn(d)(\mathbf{x} - \mathbf{m^{i'j'}}) \tag{7}$$
$$W = W + fn(d) \tag{8}$$

   where $U^{i'j'}$s are neighbors of $U^{ij}$ including $U^{ij}$ itself, $\eta$ is learning rate, $fn(d)$ is the neighborhood function which becomes 1 for d=0 and decrease with increment of d.
   (2) Set the vector $\mathbf{m^{ij}} = \mathbf{m^{ij}} + \mathbf{S}/W$.

Repeat 2. and 3. with decreasing the size of neighbors Sn for pre-defined iterations.
   As shown in step 2 of this algorithm, Pareto winner set for the integrated input vector $\mathbf{x}$ are searched for based on the concept of Pareto Optimality using the distance defined by (6) as objective function $f_h(\mathbf{x})$ for each element $\mathbf{x_h}$ in $\mathbf{x}$. Thus, the multiple units become winners. The winners and their neighboring units are modified in the update process in step 3. Overlapped neighbors are updated multiply and the overlapped region will contribute to generalization ability and integration ability of P-SOM.
   In the beginning of learning, the size of Pareto winner set became very large and it should be small in the termination of learning. But, even in the termination of learning, the size was still large in some cases. A unit can be a member of Pareto winner set if at least one of the objective function is superior to those of others in Pareto optimal set. We make this condition stronger to reduce the size of Pareto set.

**Definition 2.** *M/N Pareto Optimality*
*The M/N Pareto set P is subset of vectors in Y that is composed of the vectors which have at least M superior objective functions to those of others in P where N is number of objective functions.*

In the learning process of P-SOM, the Pareto winner set is searched for based on the concept of M/N Pareto optimality and M is updated to larger value during the learning process to reduce the size of Pareto winner set.

The map is initialized using Principal Component Analysis (PCA) to avoid the scattering of the elements in Pareto winner set. We also used batch learning method to avoid the affect from the order of the presentation of each input vector.

## 2.4   Supervised Pareto Learning SOM(SP-SOM)

P-SOM can integrate the any kind of vectors which use different distance metrics. Thus, the category vector can easily integrated. It changes the learning algorithm from unsupervised learning of conventional SOM to supervised learning to improve the accuracy of classification. The integrated input vector including category vector is composed as follows.

$$\acute{\mathbf{x}}^i = (\mathbf{x^i}, \mathbf{c^i}) \tag{9}$$

$$c^i_j = \begin{cases} 1 & \mathbf{x^i} \in C_j \\ 0 & otherwise \end{cases} \tag{10}$$

where $C_j$ is j-th category. The learning algorithm is almost same as that of P-SOM except that category vector can be used whether for finding Pareto winner set or only for labeling the units. If the integrated vector including category vector is used for finding the winners, the organization of the map is controlled by the primal integrated vector and category vector. If the primal integrated vector is used for finding the winners and the category vector is included only for updating, the category vector is considered as mere label for the unit.

In the recalling process of SP-SOM, the category vector is also used. The recalling algorithm of SP-SOM is as follows.

### SP-SOM - recalling algorithm

1. Searching for the Pareto set of units
   For given test vector $\mathbf{x^t}$, search for the pareto optimal set of the units $P = \{U^{ab}_p\}$.
2. Determination of the category
   Calculate

$$c^t_k = \sum_{U^{ij} \in P}^{m} c^{ij}_k \tag{11}$$

where $m^{ij} = (\mathbf{x}^{ij}, \mathbf{c}^{ij})$. The category of $\mathbf{x^t}$ is $C_l$ for $l = argmax_k(c^t_k)$.

As shown in this algorithm, category for a test vector is determined by the sum of the category vectors in Pareto set of units.

## 2.5   Incremental Learning of SP-SOM

As mentioned before, incremental learning will be effective for the learning of test data in the classification problem and for adaptation to the temporal change of input vectors. Two types of incremental learning mode, supervised learning and unsupervised learning are considered depending on the condition of test data.

In supervised learning, the vector for incremental learning is composed with the category vector described in the previous sub-section. In unsupervised learning, only the test vector is used for learning. The equation of the incremental learning is as follows.

$$\mathbf{m}'_{ij} = \mathbf{m}'_{ij} + \eta'(\mathbf{x}' - \mathbf{m}'_{ij}) \tag{12}$$

where $\mathbf{m}'_{ij}$ is the vector associated to $U^{ij} \in P$, $P$ is the Pareto optimal set for test vector $\mathbf{x}$, $\mathbf{x}' = (\mathbf{x}, \mathbf{c})$ for supervised learning, $\mathbf{x}' = \mathbf{x}$ for unsupervised learning, $\mathbf{c}$ is category vector of $\mathbf{x}$ and $\eta'$ is learning rate for incremental learning. This equation is equivalent to the equation for updating the winner unit in SOM except the targets are the units in Pareto winner set.

# 3   Experimental Results for Classification Problem

In this section, the experimental results for general classification problem using SP-SOM are mentioned. As the classification problem, the iris data set obtained from UCI machine learning database are used.

## 3.1   Description of Iris Data

The iris data set contains 3 classes of 50 instances of each, where each class is refers to a type of iris plant. Each data contains 4 attributes, which are sepal length(sl), sepal width(sw), petal length(pl) and petal width(pw) in cm.

## 3.2   Experimental Result for Classification Problem of Iris Data

At first, we made some experiments using SP-SOM with different combination for making integrated input vector $\mathbf{x}$. The following 4 cases are tested considering the feature of each attribute.

- Case 1: $\mathbf{x} = (sl, sw, pl, pw)$ All attributes are composed in a vector.
- Case 2: $\mathbf{x} = (\{sl, sw\}, \{pl, pw\})$ 2 vectors which represents the features of sepal and petal respectively are integrated.
- Case 3: $\mathbf{x} = (\{sl, pl\}, \{sw, pw\}$ 2 vectors which represents the features of length and width respectively are integrated.
- Case 4: $\mathbf{x} = (\{sl\}, \{sw\}, \{pl\}, \{pw\})$ All attributes are integrated as the independent vectors of length 1.

Case 1 is considered as the SOM using the category vector for supervised learning. Case 4 is the special case of SP-SOM which treats all attributes independently. This integration method will be effective for the input vectors which is composed from the attributes with unknown features. We call this Full Pareto learning SOM(FP-SOM) or Supervised Full Pareto learning SOM(SFP-SOM).

For each cases, 50 iterations of experiments are done with changing the combination of 40 data for learning and 10 data for testing. The parameters of SP-SOM are as follows.

- Map geometry: size=16x16 torus map
- Initial learning rate: 0.3
- Initial neighbor size: 4
- Pareto condition: 1/N Pareto optimality

**Table 1.** Accuracy of classification for iris data set using SP-SOM

|        | Case 1 | Case 2 | Case 3 | Case 4 |
|--------|--------|--------|--------|--------|
| Class1 | 1.000  | 1.000  | 1.000  | 0.996  |
| Class2 | 0.930  | 0.948  | 0.946  | 0.918  |
| Class3 | 0.928  | 0.928  | 0.948  | 0.910  |
| Average| 0.952  | 0.959  | 0.965  | 0.941  |

**Table 2.** Accuracy of classification for iris data set using SP-SOM

|         | SOM   | LVQ with AC | C4.5  | OC1   | LMDT  |
|---------|-------|-------------|-------|-------|-------|
| Average | 0.899 |             | 0.951 | 0.916 | 0.939 | 0.955 |

The accuracy of classification is shown in Table 1. The accuracy of the classification changes according to the composition of integrated vector. In these cases, case 3 is best and case 4, which uses FSP-SOM are worst. But, the difference is small in average. For the comparison, the results from [5] are shown in Table 2. The condition of experiments is almost same except that the iterations of experiments are 10 for these results. Except case 4, SP-SOM shows superior results. Next, we made the experiments with changing the number of the data for learning and testing. Fig.1 shows the results. X-axis of this graph denotes the number of test data and the remainder (50-x) is used for learning. Case 4' uses SFP-SOM whose Pareto condition is 2/4 Pareto optimality to improve the accuracy of case 4. With decreasing the number of the data for learning the accuracy becomes worse. But, for case 3 and case 4', the rates of decreasing is smaller than those of others.

Next, the experimental results of incremental learning are shown. During the test process, the test data are used for learning the map. For the fairness of comparison, the test data are learned by unsupervised learning method without including supervised category vectors. The result for case 3 with changing

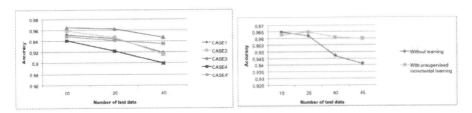

**Fig. 1.** Changes of the accuracy with changing number of test data

**Fig. 2.** Changes of the accuracy with incremental unsupervised learning

the number of test data is shown in Fig.2. Compared with the results without learning, the accuracy is improved against the decreasing of learning data with incremental learning. Even for the case of 5 learning data(45 test data), the accuracy is kept over 96% with incremental learning.

From these results, SP-SOM can perform better classification accuracy with setting the proper combination of the attributes as input vectors.

## 4 Experimental Results of Adaptive Authentication System

As mentioned before, the biometric authentication system using behavior characteristics should adapt to the change of behaviors over time. We reported the authentication method using the keystroke timing and key typing sound as the multi-modal behavior biometrics[6]. The keystroke timing data is the vector composed of the intervals of the pushing and releasing the keys. The key typing sound data is the vector composed of the maximum amplitudes of the sounds for each key typing. From each of 10 examinees, 10 data typing the same phrase "kirakira" are sampled. For the experiments using the biometric data which changes over time, we made the biometric data artificially from the sampled data because it will take very long time to wait for the change of biometrics of examinees.

At first, the map for all biometric data is organized. The size of the map is 16x16 and the iteration of the learning is 50 batch cycles for all input vectors. Next, the adaptation to the changes of the input vectors over time is examined. In the following experiment, 4 out of 15 keystroke timings and 2 out of 8 key typing sounds in the input vector are selected randomly, multiplied by 0.9 and replaced with the value before each authentication test. The case that test vectors are not learned, the case that test vector are learned by unsupervised learning and the case that test vectors are learned by supervised learning are compared. The tests are repeated 20 times. Fig.3 shows the result. FRR(False Reject Rate) is an index of authentication accuracy which means the rate of rejecting the user falsely. Without learning, FRR becomes worse with iterations. On the other hand, FRR is kept small with incremental learning. With supervised incremental learning, FRR is kept almost 0 even if the input vectors are changing time over time. Fig.4 shows

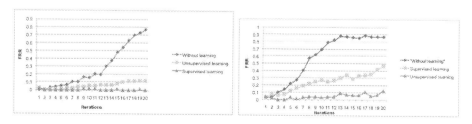

**Fig. 3.** Changes of FRR with incremental learning and without learning(1)    **Fig. 4.** Changes of FRR with incremental learning and without learning(2)

the results for the case that multiplier is set 0.8. In this case, FRR becomes worse for all case, but FRR is kept under 0.1 with supervised learning.

## 5   Conclusion

We proposed Pareto learning SOM(P-SOM) and Supervised Pareto learning SOM(SP-SOM) which based on the concept of Pareto optimality for integrating multi-modal vectors. The performance for the classification problem is confirmed in the iris classification problem. Additionally, it is shown that unsupervised incremental learning of the test data can improve the accuracy of classification and that incremental learning is also effective for the application of SP-SOM to adaptive authentication system using multi-modal behavior biometrics.

P-SOM and SP-SOM can be used for integrating more generic input vectors and the units on the map can be represented in more generic forms. For example, using the sequence data and its features of DNA sequences as input vector, the units can be represented using Hidden Markov Model for sequence data and independent vectors for each feature. As the future works, we should examine the performance of P-SOM and SP-SOM in variety of applications.

## References

1. Monrose, F., Rubin, A.D.: Keystroke Dynamics as a Biometric for Authentication. Future Generation Computer Systems (March 2000)
2. Brault, J.J., Plamondon, R.: A Complexity Measure of Handwritten Curves: Modelling of Dynamic Signature Forgery. IEEE Trans. Systems, Man and Cybernetics 23, 400–413 (1993)
3. Nakakuni, M., Dozono, H., et al.: Application of Self Organizing Maps for the Integrated Authentication using Keystroke Timings and Handwritten Symbols. Wseas Transactions on Information Science & Applications 2(4), 413–420 (2006)
4. Dozono, H., Nakakuni, M., et al.: An Integration Method of Multi-Modal Biometrics Using Supervised Pareto Learning Self Organizing Maps. In: Proc. of the Internal Joint Conference of Neural Network 2008 (2008)
5. Teh, C., Tapan, M.S.Z.: A Hybrid Supervised ANN for Classification and Data Visiualization. In: Proc. of the Internal Joint Conference of Neural Network 2008 (2008)
6. Dozono, H., Nakakuni, M., et al.: The Analysis of Key Typing Sounds using Self Organizing Maps. In: Proceedings of The 2007 International Conference on Security and Management, pp. 337–341 (2007)

# Gamma SOM for Temporal Sequence Processing

Pablo A. Estévez and Rodrigo Hernández

Dept. Electrical Engineering, University of Chile,
Casilla 412-3, Santiago, Chile
pestevez@cec.uchile.cl

**Abstract.** In this paper, we introduce the Gamma SOM model for temporal sequence processing. The standard SOM is merged with a new context descriptor based on a short term memory structure called Gamma memory. The proposed model allows increasing depth without losing resolution, by adding more contexts. When using a single stage of the Gamma filter, the Merge SOM model is recovered. The temporal quantization error is used as a performance measure. Simulation results are presented using two data sets: Mackey-Glass time series, and Bicup 2006 challenge time series. Gamma SOM surpassed Merge SOM in terms of lower temporal quantization error in these data sets.

## 1  Introduction

Self-organizing feature maps (SOMs) [1] have been recently extended for processing data sequences that are temporally or spatially connected, such as words, DNA sequences, time series, etc. [2],[3],[4]. Hammer et al. [5] presented a review of recursive self-organizing network models, and their application for processing sequential and tree-structured data. An early attempt to include temporal context is the Temporal Kohonen Map (TKM) [6], where a neuron output depends on the current input and its context of past activities. The neurons implement a local recurrence, acting as leaky integrator of signals. In the Recursive SOM model [3],[11], time is represented by feedback connections. The original SOM algorithm is used recursively on both the current input and a copy of the map at the previous time step. In addition to a weight vector, each neuron has a context vector that represents the temporal context as the activation of the entire map in the previous time step. This kind of context is computationally expensive, since the dimension of the context vectors is equal to the number of neurons in the network.

In the Merge SOM (MSOM) [2] approach, the current context is compactly described by a linear combination of the weight and the context of the last winner neuron. This context representation is more space efficient than the one used for the Recursive SOM model. MSOM represents the context in the data space, therefore the representation capability is restricted by the data dimensionality [4]. Since MSOM context does not depend on the lattice architecture, it can be combined with other self-organizing neural networks such as Neural Gas [12]. The resulting model is called Merge Neural Gas (MNG) [13].

A static neural network can be extended to a dynamical one, by adding a short term memory structure to the net. De Vries and Principe [7] introduced the Gamma neural

J.C. Príncipe and R. Miikkulainen (Eds.): WSOM 2009, LNCS 5629, pp. 63–71, 2009.

model, where the units can store their activation history in an adaptive gamma memory structure. In particular, the authors studied the focused Gamma net where the memory elements are restricted to the first layer, and then this input layer is fed to a strictly static feedforward net. Under supervised learning, the memory parameters are adapted by using a real time recurrent learning procedure.

In this paper, a new context model based on Gamma memory is added to SOM. When the order of the Gamma filter is reduced to one, the Merge SOM model is recovered. Gamma SOM and Merge SOM are compared using two time series and the temporal error quantization as a performance criterion.

## 2   Merge SOM

In this section a brief overview of Merge SOM is given. Neurons are a vector tuple $\left(\mathbf{w}^i, \mathbf{c}^i\right) \in \Re^d \times \Re^d$, where $d$ correspond to the dimensionality of the input signals, $\mathbf{w}^i$ is the weight vector, and $\mathbf{c}^i$ is the context vector associated to the $ith$ neuron. The context corresponds to the merged content of the winner neuron in the previous time step.

Given the current entry $\mathbf{x}(n)$ of a sequence, the best matching neuron (BMU), $I_n$, is the closest neuron according to the following recursive distance criterion:

$$d_i(n) = (1 - \alpha) \cdot ||\mathbf{x}(n) - \mathbf{w}^i||^2 + \alpha \cdot ||\mathbf{c}(n) - \mathbf{c}^i||^2 \tag{1}$$

where the contributions of weight and context vectors to the distance are balanced by the parameter $\alpha \in [0, 1]$. This parameter should be set as to ensure that the weight's contribution is more relevant than the context's contribution, in order to achieve a correct context representation. The current context, $\mathbf{c}(n)$, corresponds to a linear combination of the weight and context of the previous winner, $I_{n-1}$, i.e. the best matching unit in the last time step. The current context is defined as:

$$\mathbf{c}(n) = (1 - \beta) \cdot \mathbf{w}^{I_{n-1}} + \beta \cdot \mathbf{c}^{I_{n-1}}, \tag{2}$$

where the parameter $\beta$ permits to control the contribution of the weight and the context of the previous winner to the current context representation. This parameter takes values in the range $0 \leq \beta < 1$. Typically, $\mathbf{c}(0) = 0$.

### 2.1   Merge SOM Update Rules

Training takes place by adapting both weight and context vectors towards the current input and its context descriptor, respectively, using the following update rules:

$$\triangle\mathbf{w}^i = \epsilon_w(n) \cdot h_{\sigma(n)}\left(d_G(i, i*)\right) \cdot \left(\mathbf{x}(n) - \mathbf{w}^i\right) \tag{3}$$
$$\triangle\mathbf{c}^i = \epsilon_c(n) \cdot h_{\sigma(n)}\left(d_G(i, i*)\right) \cdot \left(\mathbf{c}(n) - \mathbf{c}^i\right)$$

where $n$ is the current training epoch, $\epsilon$ is the learning rate, and $h_{\sigma(n)}$ is the neighborhood function defined in the two-dimensional output grid. Typically, this neighborhood function is a Gaussian centered in the winner unit $i^*$, defined as follows:

$$h_{\sigma(n)}\left(d_G(i, i^*)\right) = exp\left(\frac{-d_G(i, i^*)}{\sigma(n)}\right) \tag{4}$$

where $d_G(i, i^*)$ is the distance in the 2D output grid between the $ith$ and the $i^*th$ neurons, defined as

$$d_G(i, i^*) = |x_i - x_{i^*}| + |y_i - y_{i^*}|. \tag{5}$$

The parameter $\sigma(n)$ is the neighborhood size and decreases exponentially during training,

$$\sigma(n) = \sigma_0 \cdot \left(\frac{\sigma_f}{\sigma_0}\right)^{\frac{n}{n_{max}}} \tag{6}$$

where $n_{max}$ is the maximum number of epochs. The learning rates $\epsilon_w(n)$ and $\epsilon_c(n)$ are annealed in the same way as (6), i.e. they decay exponentially from an initial value to a final value. It has been proved that Hebbian learning converges to the following optimal weight and context vectors at time $n$ (i.e., are stable fixed points):

$$\mathbf{w}^{opt(n)} = \mathbf{x}(n), \; \mathbf{c}^{opt(n)} = \sum_{j=1}^{n-1}(1 - \beta) \cdot \beta^{j-1} \cdot \mathbf{x}(n - j),$$

provided that there are enough neurons and neighborhood cooperation is neglected, i.e. for late stages of learning [5].

## 3 Gamma Memories

The Gamma filter is defined in the time domain as

$$y(n) = \sum_{k=0}^{K} w_k c_k(n)$$

$$c_k(n) = \beta c_k(n - 1) + (1 - \beta)c_{k-1}(n - 1) \tag{7}$$

where $c_0(n) \equiv x(n)$ is the input signal and $y(n)$ is the filter output, and $w_0, \cdots, w_K, \mu$ are the filter parameters. It has been demonstrated [10] that stability is guaranteed when $0 < \beta < 1$. The $\beta$ parameter provides a mechanism to decouple depth ($D$) and resolution ($R$) from filter order. Depth refers to how far into the past the memory stores information, a low memory depth can hold only recent information. Resolution refers to the degree to which information concerning the individual elements of the input sequence is preserved. It has be shown [9],[10] that the mean memory depth for a Gamma memory of order-K becomes

$$D = \frac{K}{(1 - \beta)} \tag{8}$$

and its resolution is

$$R = 1 - \beta.$$

Therefore, depth and resolution can be adapted in Gamma memories by changing $\beta$.

The Gamma delay operator $G(z)$ represents the transfer function using z-transform of a single filter stage

$$G(z) = \frac{(1 - \beta)}{(z - \beta)}. \tag{9}$$

Eq. (9) can be interpreted as a leaky integrator, where $\beta$ is the gain of the feedback loop. The proposed recursive rule for context descriptor of order-K can be derived directly from the transfer function (9), as follows:

$$C_k(z) = G(z)C_{k-1}(z) = \frac{1-\beta}{z-\beta}C_{k-1}(z)$$

which can be rearranged as

$$C_k(z) = (1-\beta)z^{-1}C_{k-1}(z) + \beta z^{-1}C_k(z). \tag{10}$$

By using the inverse Z-transform, the recursive expression (7) is obtained. A more detailed Gamma Filter analysis can be found in [7,8].

## 4   New Gamma Context Model

Let $\mathcal{N} = \{1,\ldots,M\}$ be a set of neurons. Each neuron has associated a weight vector $w^i \in \Re^d$, for $i = 1,\ldots,M$, obtained from a vector quantization algorithm. The Gamma context model associates to each neuron a set of contexts $\mathcal{C} = \{c_1^i, c_2^i, \ldots, c_K^i\}$, $c_k^i \in \Re^d$, $k = 1,\ldots,K$, where $K$ is the Gamma filter order. Given a sequence $s$ the context set $\mathcal{C}$ should be initialized at a fixed value, e.g. 0. By increasing the filter order, the Gamma context model can achieve an increasing memory depth without compromising resolution.

Given a sequence entry, $x(n)$, the best matching unit $I_n$ is the neuron that minimizes the following distance criterion,

$$d_i(n) = \alpha_w \left\| x(n) - w^i \right\|^2 + \sum_{k=1}^{K} \alpha_k \left\| c_k(n) - c_k^i \right\|^2 \tag{11}$$

where the parameters $\alpha_w$ and $\alpha_k$, $k \in \{1,2,\ldots,K\}$ control the contribution of the different elements. The recursive distance (11) requires calculating every context descriptor in the different filtering stages, which are built by using Gamma memories. Formally, the K context descriptors of the current unit are defined as:

$$c_k(n) = \beta c_k^{I_{n-1}} + (1-\beta)\, c_{k-1}^{I_{n-1}} \quad \forall k = 1,\ldots,K \tag{12}$$

where $c_0^{I_{n-1}} \equiv w^{I_{n-1}}$ and at $n = 0$ the initial conditions $c_k^{I_0}, \forall k = 1,\ldots,K$ are set randomly. It is easy to verify that when $K = 1$, the merge context (2) is recovered. Therefore, Merge SOM becomes a particular case of Gamma SOM when only a single Gamma filter stage is used ($K = 1$).

Because the context construction is recursive, it is recommended that $\alpha_w > \alpha_1 > \alpha_2 > \cdots > \alpha_K > 0$, otherwise errors due to poor quantization in the first filter stages would propagate through higher order contexts.

### 4.1   Gamma SOM Algorithm

The Gamma SOM algorithm is a merge between SOM and Gamma context model. A grid size of $N_y \times N_x$ is considered as in standard SOM. Neuron $ith$ has associated a weight vector, $w^i$, and a set of contexts, $c_k^i$, for $k = 1, \cdots, K$.

1. Initialize randomly weights $w^i$, and contexts, $c_k^i$, for $k = 1, \cdots, K, i = 1, \cdots, M$.
2. Present input vector, $x(n)$, to the network
3. Calculate context descriptors $c_k(n)$ using (12)
4. Find best matching unit (BMU), $I_n$, using (11)
5. Update neuron's weight and contexts using the following rule

$$\triangle \mathbf{w}^i = \epsilon_w(n) \cdot h_{\sigma(n)} \left( d_G(i, i^*) \right) \cdot \left( \mathbf{x}(n) - \mathbf{w}^i \right) \tag{13}$$

$$\triangle \mathbf{c_k}^i = \epsilon_c(n) \cdot h_{\sigma(n)} \left( d_G(i, i^*) \right) \cdot \left( \mathbf{c_k}(n) - \mathbf{c_k}^i \right)$$

6. Set $n \to n + 1$
7. If $n < L$ go back to step 2, where $L$ is the cardinality of the data set.

## 5   Experiments

Experiments were carried out with two data sets: Mackey-Glass time series and Bicup 2006 time series[1]. The parameter $\beta$ was varied from 0.1 to 0.9 with 0.1 steps. The number of filter stages K was varied from 1 to 5. The number of neurons was set to $M = \lfloor 0.15L \rfloor$, where $L$ is the length of the time series.

Training is done in two stages each one lasting 1000 epochs. In the first stage, mainly the weight (codebook) vectors are updated while the context descriptors are adjusted just a little. This is done by setting the followings parameters $\alpha_w = 0.5$, $\alpha_{c_k} = \frac{0.5}{K}$ for $k = 1, \ldots, K$, in (11). The initial and final values of parameters used in (13) and (4) were set as follows: $\sigma_0 = 0.1M$, $\sigma_f = 0.01$, $\epsilon_{w0} = 0.3$, $\epsilon_{wf} = 0.001$, $\epsilon_{c0} = 0.3$, $\epsilon_{cf} = 0.001$.

For the second training stage, parameter $\alpha$ is decayed linearly as follows:

$$\alpha_i = \frac{K + 1 - i}{\sum_{k=0}^{K}(k + 1)}, \quad i = 0 \ldots K \tag{14}$$

with $\alpha_w \equiv \alpha_0$. The initial and final values of parameters were the same as described above, except for $\epsilon_{w0} = 0.001$. The latter parameter is kept fixed in the second stage because the weight vectors have already converged during the first training stage.

After convergence of the algorithm, each neuron would have associated a receptive field defined as the mean input sequence that triggers its selection. With the aim of measuring performance, a time window of 30 past events is defined. The size of this window does not affect the function of the model, and it is used for monitoring purposes only. The temporal quantization error (TQE) [3] is used as a performance criterion. TQE measures the average standard deviation of signals within the receptive field of each neuron in the grid for a certain past input. This generates a curve of quantization error versus the index of past inputs. This curve can be integrated to yield the area under TQE curve measure.

---

[1] Available at Times Series Data Library
http://www.robjhyndman.com/TSDL/data/bicup2006.dat

### 5.1   Mackey-Glass Time Series

Figure 1 shows the Mackey-Glass time series, which is the solution of the differential equation $\frac{dx}{dt} = bx(t) \frac{ax(t-d)}{1+x(t-d)^{10}}$, for $a = -0.2$, $b = -0.1$, and $d = 17$. Five hundred points were taken from this time series. The SOM grid size was set as $9 \times 8$, with a total number of 72 neurons (about 15% of the size of the data set). Figure 2 shows the temporal quantization error for Merge SOM ($K = 1$) and Gamma SOM ($K = 3$, $K = 5$). The parameter beta was varied between 0.1 and 0.9, and the best value for Merge SOM was found ($\beta = 0.7$). This value is used in Fig. 2. It can be observed that Gamma SOM ($K = 3$) outperformed Merge SOM ($K = 1$) in terms of TQE for any value of the index of past inputs greater than 1.

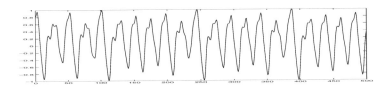

**Fig. 1.** Mackey-Glass time series

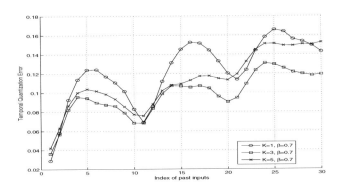

**Fig. 2.** TQE for Mackey-Glass time series using Merge SOM ($K = 1$) and Gamma SOM ($K = 3$, $K = 5$) with $\beta = 0.7$

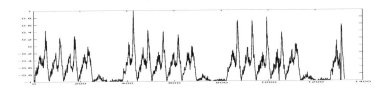

**Fig. 3.** Bicup 2006 time series

## 5.2    Bicup 2006 Time Series

Figure 3 shows the Bicup 2006 time series. It contains the number of passenger arrivals at a subway bus terminal in Santiago, Chile, for each 15-minute interval between 6:30 hours and 22:00 hours for the period March 1-21, 2005.

Fig. 4 shows the TQE performance for $K = 1, 2, 5$, filter stages, and $\beta = 0.9$, which corresponds to the best $\beta$ value found for Merge SOM. It can be observed that the TQE curve for $K = 5$ is below the curve for $K = 1$ for any index of past values greater than 1. Fig. 5 shows the TQE performance for $K = 1$ and $K = 5$ filter stages, and $\beta = 0.5; 0.7; 0.9$. It can be seen that the three curves obtained with Gamma SOM have lower error than the curve corresponding to Merge SOM ($K = 1$), independent of the $\beta$ value explored. Fig. 6 shows the area under TQE curves as a function of K values. A tendency is observed, so that the greater the K value the lower the area under TQE curve. Fig. 7 shows a two-dimensional projection of the resulting temporal vector quantization using principal component analysis (PCA) for the Bicup 2006 time series. Fig. 7 shows the projection obtained with a) Merge SOM ($K = 1$) and b) Gamma SOM ($K = 5$). The Gamma SOM projection is less noisy than the Merge SOM projection. In

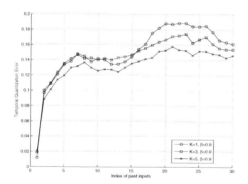

**Fig. 4.** TQE for Bicup 2006 time series using MSOM ($K = 1$, and Gamma SOM ($K = 3$, $K = 5$)

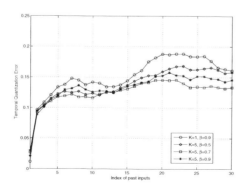

**Fig. 5.** TQE for Bicup 2006 time series using Merge SOM ($K = 1$) and Gamma SOM ($K = 5$) with different $\beta$ values

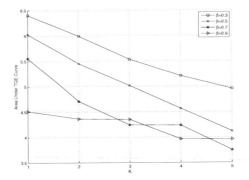

**Fig. 6.** Area under TQE curve versus number of Gamma filter stages, K, for Bicup 2006 time series

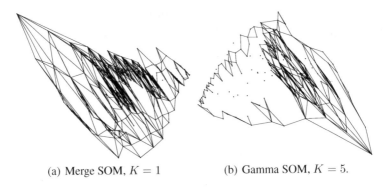

(a) Merge SOM, $K = 1$          (b) Gamma SOM, $K = 5$.

**Fig. 7.** PCA projection of temporal vector quantization result for Bicup 2006 time series, using a) Merge SOM ($K = 1$, $\beta = 0.9$), and b) Gamma SOM ($K = 5$, $\beta = 0.9$)

Fig. 7b), the cyclic nature of the Bicup time series, having different cycle lengths, can be clearly observed.

## 6    Conclusion

A new context model based on Gamma memories has been added to standard SOM with the aim of processing temporal sequences. The so-called Gamma SOM model is a generalization of Merge SOM, the latter being a particular case when a single context is used. It has been shown empirically that by adding more contexts the temporal quantization error tends to diminish, as well as the area under TQE curve. Since the proposed context model is independent of the lattice topology, it can be combined with other vector quantization algorithms such as Neural Gas, Growing Neural Gas, Learning Vector Quantization, etc, being this a promising research line.

## Acknowledgment

This research was supported by Conicyt-Chile under grant Fondecyt 1080643.

# References

1. Kohonen, T.: Self-Organizing Maps. Springer, Heidelberg (1995)
2. Strickert, M., Hammer, B.: Merge SOM for Temporal Data. Neurocomputing 64, 39–72 (2005)
3. Voegtlin, T.: Recursive Self-Organizing Maps. Neural Networks 15, 979–991 (2002)
4. Hammer, B., Micheli, A., Neubauer, N., Sperduti, A., Strickert, M.: Self Organizing Maps for Time Series. In: Proceedings of the Workshop on Self-Organizing Maps (WSOM), Paris, pp. 115–122 (2005)
5. Hammer, B., Micheli, A., Sperduti, A., Strickert, M.: Recursive Self-Organizing Network Models. Neural Networks 17, 1061–1085 (2004)
6. Chappell, G.J., Taylor, J.G.: The Temporal Kohönen Map. Neural Networks 6, 441–445 (1993)
7. De Vries, B., Principe, J.C.: The Gamma model- A New Neural Model for Temporal Processing. Neural Networks 5, 565–576 (1992)
8. Principe, J.C., Giuliano, N.R., Lefebvre, W.C.: Neural and Adaptive Systems. John Wiley & Sons, Inc., New York (1999)
9. Principe, J.C., Kuo, J.M., Celebi, S.: An Analysis of the Gamma Memory in Dynamic Neural Networks. IEEE Trans. on Neural Networks 5, 331–337 (1994)
10. Principe, J.C., De Vries, B., De Oliveira, P.G.: The Gamma Filter – A New Class of Adaptive IIR Filters with Restricted Feedback. IEEE Transactions on Signal Processing 41, 649–656 (1993)
11. Tiño, P., Farkas, I., van Mourik, J.: Dynamics and Topographic Organization of Recursive Self-Organizing Maps. Neural Computation 18, 2529–2567 (2006)
12. Martinetz, T.M., Bercovich, S.G., Schulten, K.J.: "Neural-gas" Network for Vector Quantization and its Application to Time-Series Prediction. IEEE Transactions on Neural Networks 4, 558–569 (1993)
13. Strickert, M., Hammer, B.: Neural Gas for Sequences. In: Yamakawa, T. (ed.) Proceedings of the Workshop on Self-Organizing Networks (WSOM), Kyushu, Japan, pp. 53–58 (2003)

# Fuzzy Variant of Affinity Propagation in Comparison to Median Fuzzy c-Means

T. Geweniger[1,*], D. Zühlke[2], B. Hammer[3], and Thomas Villmann[4]

[1] University of Leipzig - Med. Dep.,
Computational Intelligence Group
Semmelweisstrasse 10, 04103 Leipzig, Germany
tg@sonowin.de
[2] Fraunhofer Institute for Applied Information Technology
Schloss Birlinghoven, 53229 Sankt Augustin, Germany
[3] Clausthal University of Technology,
Institute of Computer Science,
Clausthal-Zellerfeld, Germany
[4] University of Applied Sciences Mittweida
Department of Mathematics/Physics/Informatics,
Computational Intelligence Group
Technikumplatz 17, 09648 Mittweida, Germany

**Abstract.** In this paper we extend the crisp Affinity Propagation (AP) cluster algorithm to a fuzzy variant. AP is a message passing algorithm based on the max-sum-algorithm optimization for factor graphs. Thus it is applicable also for data sets with only dissimilarities known, which may be asymmetric. The proposed Fuzzy Affinity Propagation algorithm (FAP) returns fuzzy assignments to the cluster prototypes based on a probabilistic interpretation of the usual AP. To evaluate the performance of FAP we compare the clustering results of FAP for different experimental and real world problems with solutions obtained by employing Median Fuzzy c-Means (M-FCM) and Fuzzy c-Means (FCM). As measure for cluster agreements we use a fuzzy extension of Cohen's $\kappa$ based on t-norms.

## 1 Introduction

Clustering of objects is a main task in machine learning. There exists a broad variety of different algorithms for a large range of problems. The set of algorithms includes basic ones like c-Means as well as more advanced algorithms like SOM and Neural Gas. Recently some methods have been developed to work with general similarities between the data only instead of the data points itself embedded in a metric space. Examples are Median c-means, Median and Relational Neural Gas and Affinity Propagation [5]. For recently developed affinity propagation (AP) the requirement of completely known similarity matrix is relaxed.

---

* Corresponding author.

J.C. Príncipe and R. Miikkulainen (Eds.): WSOM 2009, LNCS 5629, pp. 72–79, 2009.
© Springer-Verlag Berlin Heidelberg 2009

For some of the similarity based cluster algorithms both crisp and fuzzy variants are available, like Median Fuzzy c-Means (M-FCM) [7], which is an extension of standard Fuzzy c-means (FCM). The aim of this paper is to extend AP in a way to obtain fuzzy cluster assignments. This is possible because of the underlying statistical model of AP. After introduction of the new approach, we give exemplary applications for artificial and real world data. The results are compared with results given by recently developed M-FCM and FCM, the latter one only applicable for metric data.

## 2  Affinity Propagation and Its Fuzzy Extension

### 2.1  Affinity Propagation

Affinity propagation introduced by FREY&DUECK in [5] is a cluster algorithm based on message passing. Contrary to methods like c-means or neural maps, where the number of prototypes has to be known beforehand, affinity propagation assumes all data points as potential prototypes (exemplars) and reduces the number of respective prototypes in the course of calculation. Each data point is interpreted as a node in a network which interacts with the other nodes by exchanging real-valued messages until a set of prototypes and corresponding clusters emerges.

AP can be seen as an exemplar-dependent probability model where the given dissimilarities between $N$ data points $x_i$ (potential exemplars) are identified as log-likelihoods of the probability that the data assume each other as prototypes. More specific, the dissimilarities $s(i, k)$ between data points $i$ and $k$ may be interpreted as log-likelihoods. The goal of AP is to maximize the cost function

$$S(I) = \sum_i s\left(x_i, x_{I(i)}\right) + \sum_j \delta_j(I)$$

where $I : N \to N$ is the mapping function defining the prototypes for each data point. $\delta_j(I)$ is a penalty function

$$\delta_j(I) = \begin{cases} -\infty & \text{if } \exists j, k \ I(j) \neq j, \ I(k) = j \\ 0 & \text{otherwise} \end{cases}$$

The cost function can also be seen as proportional to log-probabilities

$$S(I) = \log\left(\Pi_i P\left(x_i, I(x_i)\right) \cdot P(I)\right)$$

with $P(x_i, I(x_i))$ as the probability that $I(x_i)$ is the prototype for $x_i$ and $P(I)$ is the probability that this assignment is valid. We notice at this point that normalization does not affect the solution.

During the computation two kind of messages are iteratively exchanged between the data until convergence: the responsibilities $r(i, k)$ and the availabilities $a(i, k)$. The responsibilities

$$r(i, k) = s(i, k) - \max_{j \neq k}\{a(i, j) + s(i, j)\}$$

reflect the accumulated evidence that point $k$ serves as prototype for data point $i$. The availabilities

$$a(i,k) = \min\left\{0, r(k,k) + \sum_{j \neq i,k} \max\{0, r(j,k)\}\right\}$$
$$a(k,k) = \max_{j \neq k}\{\max\{0, r(j,k)\}\}$$

describe the accumulated evidence how appropriately data point $k$ is seen as a potential prototype for the points $i$. Finally, the prototypes are determined by

$$I(i) = \arg\max_{j}\{a(i,k) + r(i,k)\}. \tag{1}$$

Hence, $a(i,k)$ and $r(i,k)$ can be taken as log-probability ratios [5]. The iterative alternating calculation of $a(i,k)$ and $r(i,k)$ is caused by the max-sum-algorithm applied for factor graphs [10], which can further be related to spectral clustering [9]. The number of resulting clusters is implicitly handled by the self-dissimilarities $s(k,k)$ also denoted as preferences. The larger the self-dissimilarities the finer is the granularity of clustering [5]. Common choices are the median of input similarities or the minimum thereof.

## 2.2   Fuzzy Interpretation of Affinity Propagation

We now give a fuzzy interpretation of the AP. This interpretation follows from the above probability description. We define the set of exemplars $I_E \subset N$ as before according to (1). For each $x_i, x_j \in I_E$ we define the cluster member probability $P(i,j) = 0$ iff $i \neq j$ and $P(i,i) = 1$. As previously mentioned, the log-probability ratio $r(i,k)$ measures the degree to which point $x_k \in I_E$ is suitable to be the prototype for point $x_i$. To ensure a probability description of cluster assignments we introduce the normalized responsibilities for non-exemplars

$$\hat{r}(i,k) = C\frac{r(i,k) - \max_{i|x_i \notin I_E}\{r(i,k)\}}{\max_{i|x_i \notin I_E}\{r(i,k)\} - \min_{i|x_i \notin I_E}\{r(i,k)\}}. \tag{2}$$

Then, the probabilities for mapping of data points $x_i$ onto the cluster prototype $x_k$ can be defined by

$$P(i,k) = e^{\hat{r}(i,k)}, \; P(i,k) \in [0,1],$$

if the normalization constant $C$ in (2) is chosen appropriately based on the variance of $r(i,k)$. Thus, the definition of $P(i,k)$ is compliant with the log-ratio interpretation of the responsibilities $r(i,k)$. In this way we obtain possibilistic cluster assignments, which are taken as fuzzy degrees. A probabilistic variant can be obtained by subsequent normalization of the $P(i,k)$.

We denote the fuzzy-interpreted AP by Fuzzy Affinity Propagation (FAP).

## 3   Experiments

We perform three exemplary experiments to show the abilities of the new FAP. The first experiment consists of clustering overlapping Gaussians. The distance matrix was generated using the Euclidean metric, since exact data points were known. Thus the FAP can be compared with FCM as classic fuzzy clustering for metric space data.

Subsequently, we apply the FAP algorithm to cluster text documents as examples for non-vectorial data. The first data set is from a multilangual Thesaurus of the European Union whereas the second one consists of transcripts of dialogs from psychotherapy sessions. For the latter example, additional information about meaningful clustering is available from clinical knowledge. Both data sets have in common that they are non-metric examples. The dissimilarities between the data are calculated using the Kolmogorov-complexity [1]. It is based on the minimal description length (MDL) $Z_x$ of a single document $x$ and pairwise combined documents $Z_{x,y}$. The respective normalized information distance is given by:

$$NID_{x,y} = \frac{Z_{xy} - min(Z_x, Z_y)}{max(Z_x, Z_y)}. \tag{3}$$

Usually, the MDL is estimated by the compression length $z$ according to a given standard compression algorithm (here Lempel-Ziv-Markow-Algorithm provided by 7-Zip). Due to technical reasons the estimation is non-symmetric. Therefore, the symmetrized variant $NID_{x,y}^s = \frac{(NID_{x,y}+NID_{y,x})}{2}$ is applied. Further, it has to be noted that $NID_{x,x}$ is non-vanishing in general but takes very small values [1].

For non-vectorial data the FCM is not applicable. Yet, recently a batch variant of FCM was developed, which allows to cluster data sets for which the data dissimilarities are given only [7]. It is called Median FCM (M-FCM).

## 3.1   Measures for Comparing Results

The agreement of crisp cluster solutions is typically judged in terms of Cohen's $\kappa_C \in [-1, 1]$ [2],[3]. This measure yields the statement that, if the result is greater than zero, the cluster agreements are not random but systematic. The values are usually interpreted according to the scheme given in Tab.1. For fuzzy clustering a respective method was recently described [4]. It is based on the observation that the fuzzy cluster memberships can be combined using the concept of *t-norms* [8]. Yet, due to the uncertainty in case of fuzzy sets the definition is not unique. According to the considerations in [13] we suggest to apply the fuzzy-minimum-norm.

**Table 1.** Interpretation of $\kappa$-values

| $\kappa$ - value | meaning |
|---|---|
| $\kappa < 0$ | poor agreement |
| $0 \leq \kappa \leq 0.2$ | slight agreement |
| $0.2 < \kappa \leq 0.4$ | fair agreement |
| $0.4 < \kappa \leq 0.6$ | moderate agreement |
| $0.6 < \kappa \leq 0.8$ | substantial agreement |
| $0.8 < \kappa \leq 1$ | perfect agreement |

## 3.2   Two Toy Problems

The data set for the first toy problem is created from five overlapping Gaussian distributions. We performed a fuzzy clustering based on the similarities using FAP and M-FCM [7]. Additionally we applied FCM to the original data points. According to the remark in Sec.2.1, the preferences for FAP are chosen adequately to obtain the number of the given Gaussians. Figure 1 shows the original clusters in comparison to the calculated fuzzy clustering. The agreement between the solutions and the original data as well as between the solutions of the different algorithms themselves is depicted in table Tab.2. Noticeable is the relatively low agreement to the original data, which is due to the crisp assignments of the original data.

The second toy problem is taken from the original paper by FREY&DUECK [5]. Here, AP was applied to 25 two-dimensional data points, using negative squared error as the similarity. The algorithm clusters the dataset into three groups. Applying the modified FAP yields the same three clusters, yet the cluster assignments are given as fuzzy values. Additionally, M-FCM was executed on the original data set again delivering the same three prototypes. The comparison using Cohen's $\kappa$ gives a value of 0.714 indicating a substantial agreement according to Tab.1.

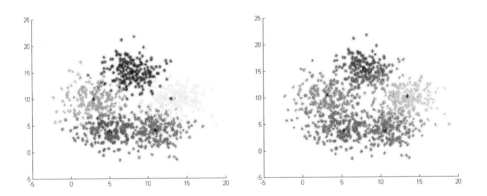

**Fig. 1.** Overlapping Gaussian distributions (left) and their color coded fuzzy cluster assignments (right) as well as cluster centers obtained by FAP. (Colored versions of these images can be obtained from the corresponding author).

**Table 2.** Cohen's $\kappa$ for different cluster algorithms for the overlapping Gaussians example

|  | Fuzzy AP |
|---|---|
| Original Data | $\kappa = 0.52$ |
| M-FCM | $\kappa = 0.79$ |
| FCM | $\kappa = 0.80$ |

## 3.3   Two Real World Problems

**Detection of phase transitions in a psycho-dynamic psychotherapy based on the narratives of the therapy sessions:** In this first real world experiment we analyze text transcripts of a series of 37 psychotherapy session dialogs from a psychodynamic therapy. It is known that the therapy was a two-phase process with the culminating point (phase transition) around session 17. This fact is based on the evaluation of several clinical therapy measures [11]. The similarity between the transcripts is determined using the universal distance description length (*Kolmogorov complexity*) estimated by the file length of the compressed texts, for details see [6]. Again, the predefined number of clusters determines the particular value of the preferences, here appropriately chosen as

$$s\left(k,k\right) = 2 * min(s(j,k)).$$

As shown in [7], cluster algorithms like Median c-Means (M-CM) and Median Neural Gas (M-NG) find similar cluster solutions separating the transcripts into two groups reflecting the break through in therapy. This concordance led to the hypothesis that narratives of the psychotherapy can be related to the therapeutic process [11].

Clustering the same data using fuzzy assignments as with FAP and M-FCM shows a smooth transition from phase one to phase two in the advised period of the therapy. The $\kappa$-value describing the concordance of these two fuzzy clustering algorithms has a value of 0.76, reflecting a substantial agreement, see Fig.2. Further, the soft phase transition is obvious that means a soft transition is triggered by the therapeutic interventions. A similar result is obtained, if psycho-physiological parameters are investigated, which were determined parallely online during the therapy seesions [12]. However, the respective processing is very time consuming. Thus, using the therapy narratives , a cheap alternative is available.

**Clustering of 'Eurovoc' documents:** The first data set, also used as an example in [6], consists of a selection of documents from the multilingual Thesaurus of the European Union 'Eurovoc'. This thesaurus contains thousands of documents, which are available in up to 21 languages each. For the experiments we selected a subset of 100 transcripts and 6 languages English, German, French, Spanish, Finnish and Dutch such that the overall database comprises 600 documents roughly classified into 6 categories: International Affairs, Social Protection, Environment, Social Questions, Education and Communications, and Employment and Working Conditions. The dissimilarities are calculated by means of the *Kolmogorov complexity* [1]. It was shown in [6] that standard AP detects 6 clusters mainly related to the language structure. Applying now FAP to this problem, the resulting agreement in terms of Cohen's $\kappa$ is just a fair agreement, see Tab. 3. However, this effect can be seen as a consequence of the rough discretization into crisp cluster assignments for standard AP. To underline this postulation we also performed M-FCM and median c-means (MCM). Both methods result also in 6 clusters as AP and FAP. And, again, there is large discrepancy between their

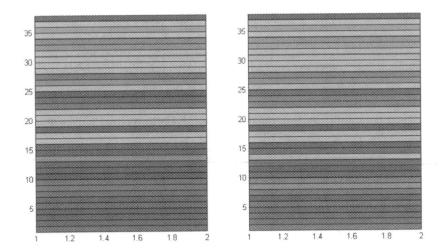

**Fig. 2.** Transition of psychotherapy sessions from phase 1 (magenta) to phase 2 (blue), the vertical axes is the number of the therapy session sequence. The clinical phase transition is identified around session 17. Left - M-FCM, right - FAP. The fuzzy assignment is coded in a blue-magenta color range. Pure color indicates a clear cluster decision whereas color shades stand for uncertain decisions. A nice agreement can be observed, also reflected by the resultig $\kappa = 0.76$. (Colored versions of these images can be obtained from the corresponding author)

**Table 3.** Cohen's $\kappa$ for the cluster solutions obtained by the different algorithms for the 'Eurovec' database

|        | AP | M-FCM | MCM |
|--------|-----|-------|-----|
| FAP    | $\kappa = 0.208$ | $\kappa = 0.810$ | $\kappa = 0.208$ |
| AP     |     | $\kappa = 0.183$ | $\kappa = 1.000$ |
| M-FCM  |     |       | $\kappa = 0.183$ |

results looking at the $\kappa$-values - only a slight agreement, see Tab. 3. However, the agreement between the fuzzy variants FAP and M-FCM on the one hand side, and between their crisp counterparts on the other hand, show both perfect agreement. Hence, one can conclude that the discretization causes the low $\kappa$-values.

## 4   Conclusion

We extended the recently proposed Affinity Propagation algorithm to a fuzzy AP. For this purpose the probabilistic interpretation of the AP is used. Following this model the responsibilities calculated during standard AP can be used in modified form for fuzzy membership assignments of the data to the cluster centers (exemplars/prototypes). Exemplary applications demonstrate the abilities

of the improved method. We compared our method with other fuzzy clustering schemes. The agreement between the methods is determined by applying
the fuzzy variant of Cohen's $\kappa$ (here based on the minimum t-norm). The FAP
shows nice agreement with the other similarity based fuzzy clustering methods in
terms of high $\kappa$-values. Yet, as shown by FREY&DUECK in [5], AP is frequently
much faster than other clustering algorithms, and - in particular - than FCM.
The same behaviour can be observed for FAP versus M-FCM. Hence, FAP can
be taken as a fast alternative to M-FCM.

# References

1. Cilibrasi, R., Vitányi, P.: Clustering by compression. IEEE Transactions on Information Theory 51(4), 1523–1545 (2005)
2. Cohen, J.: A coefficient of agreement for nominal scales. Educational and Psychological Measurement 20, 37–46 (1960)
3. Cohen, J.: Weighted chi square: An extension of the kappa method. Educational
   and Psychological Measurement 32, 61–74 (1972)
4. Dou, W., Ren, Y., Wu, Q., Ruan, S., Chen, Y., Constans, D.B.A.-M.: Fuzzy kappa
   for the agreement measure of fuzzy classifications. Neurocomputing 70, 726–734
   (2007)
5. Frey, B., Dueck, D.: Clustering by message passing between data points. Science 315, 972–976 (2007)
6. Geweniger, T., Schleif, F.-M., Hasenfuss, A., Hammer, B., Villmann, T.: Comparison of cluster algorithms for the analysis of text data using kolmogorov complexity.
   In: Proc. of the International Conferens on Neural Information Processing (ICONIP
   2008). Springer, Heidelberg (in press, 2009)
7. Geweniger, T., Zühlke, D., Hammer, B., Villmann, T.: Median variant of fuzzy c-
   means. In: Verleysen, M. (ed.) Proc. Of European Symposium on Artificial Neural
   Networks (ESANN 2009), Evere, Belgium. D-side publications (in press, 2009)
8. Hammer, B., Villmann, T.: How process uncertainty in machine learning? In: Verleysen, M. (ed.) Proc. Of European Symposium on Artificial Neural Networks
   (ESANN 2007), Brussels, Belgium, pp. 79–90. D-side publications (2007)
9. Luxburg, U.V.: A tutorial on spectral clustering. Statistics and Computing 17(4),
   395–416 (2007)
10. Pearl, J.: Probabilistic Reasoning in Intelligent System. Morgan Kaufmann, San
    Francisco (1988)
11. Villmann, T., Liebers, C., Bergmann, B., Gumz, A., Geyer, M.: Investigation of
    psycho-physiological interactions between patient and therapist during a psycho-
    dynamic therapy and their relation to speech in terms of entropy analysis using a
    neural network approach. New Ideas in Psychology 26, 309–325 (2008)
12. Villmann, T., Liebers, C., Geyer, M.: Untersuchung der psycho-physiologischen
    Interaktion von Patient und Therapeut im Rahmen für psychodynamische
    Einzeltherapien und informationstheoretische Auswertung. In: Geyer, M., Plöttner,
    G., Villmann, T. (eds.) Psychotherapeutische Reflexionen gesellschaftlichen Wan-
    dels, pp. 305–319 (2003)
13. Zühlke, D., Geweniger, T., Heimann, U., Villmann, T.: Fuzzy Fleiss-Kappa for com-
    parison of fuzzy classifiers. In: Verleysen, M. (ed.) Proc. of European Symposium
    on Artificial Neural Networks (ESANN 2009), Evere, Belgium. D-side publications
    (in press, 2009)

# Clustering with Swarm Algorithms Compared to Emergent SOM

Lutz Herrmann and Alfred Ultsch

Databionics Research Group
Department of Mathematics and Computer Science
Philipps University of Marburg
{lherrmann,ultsch}@informatik.uni-marburg.de

**Abstract.** Swarm-based methods are promising nature-inspired techniques. A swarm of stochastic agents performs the task of clustering high-dimensional data on a low-dimensional output space. Most swarm methods are derivatives of the Ant Colony Clustering (ACC) approach proposed by Lumer and Faieta. Compared to clustering on Emergent Self-Organizing Maps (ESOM) these methods usually perform poorly in terms of topographic mapping and cluster formation. A unifying representation for ACC methods and Emergent Self-Organizing Maps is presented in this paper. ACC terms are related to corresponding mechanisms of the SOM. This leads to insights on both algorithms. ACC can be considered to be first-degree relatives of the ESOM. This explains benefits and shortcomings of ACC and ESOM. Furthermore, the proposed unification allows to judge whether modifications improve an algorithm's clustering abilities or not. This is demonstrated using a set of critical clustering problems.

## 1 Introduction

Flocking behaviour of social insects has inspired various algorithms in numerous research papers over the last decade due to the ability of simple interacting entities to exhibit sophisticated self-organization abilities. A particularly interesting field of application is cluster analysis, i.e. the retrieval of groups of similar objects in high-dimensional spaces. The idea behind Ant Colony Clustering (ACC) is that autonomous stochastic agents, called ants, move data objects on a low-dimensional regular grid such that similar objects are more likely to be placed on nearby grid nodes than dissimilar ones. This task is referred to as *topographic mapping*.

In the following sections, the basic ACC algorithm by Lumer/Faieta is introduced in a notation consistent with SOM for non-vectorial data, i.e. Dissimilarity-SOM. A unifying representation for both methods is therefore derived in Section 3. Sections 4 and 5 describe how to improve topographic mapping and cluster analysis of ACC methods on basis of SOM. Finally, in Section 6 the effect of altered objective functions is empirically verified.

J.C. Príncipe and R. Miikkulainen (Eds.): WSOM 2009, LNCS 5629, pp. 80–88, 2009.

## 2    Ant Colony Clustering

The ACC method proposed by Lumer and Faieta [9] operates on a fixed regular low-dimensional grid $\mathbb{G} \subset \mathbb{N}^2$. A finite set of input samples $X$ from a vector space with norm $\|.\|$ is projected onto the grid by $m : X \rightarrow \mathbb{G}$. The mapping $m$ is altered by autonomous stochastic agents, called ants, that move input samples $x \in X$ from $m(x)$ to new location $m'(x)$. Ants move randomly on neighbouring grid nodes. Ants might pick input samples when facing occupied nodes and drop input samples when facing empty nodes. Probabilities for picking and dropping actions, respectively, are determined using objective function $\phi : \mathbb{G} \times X \rightarrow \mathbb{R}_0^+$, at which $\phi(x, i)$ denotes the average similarity between $x \in X$ and input samples located on the so-called perceptive neighbourhood around node $i \in \mathcal{G}$. Usually, the perceptive neighbourhood consists of $\sigma^2 \in \{9, 25\}$ quadratically arranged nodes at which the ant is located in the center. The set of input samples mapped onto the perceptive neighbourhood around $i \in \mathbb{G}$ is denoted with $N_x(i) = \{y \in X : y \neq x, \ m(y) \text{ neighbouring } i\}$.

$$\phi_x(i) = \frac{1}{\sigma^2} \sum_{y \in N_x(i)} \left( 1 - \frac{\|x - y\|}{\alpha} \right) \tag{1}$$

ACC methods lead to a local sorting of input samples on the grid in terms of similarities. Ants gather scattered input samples into dense piles. In literature, it has been noticed that ACC derivatives are prone to produce too many and too small clusters [1] [5]. For illustration see Figure 1.

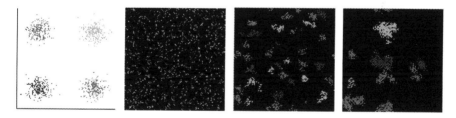

**Fig. 1.** Typical result of ACC methods. From left to right: gaussian data with 4 clusters, initial mapping of data objects, dense clusters appear, too many clusters with topological defects have finally emerged [1].

## 3    Analysis of Ant Colony Clustering by Means of Dissimilarity-SOM

The Self-Organizing Batch Maps (Batch-SOM) and its derivatives are particularly interesting for analyiss of Ant Colony Clustering (ACC) methods. Batch-SOM consist of grid $\mathbb{G}$, codebook vectors $w_i \in \mathbb{R}^n, i \in \mathbb{G}$ and a mapping function $m : X \rightarrow \mathbb{G}$ with $m(x) = \arg\min_{i \in \mathbb{G}} \|x - w_i\|$. It was shown in [5] how the objective $\phi$ of each ant is related to $m : X \rightarrow \mathbb{G}$ of Batch-SOM.

The so-called Dissimiliarity-SOM [8], often referred to as Median SOM, is a generalization of the Batch-SOM for nonvectorial input data. For the sake of simplicity, let $\|x - y\| \in \mathbb{R}_0^+$ denote the dissimiliarity of each $x, y \in X$. Codebook vectors are updated according to the generalized median, i.e. $w_i = \arg\min_{x \in X} \Phi_x(i)$. Here, $h : \mathcal{G} \times \mathcal{G} \to [0, 1]$ denotes the neighbourhood function of SOM.

$$\Phi_x(i) = \sum_{y \in X} h(m(y), i) \cdot \|x - y\| \quad \text{with} \quad \sum_{y \in X} h(m(y), i) = 1 \qquad (2)$$

In the following, the mechanism of picking and dropping ants is no longer subject of consideration. In [10] it was shown that collective intelligence can be discarded in ACC systems, i.e. same results were achieved without ants but using objective function $\phi$ directly for probabilistic cluster assignments. This simplification is evident: over a sufficient period of time, randomly moving ants may select any arbitrary subset of input samples, but re-allocation through picking and dropping depends on $\phi$ only. Probability of selection is the same on all input samples such that ants might be omitted in favor of any other subset sampling technique.

A meaningful symmetrical neighbourhood function $h : \mathbb{G} \times \mathbb{G} \to [0, 1]$ for ACC methods is defined according to the perceptive neighbourhood of ants, i.e. $h(i, j)$ is 1 if $j \in \mathbb{G}$ is located in the perceptive neighbourhood of node $i \in \mathbb{G}$ and 0 elsewhere. Equation 3 reformulates the ants' objective $\phi$ by incorporating $\Phi$ (see Equation 2).

$$\phi_x(i) = \frac{|N_x(i)|}{\sigma^2} \cdot \left(1 - \frac{\Phi_x(i)}{\alpha}\right) \quad \text{with} \quad \Phi_x(i) = \frac{\sum_{y \in X} h(m(y), i) \cdot \|x - y\|}{\sum_{y \in X} h(m(y), i)} \qquad (3)$$

The ACC method uses a fixed neighbourhood function with small radius, whereas Dissimilarity-SOM uses shrinking neighbourhood functions with large radiuses. ACC has a probabilistic update of mapping $m : X \to \mathbb{G}$, whereas Dissimilarity-SOM is deterministic. The objective function of ACC algorithms decomposes into an output density term $\frac{|N|}{\sigma^2}$ and a term $1 - \frac{\Phi}{\alpha}$ related to topographic quality. Therefore, the ACC algorithm is easily convertible into a special case of Dissimilarity-SOM, and vice versa. For a brief overview of differences see Table 1.

**Table 1.** varieties of Dissimilarity-SOM and Ant Colony Clustering

| | Dissimilarity-SOM | ACC |
|---|---|---|
| neighbourhood $h : \mathbb{G} \times \mathbb{G} \to [0, 1]$ | large shrinking | small, fixed |
| update of $m : X \to \mathbb{G}$ | deterministic | probabilistic |
| searching for update of $m : X \to \mathbb{G}$ | global $\mathbb{G}$ | local $\subset \mathbb{G}$ |
| objective function | $\Phi$ | $\frac{|N|}{\sigma^2}(1 - \frac{\Phi}{\alpha})$ |
| termination | cooling scheme | never |

## 4   Improvement of Ant Colony Clustering

From Dissimilarity-SOM, minimization of $\Phi$ is known to produce sufficiently to-pography preserving mappings $m : X \to \mathbf{G}$, e.g. when using Dissimilarity-SOM. In contrast to that, the *output density term* $\frac{|N|}{\sigma^2}$ has some major flaws. First, the output density term leads to maximization of output space densities, instead of preservation. Obtained mappings are, therefore, not related to the configuration of available clusters in the input space. Traditional ACC algorithms are not al-lowed to assign two or more objects to a single grid node (see Section 2) in order to prevent the mapped clusters from collapsing into a single grid node. Due to that, densities of input data can hardly be preserved on grid $\mathbf{G}$. In comparison with the topographic term, the output density term is much easier to maximize and, therefore, will distort the objective function $\phi$. Accounting of output den-sities is prone to distort the formation of correct topographic mappings because it is responsible for additional local optima of $\phi$.

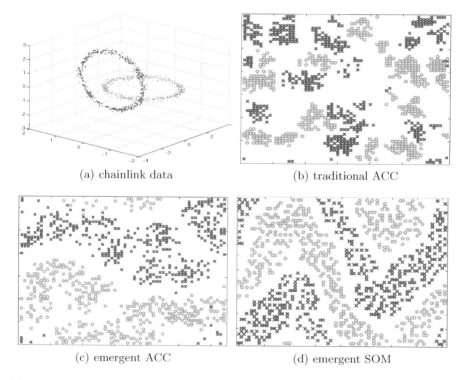

(a) chainlink data                    (b) traditional ACC

(c) emergent ACC                     (d) emergent SOM

**Fig. 2.** ACC projects looped cluster structures on a *toroid* grid. (a) Chainlink data from FCPS [11]. (b) Traditional ACC with small $\sigma$ produces too many small clusters. (c) Emergent ACC enables the formation of looped clusters. (d) Emergent SOM enables the formation of looped clusters.

The topographic term $1 - \frac{\Phi'}{\alpha}$ of the ACC objective function depends on the shape of the neighbourhood function $h : \mathbb{G} \times \mathbb{G} \to \{0, 1\}$. Usually, neighbourhoods' sizes are chosen as $\sigma^2 \in \{9, 25\}$, i.e. the immediate neighbours. From SOM it is known that the cooling scheme of the neighborhood radius vitally influences the obtained topographic mapping quality. (see [6] for details). A bigger radius enables a more continuous mapping in the sense that proximities existing in the original data are visible on the grid. This is evident because smaller neighbourhoods are more likely to exclude parts of a cluster.

In order to cope with the shortcomings mentioned above, we introduce the *Emergent Ant Colony Clustering* method. An ACC method is said be be emergent if it fulfills the following conditions:

- Ants' modifications of mapping $m : X \to \mathbb{G}$ is directed by minimization of $\Phi$
- Ants do not account for output densities.
- The perceptive neighbourhood of ants is not limited to immediate neighbours on grid $\mathbb{G}$. Instead, bigger neighbourhood radiuses are to be chosen in order to obtain SOM-like mappings.

Figure 2 illustrates the ability of emergent ACC method to preserve even looped input space clusters, which is hardly possible for traditional ACC.

## 5   Data Analysis with Emergent Ant Colony Clustering

Emergent ACC usually will provide an ESOM-like projection, i.e. input samples are uniformly mapped onto the grid. See Figure 2 for illustration. In this case, cluster retrieval cannot be achieved according to sparse regions dividing dense clusters on the grid.

A promising technique for cluster retrieval is based on so-called U-Maps [12]. Arbitrary projections from normed vector spaces onto grid $\mathbb{G} \subset \mathbb{N}^2$ are transformed into landscapes, so-called U-Maps. The U-Map technique assigns each grid node a height value that represents the averaged input space distance to its' neighbouring nodes and codebook vectors, respectively. Clusters lead to valleys on U-Maps whereas empty input space regions lead to mountains dividing the cluster valleys. This is illustrated in Figure 3 using Fisher's well-known iris data [3]. Traditional ACC produces too many valleys, whereas Emergent ACC preserves cluster structures.

The U*C cluster algorithm uses the so-called watershed transformation to retrieve cluster valleys on U-Maps. See [13] for details.

## 6   Experimental Settings and Results

In order to measure the distortion of a topographic mapping method in question, a collection of fundamental clustering problems (FCPS) is used [11]. Each data set represents a certain problem that arbitrary algorithms shall be able to handle

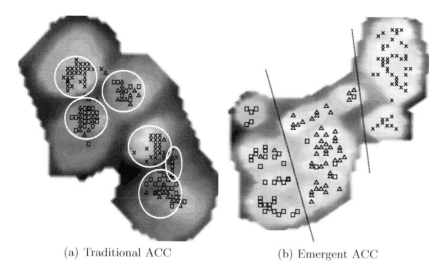

(a) Traditional ACC                    (b) Emergent ACC

**Fig. 3.** Well known iris data [3]: setosa ($\times$), versicolor ($\triangle$), virginica ($\square$). U-Maps shown as islands generated from toroid grids. Dark shades of gray indicate high inter-cluster distances. (a) Too many small clusters emerge from traditional ACC. (b) Emergent ACC preserves three clusters after the same number of learning epochs.

when facing unknown real-world data. Here, traditional and emergent ACC are tested on which one delivers the best topographic mapping.

A comprehensive overview on topographic distortion measurements can be found in [4]. Here, the so-called *minimal path length* (MPL) measurement is used.

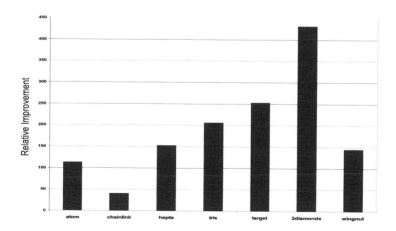

**Fig. 4.** Improvement of topographic quality measured by *minimal path length* method: percental $z$-scores of traditional over emergent ACC. Emergent ACC leads to improvements between 50% to 400% when compared to traditional ACC on different FCPS data sets.

It is an easy-to-compute measurement that sums up input space distances of grid-neighbouring data objects and codebook vectors, respectively.

$$mpl = \sum_{x \in X} \frac{1}{|N_x|} \sum_{y \in N_x} \|x - y\| \qquad (4)$$

Lower MPL values indicate less topographic distortion when moving on the grid and, therefore, a more trustworthy topographic mapping. Each algorithm is run several times with the same parametrization. MLP values indicate if accounting for output densities assists the formation of good topographic mappings, or not. All data sets from the FCPS collection were processed with the same parameters established in literature, i.e. $\alpha = 0.5$, $\sigma^2 = 25$, $k_1 = 0.3$ and $k_2 = 0.1$ on a $64 \times 64$ grid with 100 ants during 100000 iterations. The results are illustrated in Figure 4. Accounting for output densities leads to increasing MPL values on an average, i.e. worsenings of topographic mappings. Significance has been confirmed using a Kolmogorov-Smirnov test on a $\alpha = 5\%$ level. All obtained $p$-values are below $10^{-5}$.

## 7   Discussion

This work shows a previously unknown relation of two topographic mapping techniques, namely Dissimilarity-SOM and Ant Colony Clustering (ACC). It is based on the assumption [10] that stochastic agents, e.g. ants, are nothing more than an arbitrary sampling technique that is to be omitted for further analysis of formulae. This simplification is evident but may be invalid for stochastic agents guided by more than just randomness and topographic distortion, e.g. ants following pheromone trails. Our analysis of formulae does not cover algorithms that are not ACC derivatives following the Lumer/Faieta scheme. In contrast to hybrid approaches, like KohonAnts [2], our work creates a unifying basis for comprehension and creation of techniques from the fields of artificial neural networks and swarm-intelligence.

Minimal path lengths (MPL), as proposed in Section 6, are well-known topographic distortion measures. The length of input space *paths* is normalized by the cardinality $|N_x|$ of the corresponding grid neighbourhood, i.e. the number of objects mapped onto the grid neighbourhood. This is supposed to decrease error values of locally dense mappings, as produced by traditional ACC, because small radial neighbourhoods usually do not cover objects of another cluster, since locally dense mappings imply sparse dividing grid regions around clusters. Nevertheless, traditional ACC produces bigger MPL errors than emergent ACC that is not accounting for densities. We conclude that the topographic mapping quality is improved beyond our empirical evaluation.

Traditional and emergent ACC methods do not converge due to the architecture of stochastic agents. Instead, they enable perpetual machine learning. ACC methods are, therefore, to be favored over traditional methods, like Self-Organizing Maps and hierarchical clustering, when dealing with incremental learning tasks.

# 8  Summary

This work continues our last publication [5] at which the Ant Colony Clustering (ACC) method by Lumer and Faieta [9] was related to Self-Organizing Batch Maps [7]. The mechanism of picking and dropping ants was omitted in favor of a formal analysis of the underlying formulae and comparison with Kohonen's Dissimilarity-SOM. It could be shown that a unifying framwork for both methods does exist in terms of a common topographic error function. The ACC method is to be considered a probabilistic, first-class relative of Batch-SOM and, especially, Dissimilarity-SOM. The behaviour of ACC methods becomes explainable on that unifying basis.

ACC methods exhibit poor clustering abilities because of distorted topographic mappings. Improvements of topographic mapping were derived by means of SOM architecture. Perceptive areas are to be increased, and accounting for density of mapped data is futile. The novel method *Emergent ACC* does not produce dense clusters any more but uniformly distributed, SOM-like projections. Due to that, clusters are to be retrieved using U-Map technology. As predicted by our theory, an empirical evaluation showed on critical clustering problems that disregarding the density of mapped data improves the quality of topographic mapping despite of unfavorable settings.

# References

1. Aranha, C., Iba, H.: The effect of using evolutionary algorithms on ant clustering techniques. In: Pham, L., Le, H.K., Nguyen, X.H. (eds.) Proceedings of the Third Asian-Pacific workshop on Genetic Programming, Military Technical Academy, Hanoi, VietNam, pp. 24–34 (2006)

2. Fernandes, C., Mora, A., Merelo, J.-J., Ramos, V., Gimenez, J.: KANTS: Artificial Ant System for Classification. In: Ant Colony Optimization and Swarm Intelligence - Proceedings 6th Int. Conf., Brussels, Belgium (2008)

3. Fisher, R.A.: The use of multiple measurements in taxonomic problems. In: Annals of Eugenics, Part II, vol. 7, pp. 179–188. Cambridge University Press, Cambridge (1936)

4. Goodhill, G.J., Sejnowski, T.J.: Quantifying neighbourhood preservation in topographic mappings. In: Proc. 3rd Joint Symposium on Neural Computation, California Institute of Technology (1996)

5. Herrmann, L., Ultsch, A.: Explaining Ant-Based Clustering on the basis of Self-Organizing Maps. In: Proc. of the European Symposium on Artificial Neural Networks (ESANN 2008), Bruges, Belgium (2008)

6. Nybo, K., Venna, J., Kaski, S.: The self-organizing map as a visual neighbor retrieval method. In: Proc. of the Sixth Int. Workshop on Self-Organizing Maps (WSOM 2007), Bielefeld (2007)

7. Kohonen, T.: Self-Organizing Maps. Springer Series in Information Sciences, vol. 30. Springer, Heidelberg (1995, 1997, 2001)

8. Kohonen, T., Somervuo, P.: How to make large self-organizing maps for nonvectorial data. Neural Networks (15), 8–9 (2002)

9. Lumer, E., Faieta, B.: Diversity and adaption in populations of clustering ants. In: Proceedings of the Third International Conference on Simulation of Adaptive Behaviour: From Animals to Animats, vol. 3, pp. 501–508. MIT Press, Cambridge (1994)
10. Tan, S.C., Ting, K.M., Teng, S.W.: Reproducing the Results of Ant-Based Clustering without Using Ants. In: IEEE Congress on Evolutionary Computation (2006)
11. Fundamental Clustering Problem Suite, http://www.uni-marburg.de/fb12/datenbionik/data
12. Ultsch, A., Mörchen, F.: U-maps: topograpic visualization techniques for projections of high dimensional data. In: Proc. 29th Annual Conference of the German Classification Society (GfKl 2006), Berlin (2006)
13. Ultsch, A., Herrmann, L.: Automatic Clustering with U*C, Technical Report, Dept. of Mathematics and Computer Science, Philipps-University of Marburg (2006)

# Cartograms, Self-Organizing Maps, and Magnification Control

Roberto Henriques[1], Fernando Bação[1], and Victor Lobo[1,2]

[1] ISEGI-UNL, Portugal
[2] Portuguese Naval Academy, Portugal
{roberto,bacao,vlobo}@isegi.unl.pt

**Abstract.** This paper presents a simple way to compensate the magnification effect of Self-Organizing Maps (SOM) when creating cartograms using Carto-SOM. It starts with a brief explanation of what a cartogram is, how it can be used, and what sort of metrics can be used to assess its quality. The methodology for creating a cartogram with a SOM is then presented together with an explanation of how the magnification effect can be compensated in this case by pre-processing the data. Examples of cartograms produced with this method are given, concluding that Self-Organizing Maps can be used to produce high quality cartograms, even using only standard software implementations of SOM.

**Keywords:** Self-Organizing Maps, Cartograms, Magnification effect.

## 1 Introduction

Cartograms are a type of map used in various fields to convey information about data that is geo-referenced. An example of a cartogram of the population by state in the USA is given in Figure 1. The general shape of the country and of its states is recognizable, but the states with larger population are clearly identified as being "larger" than the others. The basic idea of a cartogram is to distort a geographical map by distorting the area of a region according to some variable of interest (*e.g.* population) while keeping the map, as much as possible, recognizable.

Cartograms can be a powerful way to convey information regarding characteristics of geographic regions. In the example given in Figure 1, it becomes very clear that a few states, such as California, Florida, and a few East Coast states are, population wise, more important than their geographic area would indicate. The opposite happens with most mid-western states.

An even more striking example of how useful cartograms are is presented in Figure 2 [2], where the counties where candidates John Kerry and George W. Bush won in 2004 are represented in different colours. In a standard map (*i.e.* using a common projection), Bush's victory seems overwhelming. However, in a cartogram, it can clearly be seen that it was, in fact, a close call, since John Kerry won in counties with a high population density, and the total number of votes for each candidate was very similar.

J.C. Príncipe and R. Miikkulainen (Eds.): WSOM 2009, LNCS 5629, pp. 89–97, 2009.

**Fig. 1.** Population cartogram of USA. Although the general shape of the USA is still recognizable, each state has a size proportional to it's population, not it's area.

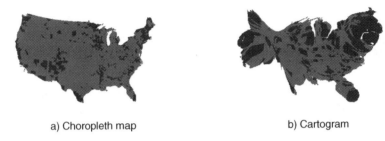

a) Choropleth map                              b) Cartogram

**Fig. 2.** USA 2004 presidential election results. Counties where John Kerry won are represented in blue, while the others are represented in red [2].

There are several algorithms for building cartograms. A good review of the work done in this area was written by Tobler [3] a few years ago. None of the algorithms has proved to be universally better than any other, since the trade-offs made to get the desired result vary. A new method called Carto-SOM, based on Self-Organizing Maps (Kohonen's Self-Organizing Maps or SOM) was recently proposed [1, 4].In this paper we present an overview of this method. One of the inconveniencies of SOM when producing cartograms is that the magnification effect [5] will introduce undesired distortions. We will show that a simple pre-processing technique can compensate this effect, for this particular case. A few examples of cartograms produced with this method are shown.

## 2   Cartogram Creation

As Keim *et.al.* [6] point out cartogram generation is a map deformation problem. The inputs of the problem are:

- a polygon map composed of a set of regions, each with an initial area given by the true geographical area (each polygon matches a region);
- a target value for the final area of each one of the polygons (regions), representing the variable of interest in that region.

The goal of the map distortion is to approximate, as much as possible, the areas of the regions to the desired target.

The cartogram creation problem may be formally described as follows:

Let $M=\{R_0, R_1, \ldots R_n\}$ be a map $M$ consisting on $n$ regions $R_i$, each of which forms
a polygon.
Let $T(M)$ be the contiguity matrix of $M$.
Let each region $R_i$ have an area of $a_i$ (area of its polygon) and a "value of interest"
$v_i$ (usually population, average income, or some other variable).
Produce a cartogram map $C=\{R'_0, R'_1, \ldots R'_n\}$ consisting of $n$ regions $R'_i$, each of
which forms a polygon with area $a'_i = k\,v_i$, with $k=\Sigma a/\Sigma v$, and $T(C)=T(M)$.

There is still another goal, loosely described as "shape similarity". Let $S$ be the
shape of $M$ and $S'$ the shape of $C$. The objective is that $S \cong S'$. The "shape similarity"
is an elusive concept that translates the ability of a reader to recognize $C$ as an
instance of $M$. This property is difficult to measure and difficult to define rigorously.
The ability to recognize $C$ is not only dependent on the ability to preserve the shape of
each $R_i$ but also on preserving certain landmark points. For instance recognizing a
certain cartogram as a cartogram of the United States is much more dependent on
preserving the shape of Florida than on preserving the shape of North Dakota.

There are many possible cartograms $C$ that achieve the desired goal, but most
mappings from $M$ to $C$ are the result of an iterative process, and only asymptotically
get the desired result (or at least some approximation to it).

## 2.1 Quantitative Evaluation of Cartograms

While it is subjective to compare the visual quality of different cartograms, it is easy
to define a numerical value that characterizes how well the cartogram shares the
available area between different regions, according to the given variable of interest.
Various such measures have been proposed. Keim $et.al.$ [6] proposes an area error
function to determine the cartogram error in each region. This relative area error $e^i_{rel}$
of a region $R_i$ is given by:

$$e^i_{rel} = \frac{\left|a'_i - a^{current}_i\right|}{a'_i + a^{current}_i} \qquad (1)$$

Where $a'_i$ is the desired area of region $R_i$ in the cartogram (to create a perfect
cartogram) and $a^{current}_i$ is the area of region $R_i$ in the cartogram.

The error measure for the whole map may be computed as the mean quadratic error,
weighed error, or simple absolute error of all regions, using the following expressions:

$$mqe(v) = \frac{\sqrt{\sum_{i=1}^{N} e(i,v)^2}}{N} \qquad (2)$$

$$we(v) = \sum_{i=1}^{N} |\,a_i \times e(i,v)\,| \qquad (3)$$

$$se(v) = \sum_{i=1}^{N} |\,e(i,v)\,| \qquad (4)$$

# 3  Building Cartograms Using SOM

Our inspiration for using SOM to build cartograms stems from the fact that it can be seen as a density estimation tool [8-10], despite the limitations imposed by the magnification effect discussed later.

The basic ideas of the Carto-SOM method are quite simple. We start by generating random points with a uniform distribution in each geographic region. The number of points in each region is proportional to the variable of interest (*e.g.* population, average income, rainfall) of that region. Those points will be characterized by two real variables that are the geographic coordinates, and a label to identify the region they belong to. A 2-dimensional SOM can then be trained with that 2 dimensional data, and labeled accordingly. The labeled SOM (seen in output space formed by the grid) can be viewed as a cartogram. If all units had the same number of points mapped to it (*i.e.* if the SOM had a uniform magnification factor of 1), it would be a perfect cartogram, from a purely quantitative point of view. The greater the number of units on the SOM, the smoother and more faithful to detail the cartogram can be.

A graphical example can make the Carto-SOM process clearer. Figure 3 shows an example of this process, using a simplified instance with only two rectangular regions. The two regions are geographically identical but have distinct values for the variable of interest, represented by *p*. The variable *p* has a high value in the dark region (*i.e.*, this is a region with a high population density), and a smaller value in the lighter region. In the following step (Figure 3.b) we randomly produce points with a uniform distribution (represented with triangles) inside each region. The number of points created is a linear function of the value of population *p*. Figure 3.c shows an initialized two-dimensional SOM (with 4 x 4 units). The SOM units are initialized so as to form a regular grid in the input space, contrary to the usual practice, in which the units are randomly initialized in the input space. We then continue with a standard SOM training phase where the units adapt to the training patterns. This means that units are moved (in the input space) in such a way that their density mimics the density of the data patterns (Figure 3.d). The algorithm continues with the labeling process (Figure 3.e). This process gives each unit a label based on the data of the

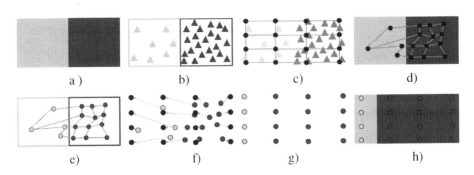

**Fig. 3.** Example of the proposed Carto-SOM method, applied to map with two rectangular regions with different population densities

region where it lies. In this case, the labels are the colors that identify each region, and thus each unit will be assigned the color of the region where it lies after training. Figure 3.g represents each unit mapped back at its initial position, and coincides with it's coordinates in the output space. Based on the units' positions and labels a final population cartogram is produced (Figure 3.h). In this cartogram the darker region (that had more population) is larger than the lighter one (that had less population), and thus the final area of the regions is proportional to their population.

While most SOMs form a regularly shaped grid (usually rectangle), the regions of a geographic map, when considered all together, don't. While it is possible to adjust the shape of the SOM to that of the geographic map, it is best to keep it rectangular and add a "outside" or "ocean" area to make the original map rectangular. The various ways of doing so are discussed in [4]. In this paper we will use a series of "outside regions" with a density equal to that of the neighboring region of interest.

There are a few minor issues, namely the possibility of having units that do not match any data (and thus have no label), and the magnification correction discussed later, that make the complete algorithm a bit more lengthily, and a complete, detailed, step by step algorithm is presented in [1].

## 4 Compensating for the Magnification Effect of SOM

The properties of the SOM as a density estimation tool have not yet been completely established, except for some very particular cases *e.g.* the one dimensional case [11, 12]. In these cases it has been found that the SOM has a bias towards low density areas *i.e.* $D_{units} = K*D_{data}{}^{\mu}$, where $K$ is a constant scaling factor, and $\mu$, known as magnification factor, is 2/3 for some known cases. A magnification effect of 1 would mean strict proportionality between the density of input patterns and the density of their assigned units. With a magnification factor of less then 1 low density areas will be proportionally over represented. Some work has been done to calculate the magnification factor in more general cases [9, 13, 14], and experimental evidence suggests that when using the original SOM algorithm the magnification factor is always less than one. Some methods have been proposed to explicitly control the magnification factor [5, 9, 15, 16]. These magnification control mechanisms require changes to the original SOM algorithm. In our proposed method, the standard SOM algorithm is used, but the original data is pre-processed to compensate for magnification.

This pre-processing consists on *boosting* the original data, so that low density areas will have even lower *surrogate* densities to compensate for the magnification. We can then use this *surrogate* dataset instead of the original one. Let us now see in detail how this dataset may be generated.

For a 1-dimensional to 1-dimensional mapping, under the special conditions described by [17], the relationship between densities of units and data points is given by,

$$D_{units} = k_1 D_{data}{}^{\mu} \tag{5}$$

where $D_{units}$ is the density of SOM units, $D_{data}$ is the density of data points, $k_1$ is a proportionality constant that results from the ratio between the total number of units and the total number of data points, and $\mu$ is the magnification factor.

To build a good cartogram, we want strict proportionality between the density of units and the density of the original data. To obtain this we must have a surrogate dataset where the density $D_{surr}$ is such that,

$$D_{units} = k_1 D_{surr}{}^\mu = k_2 D_{data} \Leftrightarrow D_{surr} = \left(\frac{k_2}{k_1}\right)^{\frac{1}{\mu}} D_{data}{}^{\frac{1}{\mu}} = k_3 D_{data}{}^{\frac{1}{\mu}} \tag{6}$$

where $k_3$ is $(k_2/k_1)^{1/\mu}$. For the sake of simplicity we will not explicitly calculate each of the constants $k_x$ involved in these calculations, since in the end we will be able to calculate the necessary variables without them.

Since we will be generating, for each region, a number of points proportional to the variable of interest $V$ and not its density, we have,

$$D_{surr} = k_3 D_{data}{}^{\frac{1}{\mu}} \Leftrightarrow \frac{V_{surr}}{A} = k_3 \frac{V_{data}{}^{\frac{1}{\mu}}}{A^{\frac{1}{\mu}}} \Leftrightarrow V_{surr} = k_3 \frac{A}{A^{\frac{1}{\mu}}} V_{data}{}^{\frac{1}{\mu}} = k_3 A^{1-\frac{1}{\mu}} V_{data}{}^{\frac{1}{\mu}} \tag{7}$$

where $V_{surr}$ is the surrogate variable of interest to be used instead of the variable of interest $V_{data}$ for each region with area $A$. The number $N$ of data points generated for each region is proportional to the variable of interest, *i.e.*, $N=k_4 V_{surr}$. Combining this with the previous equation we obtain, for each region,

$$N = k_5 A^{1-\frac{1}{\mu}} V_{data}{}^{\frac{1}{\mu}} \tag{8}$$

To compute the constant $k_5$ we only need to know how many data points we want overall $N_{total}$, since,

$$N_{total} = \sum_{all\ Re\,gions} k_5 A^{1-\frac{1}{\mu}} V_{data}{}^{\frac{1}{\mu}} \Leftrightarrow k_5 = \frac{N_{total}}{\displaystyle\sum_{all\ Re\,gions} A^{1-\frac{1}{\mu}} V_{data}{}^{\frac{1}{\mu}}} \tag{9}$$

If the original expression is valid for 2-dimensional to 2-dimensional mappings, and we knew the magnification factor exactly, then this correction would allow a proportional representation of each region, and thus a null error in the cartogram. Since that expression is just an approximation, the error will in fact be greater than 0. As for the magnification factor $\mu$, it is reasonable, from empirical experience, to assume it is approximately 2/3 [1, 9]. The final number of points for each region $i$ will thus be given by,

$$N_i = k A_i^{\frac{1}{3}} V_i^{\frac{3}{2}} \tag{10}$$

$$k = \frac{N_{total}}{\displaystyle\sum_i A_i^{\frac{1}{3}} V_i^{\frac{3}{2}}} \tag{11}$$

Since the "ocean" region outside the region of interest does not have to be faithfully represented, this correction need not be applied to that region.

## 5  Results

To test the Carto-SOM method we used Portugal's population data for 2001 (Figure 4) and USA's population data for 2000 (Figure 5). To have a comparison with other cartogram algorithms, we used Dougenik's Contiguous Area Cartogram [18], and Gastner's Diffusion Cartogram [2] approaches, since these are the most widely used.

To implement Carto-SOM method we used the MATLAB SOM-Toolbox [19]. Several training processes were performed, changing the initial SOM parameters. All the code used, including the pre-processing code and minor adjustments, is available at http://www.isegi.unl.pt/labnt/roberto. However, it should be stressed that any SOM implementation that supports labeling, together will basic data handling software (such as Ms-Excel) could be used.

For Dougenik´s method, we used an ArcGIS script file to produce the cartograms [20]. This script allows the ArcGIS user to select an input layer and to choose the number of iterations used. In this test we used 5 iterations to produce Dougenik's cartograms. For the diffusion cartogram we used an implementation produced by Michael Gastner available at his homepage [21].

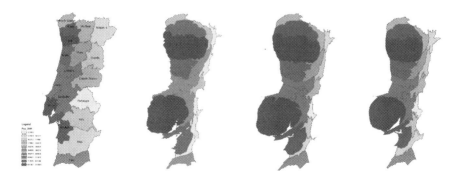

**Fig. 4.** Cartograms obtained for Portugal's population. Left to right: original map, Carto-SOM, Dougenik, and Gestner's method.

**Fig. 5.** Cartograms obtained for the population of the USA. Left to right: original map, Carto-SOM, Dougenik, and Gestner's method.

In Table 1 we present the cartogram errors obtained. Carto-SOM performance measured by this error is significantly better than the others for the USA, and is still quite competitive for Portugal.

**Table 1.** Keim error evaluation using different criteria on the various datasets

|  | USA | | | Portugal | | |
|---|---|---|---|---|---|---|
|  | se (%) | mqe (%) | we (%) | se (%) | mqe (%) | we (%) |
| Dougenik | 16.02 | 3.08 | 11.09 | 12.29 | 3.80 | 8.65 |
| Diffusion | 5.33 | 1.45 | 4.71 | 1.57 | 0.46 | 1.18 |
| Carto-SOM | 3.70 | 0.74 | 0.65 | 3.97 | 1.07 | 1.29 |

## 6 Conclusions

In this paper we presented a method for building cartograms, called Carto-SOM, which is based on the Self-Organizing Map (SOM) algorithm. The SOM is used to perform a 2-dimensional to 2-dimensional mapping, and input data is preprocessed to compensate for the magnification effect. Tests show that cartograms created using the Carto-SOM are good and accurate representations of the variables of interest. Visually we can see that Carto-SOM is an efficient cartogram building algorithm. Finally, it must be emphasized that, using the Carto-SOM method, the only software necessary to create a cartogram is a standard implementation of the SOM algorithm. Such implementations are widely available, both in commercial data analysis programs and public domain packages.

## References

1. Henriques, R., Bação, F., Lobo, V.: Carto-SOM - Cartogram creation using self-organizing maps. International Journal of Geographic Information Science (to be published) (2009)
2. Gastner, M., Newman, M.E.J.: Diffusion-based method for producing density-equalizing maps. Proceedings of the National Academy of Sciences of the United States of America 101(20), 7499–7504 (2004)
3. Tobler, W.: Thirty-five years of computer cartograms. In: Annals, Assoc. American Geographers, pp. 58–73 (2004)
4. Henriques, R.: Cartogram creation using self-organizing maps. In: ISEGI 2006, p. 144. New University of Lisbon, Lisbon (2006)
5. Bauer, H.-U., Der, R., Herrmann, M.: Controlling the Magnification Factor of Self-Organizing Feature Maps. Neural Computation 8, 757–771 (1996)
6. Keim, D.A., North, S.C., Panse, C.: Medial-axes based cartograms. In: IEEE Computer Graphics and Applications, pp. 60–68 (2005)
7. Heilmann, R., et al.: RecMap: rectangular map approximations. In: IEEE Symposium on Information Visualization. IEEE, Austin (2004)
8. Kohonen, T.: Self-Organizing Maps, 3rd edn. Information Sciences, p. 501. Springer, Heidelberg (2001)

9.  Merenyi, E., Jain, A., Villmann, T.: Explicit Magnification Control of Self-Organizing Maps for "Forbidden" Data. IEEE Transactions on Neural Networks 8(3), 786–797 (2007)
10. Sarajedini, B.A., Chau, P.M.: Cumulative Distribution Estimation With Neural Networks. In: WCNN 1996. Laurence Erlbaum, California (1996)
11. Ritter, H., Schulten, K.: Extending Kohonens Self-organizing Mapping Algorithm to Learn Ballistic Movements. In: NATO ASI Series. Neural Computers, vol. F41. Springer, Heidelberg (1988)
12. Cottrell, M., Fort, J.C., Pages, G.: Theoretical Aspects of the SOM algorithm. Neurocomputing 21, 119–138 (1998)
13. Fort, J.C.: SOM's mathematics. Neural Networks 19(6-7), 812–816 (2006)
14. Dersch, D.R., Tavan, P.: Asymptotic level density in topological feature maps. IEEE Transactions on Neural Networks 6(1), 230–236 (1996)
15. DeSieno, D.: Adding a conscience to competitive learning. In: Proceedings of ICNN 1988, International Conference on Neural Networks. IEEE Service Center, Los Alamitos (1988)
16. Hammer, B., Hasenfuss, A., Villmann, T.: Magnification control for batch neural gas. Neurocomputing 70(7-9), 1225–1234 (2007)
17. Ritter, H.: Asymptotic level density for a class of vector quantization processes. IEEE Transactions on Neural Networks 2(1), 173–175 (1991)
18. Dougenik, J., Chrisman, N., Niemeyer, D.: An algorithm to construct continuous area cartograms. In: Professional Geographer, pp. 75–81 (1985)
19. Vesanto, J., et al.: SOM Toolbox for Matlab, vol. 5, p. 59. Helsinki University of Technology, Espoo (2000)
20. Schmid, S. CartoCreator: cartogram creator (2005), http://arcscripts.esri.com/details.asp?dbid=14226 (August 10, 2005)
21. Gastner, M.: Diffusion cartogram - cartogram code (2005), http://www.santafe.edu/~mgastner/

# Concept Mining with Self-Organizing Maps for the Semantic Web

Timo Honkela and Matti Pöllä

Helsinki University of Technology
Department of Information and Computer Science
Adaptive Informatics Research Centre
P.O. Box 5400 FI-02015 TKK, Finland
{timo.honkela,matti.polla}@tkk.fi

**Abstract.** In this paper, we discuss problems related to the basic Semantic Web methodologies that are based on predicate logic and related formalisms. We discuss complementary and alternative approaches. In particular, we suggest how the Self-Organizing Map can be a basis for making the Semantic Web more semantic.

## 1 Introduction

It is clear that the use of standardized formats within computer science is beneficial. For instance, the widespread use of the World Wide Web would not have been possible without the adoption of HTML–a standardized markup language for describing the structure of hypertext documents and link relationships between them. However, the approach of using metadata to describe the contents of documents has been found problematic and, for example, Google has abandoned the use of keywords defined in HTML META elements due to their poor quality.

### 1.1 Standardization of Content

There are serious attempts to create standards for metadata. In the Semantic Web, Resource Description Framework (RDF[1]), is a framework for describing metadata of Web resources, such as the title, author, modification date, content, and copyright information of a Web page. However, it is to be noted that formalizations such as RDF do not solve all interoperability and consistency problems. A W3C Recommendation states that "RDF does not prevent anyone from making assertions that are nonsensical or inconsistent with other statements, or the world as people see it. Designers of applications that use RDF should be aware of this and may design their applications to tolerate incomplete or inconsistent sources of information."[2]

---

[1] Resource Description Framework, http://www.w3.org/RDF/
[2] Resource Description Framework (RDF): Concepts and Abstract Syntax, http://www.w3.org/TR/rdf-concepts/#section-anyone

J.C. Príncipe and R. Miikkulainen (Eds.): WSOM 2009, LNCS 5629, pp. 98–106, 2009.
© Springer-Verlag Berlin Heidelberg 2009

In information science, an ontology is a formal representation of a set of concepts within a domain and the relationships between those concepts. It is used to reason about the properties of that domain, and may be used to define the domain. In general, the idea is that an ontology is a shared conceptual model of the domain. The Web Ontology Language (OWL) is a family of knowledge representation languages for authoring ontologies. eXtensible Markup Language (XML) is often used to implement these descriptions.

In traditional artificial intelligence, it was assumed that one conceptual system could be built to serve all purposes [1]. Later, with the Semantic Web developments, it was recognized that different, potentially mutually incompatible ontologies can be built. To facilitate this, a specialized language called eXtensible Stylesheet Language Transformations (XSLT) has been developed for the transformation of XML documents into other XML documents. Such formalisms do not automate the conceptual mapping process. For instance, the developers of these methodologies themselves describe the situation as follows: "If all the applications are changed to use XML, the programmer only has to learn to handle XML data, not the full range of weird internal formats in which data could otherwise be stored and transferred. This means that some of the application glue can be constructed using XML tools such as XSLT, the transformation language (http://www.w3.org/TR/xslt). The bad news is that the problem of effectively exchanging data doesn't go away. For every pair of applications, in fact for each way in which they need to be linked, someone has to create an 'XML to XML bridge.' That is, if you take XML files from two different applications, you can't just merge them. To make a (XML) query on an XML document, but add in some constraints from another document, you can't just merge the two queries. It's not as though everything is in relational databases where common elements can be used so that data is joined together."[2]

## 1.2  Subjective and Complex Paths from Data to Metadata

When pictorial or sound data is considered, an associated metadata description may be based on a pre-defined classification or framework, for instance, like an ontology within the current Semantic Web technologies. However, even if something like the identity of the author or the place of publishing can usually be determined unambiguously, the same is not true for the description of the contents. In the domain of information retrieval, it was found that in spontaneous word choice for objects in five domains, two people favored the same term with less than 20% probability [3]. It has also been shown that different indexers, well trained in an indexing scheme, might assign index terms for a given document differently [4]. It has been observed that an indexer might use different terms for the same document at different times. The meaning of an expression (queries, descriptions) in any domain is graded and changing, biased by the particular context.

Vygotski has stated that the world of experience must be greatly simplified and generalized before it can be translated into symbols. This is the way how communication becomes possible as the experience of an individual resides only in his own mind and is not communicable as such [5]. Conceptualization is a

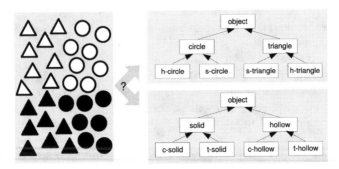

**Fig. 1.** An illustration of two conflicting conceptual systems on the surface level. These systems prioritize the distinctive features differently.

complex process that takes place in a socio-cultural context, i.e., within a community of interacting individuals whose activities result into various kinds of cultural artifacts such as written texts.

It is a very basic problem in information retrieval and knowledge management that different words and phrases are used for expressing similar objects of interest. Natural languages are used for the communication between human beings, i.e., individuals with varying background, knowledge, and ways to express themselves. When rich contents are considered this phenomenon should be more than evident. Therefore, if the content description is based on a formalized and rigid framework of a classification system, problems are likely to arise. Fig. 1 shows a simple example of two conflicting formalizations at the intermediate level of a hierarchical conceptual representation.

If a word or an expression is seen without the context there are more possibilities for misunderstanding. Thus, for a human reader, contextual information is often very beneficial. The same need for disambiguation can also be relevant for information systems. As the development of ontologies and other similar formalizations are, in practice, grounded in the individual understanding and experience of the developers and their socio-cultural context, the status of individual items in a symbolic description may be unclear or ambiguous. Even if an ontology is ambiguously defined and would, in principle, provide a machine readable representation for content description, there is always a man-machine interface needed at a conceptual level. This is because ontology-based information systems do not deal with symbol grounding [6] and therefore they are finally always dependent on human interpretation of the representations.

## 2    Semantic Processing with the SOM

There has been a lot of research attempting to create hybrid systems that would combine the benefits of symbolic knowledge representation and reasoning methods in one hand and the learning and symbol grounding abilities of artificial neural network and pattern recognition methods on the other [7,8,9]. Soft computing methods have also reached some popularity among Semantic Web researchers [10].

The methods that are used to manage data should be able to deal with contextual information, or even provide the necessary context needed in the interpretation of the data. In this article, the Self-Organizing Map (SOM) by [11] is considered as an example of such a method.

## 2.1  Word Category Maps

The *word category map* is a self-organizing semantic map [12] that describes relations of words based on their averaged short contexts. The $i$th word in the sequence of words is represented by an $n$-dimensional real vector $x_i$ with random-number components. The averaged context vector of this word reads

$$X(i) = \begin{bmatrix} \mathrm{E}\{x_{i-1}|x_i\} \\ \varepsilon x_i \\ \mathrm{E}\{x_{i+1}|x_i\} \end{bmatrix} , \tag{1}$$

where E denotes the estimate of the expectation value evaluated over the text corpus, and $\varepsilon$ is a small scalar number. Now the $X(i) \in \Re^{3n}$ constitute the input vectors to the word category map. In our experiments $\varepsilon = 0.2$ and $n = 90$. The training set consists of all the $X(i)$ with different $x_i$.

The SOM is labeled after the training process by inputting the $X(i)$ once again to the word category map and labeling the best-matching units according to symbols corresponding to the $x_i$ parts of the $X(i)$. In this method a unit may become labeled by several symbols, often synonymous or forming a closed attribute set. Usually interrelated words that have similar contexts appear close to each other on the map.

Ritter and Kohonen used artificially generated three-word sentences in the generation of the word category maps [12]. Honkela et al. later used sentence fragments from a natural corpus for the first time to create a meaningful two-dimensional word category map [13]. In these experiments, short contexts were used. This resulted into maps in which the overall order corresponded with linguistic syntactic categorization and semantic categories emerged in the local structures of the maps. If the context of a word consists of the whole document, i.e. document-word matrices are used, the overall order on the map corresponds more closely to some semantic topic structure. In such analysis it is advisable to limit the selection of words to those for which the entropy over the document collection is not high.

## 2.2  Conceptual Hierarchies

A SOM may span a conceptual hierarchy. This can be made explicit by clustering of the SOM (see e.g. [14,15,16]). Vesanto and Alhoniemi explain the motivation for clustering in which similar map units are grouped by the facilitation of quantitative analysis of the map and the data [14]. They have use both hierarchical agglomerative and partitive clustering.

To demonstrate how the SOM can be used to analyze the hierarchical structure of data, we use a non-hierarchical data set describing well known composers, violinists, and pianists with a 11-dimensional feature vector (see Table 1).

**Table 1.** An excerpt of the data used in the experiment. The first six variables refer to the persons status as composer, violinist and/or pianist and the next four variables refer to the style period of composers (reneissance, baroque, classical or romantic). The last variable gives the year of birth.

| | | | | | | | | | | | |
|---|---|---|---|---|---|---|---|---|---|---|---|
| 1 | 1 | 0 | 0 | 0 | 0 | 1 | 0 | 0 | 0 | 1520 | Galilei |
| 1 | 1 | 0 | 0 | 0 | 0 | 0 | 1 | 0 | 0 | 1685 | Bach |
| 1 | 1 | 0 | 0 | 0 | 0 | 0 | 0 | 1 | 0 | 1743 | Boccherini |
| 1 | 1 | 0 | 0 | 0 | 0 | 0 | 0 | 0 | 1 | 1822 | Franck |
| 0 | 0 | 1 | 1 | 0 | 0 | 0 | 0 | 0 | 0 | 1945 | Perlman |
| 0 | 0 | 0 | 0 | 1 | 1 | 0 | 0 | 0 | 0 | 1829 | Rubinstein |

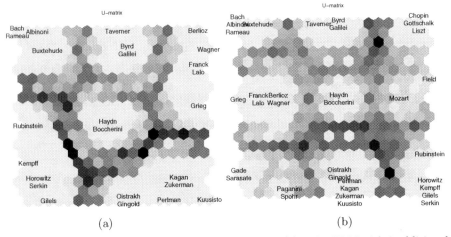

(a)                    (b)

**Fig. 2.** 25 composers, pianists and violinists on a SOM (a) and a SOM with 9 additional composers, pianists and violinists (b)

In the first phase, a 10×10 SOM is trained using the data in Table 1 and the U-matrix visualization of the resulting map is presented in Figure 2a. The birth year column was scaled into [0, 1] for normalization purposes.

Then, the hierarchical structure of the clusters was analyzed using the unweighted average distance between the BMU units. The result of the cluster analysis is presented in a dendrogram in Figure 3. Comparing the cluster hierarchy with the corresponding regions in Figure 2a we can find the cluster of Perlman, Kagan, Kuusisto and Oistrakh, who are all violinists. This cluster is reflected in the hierarchy as a separate branch. The cluster in the upper left corner of the map consists of composers from the baroque era (Bach, Rameau, Albinoni) and is also found as a separate branch in the dendrogram.

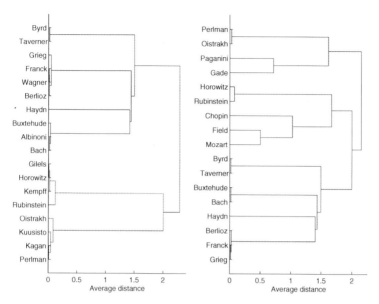

**Fig. 3.** Dendrograms based on the two SOMs

This analysis was then conducted with an extended set of data. Persons who are both composers and violinists, or composers and pianists were added. The map is shown in Figure 2b. In the new map, a cluster of composer-pianists (Chopin, Gottschalk and Liszt) appears in the upper right corner and a cluster of composer-violinists (Paganini and Spohr) appears in the lower left section of the map. If the hierarchy had been fixed for the original data set the new clusters would not be visible as the new samples would be categorized according to the original hierarchy. By generating the hierarchy from the data itself, we can find a meaningful categorization scheme without the need for fixed categories.

An interesting case from the point of view of traditional conceptual hierarchies is the question of multiple inheritance. In these experiments, the first data set was constructed so that there were no instances of multiple inheritance whereas in the second experiment there were several including cases in which a person was both a composer and a pianist or a composer and a violinist (consider e.g. Paganini as a famous case).

As the SOM provides a convenient means to model concept formation and symbol grounding, and can be used for implicit representation of conceptual hierarchies, one can ask how would it be possible to conduct basic logical reasoning within the SOM framework. Frank has presented a model in which basic logical operations can be represented as some kind of Venn diagram manipulations on the SOM surface [17].

## 2.3   Semantic Spaces Based on Documents and Images

The WEBSOM method was developed to facilitate an automatic organization of text collections into visual and browsable document maps [18,19]. Based on the SOM algorithm [11], the system organizes documents into a two-dimensional plane in which two documents tend to be close to each other if their contents are similar. The similarity assessment is based on the full-text contents of the documents. In the original WEBSOM method [20] the similarity assessment consisted of two phases. In the first phase, a word-category map [12,13] was formed to detect similarities of words based on the contexts in which they are used. The Latent Semantic Indexing (LSI) method [21] is nowadays often used for similar purpose. In the second phase, the document contents were mapped on the word-category map (WCM). The distribution of the words in a document over the WCM was used as the feature vector used as an input for the document SOM. Later, the WEBSOM method was streamlined to facilitate processing of very large document collections [22] and the use of the WCM as a preprocessing step was abandoned.

The PicSOM method [23,24,25] was developed for similar purposes than the WEBSOM method for content-based image retrieval, rather than for text retrieval. Also the PicSOM method is based on the SOM algorithm [11]. The SOM is used to organize images into map units in a two-dimensional grid so that similar images are located near each other. The PicSOM method brings three advanced features in comparison with the WEBSOM method. First, the PicSOM uses a tree-structured version of the SOM algorithm (Tree Structured SOM, TS-SOM) [26] to create a hierarchical representation of the image database. Second, the PicSOM system uses a combination of several types of statistical features. For the image contents, separate feature vectors have been formed for describing colors, textures, and shapes found in the images. A distinct TS-SOM is constructed for each feature vector set and these maps are used in parallel to select the returned images. Third, the retrieval process with the PicSOM system is an iterative process utilizing relevance feedback from the user. A retrieval session begins with an initial set of different images uniformly selected from the database. On subsequent rounds, the query focuses more accurately on the user's needs based on their selections. This is achieved as the system learns the user's preferences from the selections made on previous rounds.

This basic principle can be applied in multiple ways to provide a bridge between the raw data directly linked with some phenomenon and the linguistic and symbolic description of its conceptual structure.

## 3   Conclusions

We have presented some motivation why the core technologies in Semantic Web should not solely rely on predicate logic and related formalisms. In an early account on similar concerns, Winograd and Flores emphasized context in understanding communication and information [27]. They criticized the notion of context-independent knowledge that has been underlying traditional AI efforts. In the Pragmatic Web manifesto, it is pointed out that most of the ontologies used in practice assume a certain context and the perspective of some community [28]. Ontologies are not fixed, but co-evolve with their communities of use.

Communication partners have to agree continuously on what they can assume to be the shared background. Often parties from different professional, social, and cultural backgrounds need to understand each other [28]. It means that there is a need for meaning negotiations and that there is a cost related to the use of ontologies, not only to the development [29].

In this paper, we have argued for a certain data-driven approach in which the original data is analyzed automatically rather than relying on hand-crafted ontologies and their use as a basis for choosing descriptors in the metadata. In summary, the interplay between skeletal knowledge structures and the actual use of knowledge is a highly complex process. In this area, dynamic and adaptive unsupervised learning methods such as the SOM can be very useful.

# References

1. Lenat, D.B.: CYC: A large-scale investment in knowledge infrastructure. Communications of the ACM 38(11), 33–38 (1995)
2. Hendler, J., Berners-Lee, T., Miller, E.: Integrating applications on the semantic web. Journal of the Institute of Electrical Engineers of Japan 122(10), 676–680 (2002)
3. Furnas, G.W., Landauer, T.K., Gomez, L.M., Dumais, S.T.: The vocabulary problem in human-system communication. Communications of the ACM 30(11), 964–971 (1987)
4. Bates, M.J.: Subject access in online catalog: a design model. Journal of the American Society of Information Science 37(6), 357–376 (1986)
5. Vygotsky, L.: Thought and language. MIT Press, Cambridge (1986) (originally published in 1934)
6. Harnad, S.: The symbol grounding problem. Phys. D 42(1-3), 335–346 (1990)
7. Orponen, P., Florén, P., Myllymäki, P., Tirri, H.: A neural implementation of conceptual hierarchies with bayesian reasoning. In: Proc. of ICJNN, International Joint Conference on Neural Networks, pp. 297–303. IEEE Computer Society Press, Los Alamitos (1990)
8. Tirri, H.: Implementing expert system rule conditions by neural networks. New Generation Computing 10(1), 55–71 (1991)
9. Sun, R., Wermter, S. (eds.): Hybrid Neural Systems 1998. LNCS, vol. 1778. Springer, Heidelberg (2000)
10. Ma, Z.: Soft Computing in Ontologies and Semantic Web. Springer, Heidelberg (2006)
11. Kohonen, T.: Self-Organizing Maps. Springer, Heidelberg (2001)
12. Ritter, H., Kohonen, T.: Self-organizing semantic maps. Biological Cybernetics 61(4), 241–254 (1989)
13. Honkela, T., Pulkki, V., Kohonen, T.: Contextual relations of words in Grimm tales analyzed by self-organizing map. In: Fogelman-Soulié, F., Gallinari, P. (eds.) Proceedings of ICANN 1995, International Conference on Artificial Neural Networks, Paris, France, EC2 et Cie, Paris, pp. 3–7 (October 1995)
14. Vesanto, J., Alhoniemi, E.: Clustering of the self-organizing map. IEEE Transactions on Neural Networks 11(3), 586–600 (2000)
15. Siponen, M., Vesanto, J., Simula, O., Vasara, P.: An approach to automated interpretation of SOM. In: Allinson, N., Yin, H., Allinson, L., Slack, J. (eds.) Advances in Self-Organizing Maps, pp. 89–94. Springer, London (2001)

16. Kaski, S., Venna, J., Kohonen, T.: Coloring that reveals cluster structures in multivariate data. Australian Journal of Intelligent Information Processing Systems 6(2), 82–88 (2000)
17. Frank, S.: Sentence comprehension as the construction of a situational representation: A connectionist model. In: Proceedings of AMKLC 2005, International Symposium on Adaptive Models of Knowledge, Language and Cognition, Espoo, Finland, Helsinki University of Technology, pp. 27–33 (2005)
18. Honkela, T., Kaski, S., Lagus, K., Kohonen, T.: WEBSOM—self-organizing maps of document collections. In: Proceedings of WSOM 1997, Workshop on Self-Organizing Maps, pp. 310–315. Helsinki University of Technology, Espoo (1996)
19. Lagus, K., Kaski, S., Kohonen, T.: Mining massive document collections by the WEBSOM method. Information Sciences 163, 135–156 (2004)
20. Honkela, T., Kaski, S., Lagus, K., Kohonen, T.: Newsgroup exploration with WEBSOM method and browsing interface. Technical Report A32, Helsinki University of Technology, Laboratory of Computer and Information Science, Espoo, Finland (1996)
21. Deerwester, S.C., Dumais, S.T., Landauer, T.K., Furnas, G.W., Harshman, R.A.: Indexing by latent semantic analysis. Journal of the American Society of Information Science 41, 391–407 (1990)
22. Kohonen, T., Kaski, S., Lagus, K., Salojärvi, J., Honkela, J., Paatero, V., Saarela, A.: Self organization of a massive text document collection. In: Kohonen Maps, pp. 171–182. Elsevier, Amsterdam (1999)
23. Laaksonen, J., Koskela, M., Oja, E.: PicSOM: Self-organizing maps for content-based image retrieval. In: Proceedings of IEEE International Joint Conference on Neural Networks (IJCNN 1999), pp. 2470–2473 (1999)
24. Laaksonen, J., Koskela, M., Oja, E.: PicSOM - self-organizing image retrieval with MPEG-7 content descriptions. IEEE Transactions on Neural Networks, Special Issue on Intelligent Multimedia Processing 13(4), 841–853 (2002)
25. Koskela, M., Laaksonen, J., Sjöberg, M., Muurinen, H.: PicSOM experiments in TRECVID 2005. In: Proceedings of the TRECVID 2005 Workshop, pp. 267–270 (2005)
26. Koikkalainen, P., Oja, E.: Self-organizing hierarchical feature maps. In: Proc. IJCNN 1990, International Joint Conference on Neural Networks, Washington, DC, vol. II, pp. 279–285. IEEE Service Center, Piscataway (1990)
27. Winograd, T., Flores, F.: Understanding Computers and Cognition: A New Foundation for Design. Ablex, New Jersey (1986)
28. Schoop, M., de Moor, A., Dietz, J.L.: The pragmatic web: a manifesto. Commun. ACM 49(5), 75–76 (2006)
29. Honkela, T., Kononen, V., Lindh-Knuutila, T., Paukkeri, M.S.: Simulating processes of concept formation and communication. Journal of Economic Methodology 15(3), 245–259 (2008)

# ViSOM for Dimensionality Reduction in Face Recognition

Weilin Huang and Hujun Yin

School of Electrical and Electronic Engineering
The University of Manchester
Manchester, M60 1QD, UK
`weilin.huang@postgrad.manchester.ac.uk`, `h.yin@manchester.ac.uk`

**Abstract.** The self-organizing map (SOM) is a classical neural network method for dimensionality reduction and data visualization. Visualization induced SOM (ViSOM) and growing ViSOM (gViSOM) are two recently proposed variants for a more faithful, metric-based and direct data representation. They learn local quantitative distances of data by regularizing the inter-neuron contraction force while capturing the topology and minimizing the quantization error. In this paper we first review related dimension reduction methods, and then examine their capabilities for face recognition. The experiments were conducted on the ORL face database and the results show that both ViSOM and gViSOM significantly outperform SOM, PCA and related methods in terms of recognition error rate. In the training with five faces, the error rate of gViSOM dimension reduction followed by a soft $k$-NN classifier reaches as low as 2.1%, making ViSOM an efficient approach for data representation and dimensionality reduction.

## 1  Introduction

Dimensionality reduction techniques have been widely used for data preprocessing, which provide basis for further analysis, management and storage of the data. It extracts meaningful information from high-dimensional data and represents the data by fewer dimensions, and thus can greatly facilitate data analysis, clustering and classification. For instance, in face recognition, each face is represented by a large number of pixel values. It is difficult or inefficient to directly operate on such high-dimensional data. Thus reducing dimensionality has become an important issue in data intensive pattern recognition.

Principal component analysis (PCA) is a primary technique and is regarded as the foundation for many dimensionality reduction techniques. PCA seeks a linear projection that best represents the data in the least-squares sense. It has been widely used in data analysis due to its computational simplicity and analytical tractability. *Eigenface* [1] is a famous application of PCA in face recognition. However, the linearity of PCA limits its power for complex and increasingly large data sets, as it is not capable of revealing nonlinear structure of the data defined by beyond second order statistics. Several PCA-based nonlinear techniques for

J.C. Príncipe and R. Miikkulainen (Eds.): WSOM 2009, LNCS 5629, pp. 107–115, 2009.
© Springer-Verlag Berlin Heidelberg 2009

dimensionality reduction have been proposed recently. For example, kernel-PCA [2] extends PCA to nonlinear by using a kernel function in the input space. Local linear embedding (LLE)[3], a local PCA method, learns the underlying manifold of the data by minimizing an embedding function; while Isomap [4] captures the topology structure by computing the geodesic manifold distances between data points. Meanwhile, there has been previous work on applying these nonlinear techniques for face recognition [5,6,7,8].

Neural networks provide alternative approaches to nonlinear data projection and dimension reduction. The SOM [9] is one of the classical methods for clustering, dimension reduction and data visualization. Dimension reduction is achieved by establishing a topological order of the projection between input data and their corresponding neurons on the map. The applications of SOM in face recognition and comparisons with PCA-based methods can be found in [10,11]. For a more natural and direct display of data structure, ViSOM [12] has been proposed and improved recently by a growing variant, gViSOM [13]. The inter-point distances are locally preserved on the map along with the topology. It has been shown that ViSOM provides a better visual exhibition of data points and their distribution on the map than SOM [12,13]. ViSOM (or gViSOM) represents a metric scaling of the input space and has comparable capability for highly nonlinear manifold learning with other nonlinear PCA methods, such as LLE and Isomap [13]. A review on nonlinear dimensionality reduction is given in [14]. In this paper, we apply the ViSOM and gViSOM for dimensionality reduction in face recognition. We examine their performances and compare them with SOM and several PCA-based methods.

The rest of the paper is organized as follows. PCA-based algorithms, both linear and nonlinear, are briefly reviewed in Section 2, followed by the introduction of SOM-based methods in Section 3. Section 4 presents the classifiers used in the experiment, and the experimental details and results are shown in Section 5. Finally, Section 6 concludes the paper.

## 2   PCA-Based Methods

PCA [1] is a classical linear dimension reduction method aiming at finding principal orthogonal directions from a data set by solving an eigenvalue problem. While discarding a large number of minor components, a small number of principal components are retained to form a linear, low-dimensional subspace, known as *eigenface* in face recognition. Raw face images are projected onto the *eigenface* subspace first, and the classification is carried out in the reduced space.

Kernel-PCA [2] projects input data onto a high-dimensional feature space by using a hypothetic nonlinear function. Then the standard PCA is performed on the high-dimensional data set via a kernel function. There are two commonly used kernel functions: *polynomial* and *Gaussian radial basis*.

LLE [3] is a local PCA method and is capable of mapping high-dimensional nonlinear data onto a single global coordinate system of lower dimensionality. The neighborhood or topology is preserved in the embedding space by minimizing the cost functions in the input space and output space respectively. The optimal weights of input data can be found by solving a least squares problem of the cost function in the input space, while the embedding vectors in the output space are computed as an eigenvalue problem.

Isomap [4] seeks to learn the underlying manifold structure of a data set by computing the geodesic manifold distances between all pairs of data points. It first defines a neighborhood graph, over which each point is connected to all its neighbors in the input space. Then the geodesic distances of all pairs of points are computed via the shortest path on the neighborhood graph (using Floyd's algorithm). Finally multidimensional scaling is applied to the distance matrix to construct the embedding of the data to preserve intrinsic geometry structure of the data.

Curvilinear component analysis (CCA) [15] is another method to represent nonlinear data structure in a lower-dimensional space. The intrinsic geometric property of the data is revealed by preserving local distance relationships via an error function. A neighborhood function is used for local topology preservation and emphases on maintaining shorter distances than longer ones.

## 3   SOM-Based Methods

### 3.1   SOM

SOM is an unsupervised learning loosely based on the retinatopic mapping: an ordered projection of visual retina to visual cortex [16]. It uses a set of neurons ranged often in a 2-D lattice to form a topological mapping of the input space. The SOM learns the topological structure of the input by updating the weight vectors of a neighborhood of the winning neurons when being presented with an input. In the case of dimensionality reduction and data visualization, high-dimensional data are projected onto a low-dimensional SOM, represented by the order, location or index of the neuron on the map. The data structure learned by SOM reveals the relative or ordinal relationships among input data. However, it is unable to reproduce the quantitative distances between the input points on the reduced space. In many applications, a more faithful and metric scaling of the input space is more desirable in data visualization and dimensionality reduction [12,13].

### 3.2   ViSOM

The visualization induced SOM (ViSOM) [12] has been proposed to extend the SOM for faithful (local) distance preservation on the map. The ViSOM preserves the distance quantities along with the topology of data set. The updating force of SOM, $[x(t) - w_k(t)]$, can be decomposed into two components: $[x(t) - w_v(t)] + [w_v(t) - w_k(t)]$. The first term is the updating force from the winner $v$ to the input $x(t)$, which is the same to the updating force of the winner. The second term is a lateral force that brings the neighboring neurons to the winner. This lateral contraction force is regulated in order to maintain a uniform inter-neuron distance locally on the map in ViSOM [12]. The ViSOM algorithm is briefly described below.

1. Select the winner $v$ when an input $x(t)$ is presented, and update its weight

$$\Delta w_v(t) = \alpha(t)[x(t) - w_v(t)] \tag{1}$$

2. Update the weights of the nodes in the neighborhood according to

$$\Delta w_l(t) = \alpha(t)\eta(\varphi, l, t)[x(t) - w_v(t)] + \beta[w_v(t) - w_l(t)] \tag{2}$$

where $\beta = d_{vl}/(\delta D_{vl}) - 1$, $d_{vl}$ is the distance of neuron weights in the input space, $D_{vl}$ is the distance of neuron indexes on the map, and $\delta$ is the resolution parameter. The neighborhood function $\eta$ is similar to that in the SOM, the width of the neighborhood decreases from an initially large value to a final small value but not just to one as in the SOM.

3. Refresh the map by using the weights of randomly chosen neurons as the input at a small percentage of updating times.

The ViSOM learns the local quantitative distances of the input data by regularizing the inter-neuron contraction force, while it captures the ordering and minimizes the quantization error. The distance of two (local) projected points on the map is proportional to the distance of these two points in the input space, making feature representation and data visualization more faithful and quantitatively measurable. The resolution of the map can be enhanced by incorporating the local linear projection (LLP) method [17], which projects a data point onto the sub plane spanned by the two closest edges instead of to the winner.

## 3.3  gViSOM

It has been shown that SOMs of prefixed size are difficult to converge to highly nonlinear manifolds [13]. For improving the local distance-preserving capability of ViSOM, an incremental or growing ViSOM (gViSOM) has been proposed [13] for embedding and metric-scaling nonlinear manifolds. Details of the gViSOM algorithm are as follows,

1. Start with a small initial map (e.g. $5 \times 5$) of either rectangular or hexagonal. Place the initial map onto a linear subspace of either the entire or a local region of the data space. Set the desired resolution and the neighborhood size.

2. Randomly draw a data sample from the data space and find the winning neuron with the shortest distance.

3. If the sample falls within the neighborhood, update the weights of the neurons of the neighborhood using the ViSOM algorithm; otherwise go back to Step 2.

4. At regular iteration intervals (e.g. 2000 iterations), if the growing condition is met (that is, the data is underrepresented by the existing map), grow the map by adding a column or row to the side with the highest activities (measured by the winning frequencies). The added column or row is a linear extrapolation of the existing map. Other growing structures can be used, such as incrementing polygons instead of entire column or row for a free structure of the map and efficient use of neurons.

5. As in the ViSOM, at regular intervals (every certain number of iterations), refresh the map (neurons) probabilistically.
6. Check if the map has converged. If not go back to Step 2.
7. Project all data samples onto the map, either to the neurons or by the LLP resolution enhancement.

## 4    Classifiers

For classification, three common classifiers were used in our system: the Nearest-Neighbor (NN), soft $k$-Nearest Neighbor (soft $k$-NN) and the Linear Discriminant Analysis (LDA). NN is the simplest classifier assigning a test sample to the class of the most similar example in the training set. In soft $k$-NN classifier [11], each principal component outputs a confidence value, which gives the degree of support for each component in every face representation, and then the final decision is given by considering all of these confidence values. LDA [18] is an efficient and widely used linear classifier. It tries to find the linear projection of the data set that minimizes within-class scatter while maximizes between-class separation. The ratio of the determinant of the between-class scatter matrix and the within-class scatter matrix in the projected space is maximized by solving an eigenvalue problem.

## 5    Experiments and Results

In the experiment, the described methods were used for reducing data dimensions in the preprocessing of raw face images, and then one of the classifiers was used for classification. The performances of the dimension reduction methods were evaluated based on the same classifier. The experiment was conducted on a publicly available database, the ORL database (of Olivetti Research Laboratory), which consists of 40 subjects with 10 different face images for each subject. All images in the database were taken against a dark homogeneous background with an up-right, frontal position and have the same size of $92 \times 112$. Face images vary slightly in term of lighting conditions, facial expressions or facial details. Examples of two subjects are shown in Fig.1.

**Fig. 1.** Examples of ORL face images

**PCA-Based Methods.** In PCA-based methods, the number of dimensions ($92 \times 112$) of face image is reduced to 60. Two types of kernel-PCA, *polynomial* (KPCA1) and *Gaussian Radial Basis* (KPCA2), were used with degree of 2 and radius of 30, respectively. The sizes of neighborhood used by LLE and Isomap were set to 30 and 120 respectively. The 60-dimensional face representations are then used for training and testing by a NN, soft $k$-NN or LDA classifier.

**SOM-Based Dimensionality Reduction.** The face images are first locally sampled by moving a window of size $5 \times 5$ over the entire image by 4 pixels each time. The sampled images are reconstructed to the sizes of $25 \times 23 \times 28$ after sampling. That is, each sampled face image contains $23 \times 28$ 25-dimensional subsamples. These 25-dimensional samples are used as the inputs for SOM-based maps, which are trained by implementing the SOM-based algorithms with 50000 updates. For each method, its size and parameters have been optimized to its best performance. For example, the sizes of SOM, ViSOM and gViSOM varied from $5 \times 5$ to $30 \times 30$, and the chosen sizes represent the cases with the best performances (i.e. $10 \times 10$ for SOM, $30 \times 30$ for ViSOM with $\delta$ of 0.5 and $16 \times 19$ for gViSOM starting at $5 \times 5$ with $\delta$ of 0.6).

Then all 25-dimensional samples in each face image are passed through the trained SOM, ViSOM and gViSOM, and represented by the 2-D index values of the corresponding winners on the maps. Thereby, on the trained map, each face image has a corresponding 2-D face projection (as shown in Fig. 2), which is used for further classification. Each dimension of the face projection can be reconstructed as a feature face (with size of $23 \times 28$, examples of two subjects are shown in Fig. 3), which resemble features of the original face images. As can be seen, ViSOM methods resemble better the original features due to its the metric preserving property in feature extraction. For a full and objective evaluation, the performances of SOM-based and PCA-based methods were investigated on the same classifier for each experiment on all subjects of the ORL database. The number of training images was varied from 3, 4, 5 to 6 per subject and the remaining 7, 6, 5, and 4 were used as test images respectively. The results reported are the average results of 10 independent implementations with different

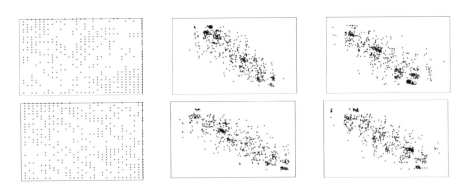

**Fig. 2.** Face projections of SOM (left), ViSOM (center) and gViSOM (right)

**Fig. 3.** Feature faces of SOM (left), ViSOM (center) and gViSOM (right)

**Table 1.** Error rates of PCA-based methods followed by a NN, soft $k$-NN or LDA classifier

| No.of training faces | Error rates(%) | | | | | |
|---|---|---|---|---|---|---|
| | PCA | KPCA1 | KPCA2 | LLE | ISOMAP | CCA |
| | NN Classifier | | | | | |
| 3 | 13.25 | 13.54 | 12.25 | **11.25** | 14.21 | 12.25 |
| 4 | 8.08 | 8.64 | **7.17** | 7.29 | 8.54 | 8.17 |
| 5 | **5.65** | 5.75 | 5.80 | 5.70 | 6.85 | 5.80 |
| 6 | **3.56** | **3.56** | 4.06 | 4.13 | 4.13 | 4.06 |
| | soft $k$-NN Classifier | | | | | |
| 3 | 13.43 | 15.50 | 11.75 | **11.29** | 14.14 | 12.68 |
| 4 | 8.69 | 9.42 | 9.08 | **7.25** | 8.79 | 8.46 |
| 5 | 6.15 | 6.45 | 7.00 | **5.50** | 6.90 | 5.85 |
| 6 | 4.26 | 4.50 | 5.87 | 3.94 | 4.75 | **3.81** |
| | LDA Classifier | | | | | |
| 3 | **9.36** | 11.07 | 10.07 | 9.71 | 13.46 | 12.15 |
| 4 | **5.08** | 6.00 | 5.96 | 6.75 | 8.37 | 7.21 |
| 5 | 3.80 | 4.15 | 4.95 | **3.75** | 6.90 | 5.40 |
| 6 | 3.12 | 3.31 | 3.19 | **2.69** | 4.31 | 4.31 |

randomly chosen training images. Meanwhile, the same choices of training (and test) images were used by all the methods to ensure an unbiased comparison. The results of PCA-based methods followed by the NN, soft $k$-NN or LDA classifier are shown in Tables 1. The performances of SOM-based methods with the NN or soft $k$-NN classifier are listed in Table 2, $ViSOM^*$ and $gViSOM^*$ denote the projections with the LLP resolution enhancement.

The tables show that with more training samples, error rates decrease in all methods as expected. The SOM have the similar performances to PCA-based methods with the NN classifier; ViSOM and gViSOM yield markedly improved performances than SOM and PCA-based methods with about 2% lower error rates in every implementation. With soft $k$-NN classifier, LLE has slightly lower error rates than other PCA-based methods; while SOM-based methods have about 2-3% improvements over the LLE, and gViSOM with LLP has even

**Table 2.** Error rates of SOM-based methods followed by a NN or soft $k$-NN classifier

| No. of training faces | Error rates(%) | | | | |
|---|---|---|---|---|---|
| | SOM | ViSOM | ViSOM* | gViSOM | gViSOM* |
| | NN Classifier | | | | |
| 3 | 11.57 | 10.61 | **10.50** | 10.86 | 10.79 |
| 4 | 7.50 | 6.42 | **6.37** | 6.46 | 6.54 |
| 5 | 5.85 | **4.30** | 4.35 | 4.40 | 4.50 |
| 6 | 3.81 | 2.69 | **2.63** | 2.75 | 2.88 |
| | soft $k$-NN Classifier | | | | |
| 3 | 8.04 | 7.75 | 7.32 | 7.21 | **6.71** |
| 4 | 4.46 | 3.88 | 3.79 | **3.67** | **3.67** |
| 5 | 3.20 | 2.80 | 2.40 | 2.55 | **2.10** |
| 6 | 1.88 | 1.25 | 1.19 | 0.81 | **0.75** |

better performances with more than 1% further improvement. The error rates of gViSOM with LLP in training five and six faces are as low as 2.1% and 0.75% respectively. These results are also better than the results of the PCA-based methods followed by a LDA classifier, with performance improvements of 2-6% in error rate.

## 6    Conclusions

In this paper, we have applied the recently proposed ViSOM and gViSOM for face recognition. The capabilities of them for dimensionality reduction and feature extraction are compared with the standard SOM and several nonlinear PCA methods. The experimental results on a real-world face database show that SOM-based methods achieve a comparable or better performance than widely used PCA-based methods for feature extraction and dimensionality reduction; while metric preserving ViSOM and gViSOM outperform the SOM and these PCA-based methods with significant 2-6% improvement in the error rate. The gViSOM followed by a soft $k$-NN classifier gives the lowest error rate. This demonstrates that faithful representation of high dimensional data is important in pattern recognition and ViSOM offers an effective nonlinear data projection for face recognition.

## References

1. Turk, M., Pentland, A.: Eigenfaces for recognition. Journal of Cognitive Neuroscience 3, 71–86 (1991)
2. Scholköpf, B., Smola, A., Müller, K.-R.: Nonlinear component analysis as a kernel eigenvalue problem. Neural Computation 10, 1299–1319 (1998)
3. Roweis, S.T., Saul, L.K.: Nonlinear dimensionality reduction by locally linear embedding. Science 290, 2323–2326 (2000)

4. Tenenbaum, J.B., de Silva, V., Langford, J.C.: A global geometric framework for nonlinear dimensionality reduction. Science 290, 2319–2323 (2000)
5. Pang, Y.H., Teoh, A.B.J., Wong, E.K., Abas, F.S.: Supervised locally linear embedding in face recognition. In: Int. Symp. on Biometrics and Security Technologies, pp. 1–6 (2008)
6. Yang, M.H.: Kernel eigenfaces vs. kernel Fisherfaces: face recognition using kernel methods. In: Proc. 5th IEEE Int. Conf. on Automatic Face and Gesture Recognition (RGR 2002), Washington DC, pp. 215–220 (2002)
7. Yang, M.H., Ahuja, N., Kriegman, D.: Face recognition using kernel eigenfaces. IEEE Int. Conf. on Image Processing 1, 37–40 (2000)
8. Yang, M.H.: Extended Isomap for pattern classification. In: Proc. National Conference on Artificial Intelligence, pp. 224–229 (2002)
9. Kohonen, T.: Self-Organizing Maps, 2nd edn. Springer, Heidelberg (1997)
10. Lawrence, S., Giles, C.L., Tsoi, A., Back, A.: Face recognition: Aconvolutional neural-network approach. IEEE Trans. Neural Networks 8, 98–113 (1997)
11. Tan, X., Chen, S., Zhou, Z., Zhang, F.: Recognizing partially occluded, expression variant faces from single training image per person with SOM and soft k-NN ensemble. IEEE Trans. on Neural Networks 16, 875–886 (2005)
12. Yin, H.: ViSOM-a novel method for multivariate data projection and structure visualization. IEEE Trans. on Neural Networks 13, 237–243 (2002)
13. Yin, H.: On multidimensional scaling and the embedding of self-organizing maps. Neural Networks 21, 160–169 (2008)
14. Yin, H.: Nonlinear dimensionality reduction and data visualization: A review. Int. Journal of Automation and Computing 3, 294–303 (2007)
15. Demartines, P., Hérault, J.: Curvilinear component analysis: a self-organizing neural network for nonlinear mapping of data sets. IEEE Trans. on Neural Networks 8, 148–154 (1997)
16. Von der Malsburg, C., Willshaw, D.J.: Self-organization of orientation sensitive cells in the striate cortex. Biological Cybernetic 14, 85–100 (1973)
17. Yin, H.: Resolution enhancement for the ViSOM. In: Proc. Workshop on Self-Organizing Maps (WSOM 2003), pp. 208–212 (2003)
18. Belhumeur, P.N., Hespanha, J.P., Kriegman, D.J.: Eigenfaces vs. Fisherfaces: Recognition using class specific linear projection. IEEE Trans. on Pattern Analysis and Machine Intelligence 19, 711–720 (1997)

# Early Recognition of Gesture Patterns Using Sparse Code of Self-Organizing Map

Manabu Kawashima, Atsushi Shimada, and Rin-ichiro Taniguchi

Department of Advanced Information Technology, Kyushu University
744, Motooka, Nishi-ku, Fukuoka 819–0395 Japan
{kawashima,atsushi,rin}@limu.ait.kyushu-u.ac.jp
http://limu.ait.kyushu-u.ac.jp

**Abstract.** We propose a new gesture recognition method which is called "early recognition". Early recognition is a method to recognize sequential patterns at their beginning parts. Therefore, in the case of gesture recognition, we can get a recognition result of human gestures before the gestures have finished. We realize early recognition by using sparse codes of Self-Organizing Map.

**Keywords:** Self-Organizing Map, Gesture Recognition, Early Recognition.

## 1   Introduction

Man-machine seamless 3-D interaction is an important tool for various interactive systems such as virtual reality systems, video game consoles, etc. To realize such interaction, the system has to estimate human gestures in real-time. Generally, gesture recognition is achieved after all sequential postures, which make up a gesture, are input to a system. Therefore, if a long gesture is input to the system, we have to wait for the response for a certain time until the system output the recognition result. This is a problem to realize "real-time" interaction between a human and a machine.

Recent years, a new approach called "early recognition" has been proposed for gesture recognition[1,2]. The early recognition means that a system output a recognition result before a gesture has finished. It is very useful technique to realize real-time interaction. In this paper, we also propose another approach to realize early recognition. Our approach uses Self-Organizing Map (SOM) and it is improved approach of gesture recognition proposed by Shimada *et al.*[3]. In their approach, all postures which make up gestures are learned by the SOM. Therefore, each gesture is represented by an activated pattern of neurons called "Sparse Code". They let another SOM learn the sparse codes of each gesture.

In our approach, we also use the sparse codes which represent each gesture pattern. When a posture is input to the map, a neuron will activate for the posture. We estimate which gesture will be observed when the neuron is selected as winner (activated neuron) based on Bayesian estimation.

J.C. Príncipe and R. Miikkulainen (Eds.): WSOM 2009, LNCS 5629, pp. 116–123, 2009.

## 2  Sparse Code for Gesture Recognition

### 2.1  Self-Organizing Map

Self-Organizing Map (SOM) is one of the most widely used artificial neural network algorithms proposed by Kohonen[4,5]. An input vector $I$ has connections with all of neurons $(1, \ldots, u, \ldots, N)$ on the map. Each neuron has a weight vector $W_u$. In the training phase, a neuron $c$ that satisfies equation (1) is selected as the winner neuron and then weight vectors are updated by equation (2).

$$c = \underset{u}{\mathrm{argmin}} \; (\|I - W_u\|) \tag{1}$$

$$W_u(t + 1) = W_u(t) + h_{c,u}(I(t) - W_u(t)) \tag{2}$$

$h_{cu}$ is a neighborhood kernel defined by equation (3).

$$h_{c,u} = \alpha(t) \cdot \exp\left(-\frac{\|r_c - r_u\|^2}{2\sigma^2(t)}\right) \tag{3}$$

With increasing $\|r_c - r_u\|$ and $t$, $h_{cu}$ converges to zero. $\alpha(t)$ is a monotonically decreasing function of $t$ $(0 < \alpha(t) < 1)$, and $\sigma^2(t)$ defines the width of the kernel.

On the other hand, a neuron $c$ is also selected as the winner in the recognition phase. Generally, the category to which the $c$ belongs is regarded as the recognition result. Meanwhile, the weight vector $W_c$ or the coordinate of the neuron $c$ are also used as the recognition result. In this paper, we define a vector $O$ as an output for an input vector $I$.

### 2.2  Sparse Code

We can represent a posture as $I(t), (1, \ldots, t, \ldots, T)$ which is an unit of a gesture, where $T$ is the time length of the gesture. Under defining the winner for the input $I(t)$ as $c(t)$, we can get the $c(t)$ for all of the postures which make up a gesture as shown in Figure 1. In this figure, each circle shows a neuron of the SOM, and the bottom part of this figure shows a gesture. The gray-colored neurons are winners for the sequential postures $(I(1), I(2), I(3))$. Our approach uses not only these winners but also the other neurons as the upper layer's input. In other words, each neuron $u$ has a state $s_u$ that it is winner for a gesture or not, and whole state of the neurons are used for the upper layer's input. The $s_u$ is defined as follows.

$$s_u = \begin{cases} 1, & \text{if } u = c(t)|_{t \in T} \\ 0, & \text{otherwise} \end{cases} \tag{4}$$

Therefore, the output vector of the first layer is represented as $S = (s_1, \ldots, s_N)$. This output can be regarded as a "Sparse Code" which represents an activated pattern of winner neurons for a gesture. There is a possibility that the same neuron is selected as the winner more than once. In such a case, we set the neuron state $s_u = 1$. This helps the upper layer's SOM to achieve time invariant recognition since the sparse code is similar

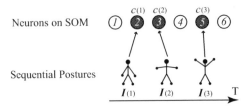

**Fig. 1.** Winner Neurons for a Gesture

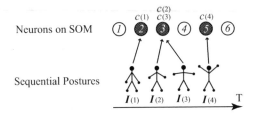

**Fig. 2.** Time Invariant Recognition with Sparse Code

between gestures whose time length is different from each other as shown in Figure 2. In this figure, though the time length of the gesture is longer than the one in Figure 1, the sparse code is the same.

## 3    Early Recognition

Shimada *et al.* directly utilized the sparse code *S* for training of gestures in the next layer's SOM. In contrast, we utilize a part of the sparse code to achieve early recognition of human gestures. We introduce Bayesian estimation method to acquire posterior probability. At time $t$, a posture $I(t)$ is observed by the motion capture system. What we want to know is which gesture includes the posture $I(t)$. In our approach, we acquire posterior probability based on the sparse code of the map. In the SOM, only one neuron will be activated for a posture. Therefore, we need a probability that a gesture will be observed under the condition that a neuron is activated for the current posture. In the following section, we expressly define the probability.

### 3.1    Representation of Probability

We denote $m$ the class of gesture, $A_m$ the event of $m$ and $X_{c(t)}$ the event that the neuron $c$ is activated at time $t$. Our aim is to acquire $P_t(A_m|X_u)$ which is calculated by the prior probability and likelihood.

$$P_t(A_m|X_{c(t)}) = \frac{P(X_{c(t)}|A_m)P_t(A_m)}{P_t(X_{c(t)})} \qquad (5)$$

$$P_t(X_{c(t)}) = \sum_A P(X_{c(t)}|A_m)P_t(A_m) \qquad (6)$$

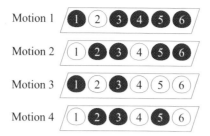

**Fig. 3.** Example of Sparse Code for Each Gesture

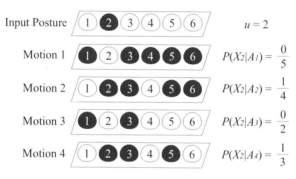

**Fig. 4.** Example of Likelihood $P_t(X_u|A_m)$

where $P(X_{c(t)}|A_m))$ is the likelihood and $P_t(A_m)$ is the prior probability. The likelihood is calculated by following equation.

$$P(X_u|A_m) = \frac{s_{u,m}}{\sum_u s_{u,m}} \qquad (7)$$

where $s_{u,m}$ is a state of a neuron $u$ for a sparse code of gesture $m$. Fig. 3 shows an example of sparse code for each gesture. After training of gestures, neurons are activated for each gesture. The activated neurons are painted in black. If $s_{u,m} > 0$, it denotes that the neuron $u$ is activated for the gesture $m$. From the other viewpoint, the neuron $u$ will be activated when the gesture $m$ is observed. Therefore, the likelihood can be described by the ratio of $s_{u,m}$. Fig. 4 shows an example of calculation of the likelihood. When the neuron $u = 2$ is activated for the current posture, the likelihood of $P(X_2|A_2) = 1/4$ since the neuron $u = 2$ is one of four neurons which are activated for the gesture $m = 2$. On the other hand, $P(X_2|A_3) = 0$ since the neuron $u = 2$ is not activated for the gesture $m = 3$.

On the other hand, $P_t(A_m)$ represents the probability of observation of each gesture. Generally, we use $P_t(A_m) = 1/M$ ($M$ is the number of gestures) for the prior probability. However, a gesture actually consists of different length of postures from the other gestures. Therefore, we use the ratio of activated neurons for each gesture as the prior probability. And the ratio can be calculated from the sparse code of each gesture. The

$P_t(A_m)$ becomes larger with increasing the number of activated neurons for the gesture $m$. Actually, the $P_t(A_m)$ is calculated as follows.

$$P_t(A_m) = \begin{cases} \dfrac{\sum\limits_u s_{u,m}}{\sum\limits_u \sum\limits_m s_{u,m}}, & \text{if } t = 1 \\[2em] \dfrac{P_{t-1}(A_m|X_{c(t)}) \sum\limits_u s_{u,m}}{\sum\limits_m \left\{ P_{t-1}(A_m|X_{c(t)}) \sum\limits_u s_{u,m} \right\}}, & \text{otherwise} \end{cases} \tag{8}$$

We use the previous posterior probability to update the prior probability.

## 3.2   Algorithm of Early Recognition

The algorithm of early recognition consists of two steps; preprocessing to generate the sparse code and online processing to estimate the gesture.

### Generation of Sparse Code

**Step 1.** Let SOM learn all postures which are element of gestures.
**Step 2.** Acquire sparse code $S_m$ for each gesture $m(1, ..., m, ..., M)$.

$$S_m = (s_{1,m}, ..., s_{u,m}, ..., s_{N,m}) \tag{9}$$

$$s_{u,m} = \begin{cases} 1, & \text{if } u = c(t)|_{t \in T, m \in M} \\ 0, & \text{otherwise} \end{cases} \tag{10}$$

We denote $s_{u,m}$ the state of neuron $u$ whether it activates for gesture $m$.
**Step 3.** Gradate sparse codes by following Gaussian filter. This step is effective for generation of shift invariant sparse code.

$$s_u = \sum_v \left\{ s_v \exp\left( -\frac{||r_v - r_u||^2}{2\sigma^2} \right) \right\} \tag{11}$$

### Recognition

**Step 1.** Calculate the likelihood $P(X_u|A_m)$ and the initial prior probability at time $P_1(A_m)$ for all neurons and for all gestures by equation (7) and (8) respectively.
**Step 2.** Acquire the winner neuron $c(t)$ which activates for current posture $T(t)$ by equation (1).
**Step 3.** Calculate the posterior probability $P_t(A_m|X_{c(t)})$ for all gestures by equation (5).
**Step 4.** If $P_t(A_m|X_{c(t)}) > TH$ (TH is a threshold ($\frac{1}{M} < th \leq 1$)), output the gesture $m$ as the result of early recognition. Otherwise, update the prior probability according to equation (8) and jump to Step 1.

# 4 Experimental Results

## 4.1 Conditions

We demonstrate the gesture recognition using motion data from Carnegie Mellon University's Graphics Lab motion-capture database (http://mocap.cs.cmu.edu/). The motion data is composed of 25 measured markers. Each marker is composed of data of (x, y, z)-axis. We used seven of all 3-D positions as the feature vector which represents posture information. We extracted six kinds of gestures shown in Table 1. We used 120 training samples (20 samples for each gesture) and 60 test patterns (10 patterns for each gesture).

## 4.2 Results

At first, we let the SOM learn the training samples. The initial size of the map was $50 \times 50$. After the training, we let the SOM generate the sparse code for each gesture. Next, we input each test pattern into the map and investigated the posterior probabilities which were calculated based on the sparse code. We set the threshold $TH = 0.95(95\%)$ to output the result of early recognition. Table 2 shows the result. Each cell shows which gesture was output as the result of early recognition when a gesture $m$ had been input to the map. For example, when test samples of gesture 5 had been input to the map, 9 of 10 samples were output correctly, and one of 10 samples was erroneously output as gesture 6. Therefore, the accuracy of gesture 5 was 90%.

**Table 1.** Details of Each Gesture

| Gesture | Length | Detail |
|---|---|---|
| 1 | 9 ~ 100 | spread his hands and put them on his shoulder |
| 2 | 7 ~ 90 | hold up his hands |
| 3 | 7 ~ 95 | make a posture of Arm Shoulder Throw |
| 4 | 12 ~ 127 | swirl his hands in front of his body |
| 5 | 19 ~ 259 | move his hands up and down |
| 6 | 18 ~ 320 | twist his right hand |

**Table 2.** Result of Early Recognition

| Output\Input | Gesture 1 | Gesture 2 | Gesture 3 | Gesture 4 | Gesture 5 | Gesture 6 |
|---|---|---|---|---|---|---|
| Gesture 1 | 10 | 8 | 0 | 0 | 0 | 0 |
| Gesture 2 | 0 | 2 | 0 | 0 | 0 | 0 |
| Gesture 3 | 0 | 0 | 10 | 0 | 0 | 0 |
| Gesture 4 | 0 | 0 | 0 | 10 | 0 | 0 |
| Gesture 5 | 0 | 0 | 0 | 0 | 9 | 3 |
| Gesture 6 | 0 | 0 | 0 | 0 | 1 | 7 |
| Accuracy(%) | 100 | 20 | 100 | 100 | 90 | 70 |

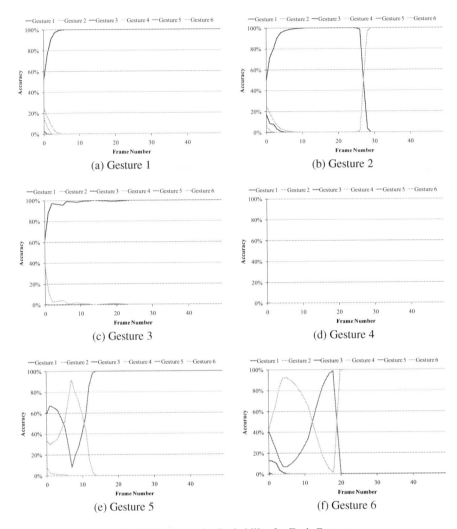

**Fig. 5.** The Posterior Probability for Each Gesture

Finally, Fig. 5 shows a graph which represents the probability transition at each frame when the longest test pattern was input to the map. The horizontal line is the frame number and the vertical line is probability. We show the result of first 50 frame only because of comparison among test patterns comprehensibly.

### 4.3   Discussion

We can see that the SOM could output the early recognition result about gesture 1, 3 and 4 successfully. On the other hand, there were a few mistake about gesture 5 and 6. Especially, most of gesture 2 were recognized falsely. We investigated why the SOM failed to recognize the test patterns of gesture 2, 5 and 6. All gestures have similar postures in the first few frame since each gesture starts from standing posture. In other words, some postures in a gesture share the neuron with the other gesture which has similar postures. For briefly, we define two kinds of gestures; gesture A and gesture B which partially have common postures. When the common posture of gesture A and B was input to the map, the likelihoods for each gesture were calculated. In this case, the difference of likelihoods is not so large between gesture A and B. At this point, the posterior probability of gesture A is also not different from gesture B. However, if such a common posture continues for a while, the difference of posterior becomes larger. This results in false recognition. Namely, there is a possibility of false recognition if the common posture continues for a long time.

## 5  Conclusion

We have proposed a new framework of early recognition of human gestures. We have used Self-Organizing Map (SOM) to learn human gestures. The SOM outputs sparse codes for each gesture. We estimated a human gesture based on Bayesian estimation using the sparse code. We got positive results of early recognition in the experiment. We are now researching to tackle the problem of common postures which are included in some gestures.

## References

1. Mori, A., Uchida, S., Kurazume, R., Taniguchi, R., Hasegawa, T., Sakoe, H.: Early recognition and prediction of gestures. In: Proc. of International Conference on Pattern Recognition, vol. 3(4), pp. 560–563 (2006)
2. Uchida, S., Amamoto, K.: Early Recognition of Sequential Patterns by Classifier Combination. In: Proc. of International Conference on Pattern Recognition (2008) (CD-ROM)
3. Shimada, A., Taniguchi, R.: Gesture Recognition Using Sparse Code of Hierarchical SOM. In: Proc. of International Conference on Pattern Recognition (2008) (CD-ROM)
4. Kohonen, T.: Self-Organization and Associative Memory. Springer, Heidelberg (1989)
5. Kohonen, T.: Self-Organizing Maps. Springer Series in Information Science (1995)

# Bag-of-Features Codebook Generation by Self-Organisation

Teemu Kinnunen, Joni-Kristian Kamarainen*, Lasse Lensu,
and Heikki Kälviäinen

Machine Vision and Pattern Recognition Laboratory (MVPR),
*MVPR/Computational Vision Group, Kouvola
Department of Information Technology
Lappeenranta University of Technology
Finland

**Abstract.** Bag of features is a well established technique for the visual categorisation of objects, categories of objects and textures. One of the most important part of this technique is codebook generation since its within-class and between-class discrimination power is the main factor in the categorisation accuracy. A codebook is generated from regions of interest extracted automatically from a set of labeled (supervised/semi-supervised) or unlabeled (unsupervised) images. A standard tool for the codebook generation is the c-means clustering algorithm, and the state-of-the-art results have been reported using generation schemes based on the c-means. In this work, we challenge this mainstream approach by demonstrating how the competitive learning principle in the self-organising map (SOM) is able to provide similar and often superior results to the c-means. Therefore, we claim that exploiting the self-organisation principle is an alternative research direction to the mainstream research in visual object categorisation and its importance for the ultimate challenge, unsupervised visual object categorisation, needs to be investigated.

## 1 Introduction

Visual object categorisation (VOC) means automatic detection of categories (e.g., "face", "motorbike", etc.) of objects in images. During the last decade, VOC has become an important and active research topic in computer vision. The motivation originates from the desire to automatically search the vast amount of digital image and video data distributed on the Internet. Researchers in this field have accepted the "Bag-of-Features" (BoF) approach (see, e.g., [1,2,3] and Fig. 1) as the main processing principle and it has achieved the mainstream status. In this work, we accept the main principle, but want also to revise one of its intrinsic parts: the visual feature codebook generation. A standard tool for the inter-category codebook generation is the c-means clustering. The state-of-the-art results have been achieved by enhancing the standard c-means with more sophisticated processing and optimisation.

J.C. Príncipe and R. Miikkulainen (Eds.): WSOM 2009, LNCS 5629, pp. 124–132, 2009.

Very recently, an ultimate challenge of visual object categorisation has been proposed [4,5]: unsupervised visual object categorisation. In the unsupervised problem, there is no training or validation sets with manually labeled ground truth, which, on the other hand, prevents using the most effective enhancements in the codebook generation. Now we need to revisit and revise the standard parts of the BoF approach. In this work we revisit the codebook generation part and investigate whether a self-organisation principle, especially self-organising map (SOM) [16], can provide novel or superior characteristics to the c-means.

## 2 Related Work

Due to the active past and current work in the field of supervised VOC, the reported results are now very incremental. For example, there are two main directions in the codebook generation algorithms: replacement of the c-means with another "more tailored" clustering method and enhancement of the c-means with application-specific parts. The latter one has been more successful.

Jurie and Triggs [6] have developed a clustering method which is more robust than the c-means. Their method avoids setting all cluster centres into high density areas, which is typical to the c-means. Their algorithm first chooses N samples randomly and then computes maximal density of the samples using mean-shift estimator. Then it assigns a cluster centre point to the maximal density and eliminates all samples that are within a certain radius from the cluster centre. Then the algorithm repeats these steps with remaining samples as long as there are too many samples left or the number of clusters is too low. Interestingly, this "topology preserving" enforcement is very similar to the main characteristic of self-organisation.

Gemert et al. [2] have developed a method based on the c-means. They replace the simple learning rule, which assigns a sample to the closest cluster, with uncertainty, plausibility and distance values. These values are used in the codebook generation. For example, if a data point is in the middle of two clusters, it will be assigned with the proportion of 50% to the both clusters.

Problem-specific clustering approaches have been developed as well. Leibe et al. [7] use hierarchical clustering to generate the codebook. Many other successful methods, however, use directly the c-means [8,9]. The main property in these enhancements is in locating the cluster centres to spread in a more intelligent manner than converging to few high density regions of the input samples.

One problem-specific enhancement outside clustering is to utilise the spatial information in the codebook generation or probing. For example, Lazebnik et al. [10] reported a method which uses a spatial pyramid to organise descriptors based on their appearance and location. These enhancements, however, are particularly unsuitable for unsupervised methods.

In the recent work on unsupervised visual object categorisation, Sivic et al. [4] presented an unsupervised method utilising Latent Dirichlet Allocation (LDA) model. They improved the original LDA by introducing hierarchical LDA (hLDA). With the hierarchy, they were able to improve the categorisation performance, but the results were reported only for a small number of categories and it is not clear if the approach generalises well.

## 3   Bag-of-Features Framework and Self-organisation in Codebook Generation

The general principle in the bag-of-features approach is very simple. First, interest points are automatically detected from the images, e.g., by using the SIFT [11], Maximal Stable Extremal Regions (MSER) [12] or salient region detector [13]. Then, invariant region descriptors are formed around these interest points (included to, e.g., the SIFT, Speeded Up Robust Features (SURF) [14] and Gradient Location and Orientation Histogram (GLOH) [15]). Then comes an important part: the descriptors are used to form a compact codebook. From any observed image, the interest point detection and descriptor formation parts are exactly the same, but then the contents in the image should be classified according to the "loads" in the codebook. Prior to the categorisation, spatial processing, such as segmentation, can be performed, but generally the main structure is obeyed. Now, it is clear that the codebook plays an essential role in this kind of system. The system is depicted in Fig. 1.

We are using bag-of-features approach, which is similar to the system which was presented by Dance et al. [1], to generate feature histograms for the images. These feature histograms are used to describe images. Let $D$ be a set of descriptors which are extracted from an image using a local feature extractor such as SIFT, and $CB$ be a codebook which contains $M$ words. In practice, words in the $CB$ are clusters' centre points. Let $N$ be the number of descriptors extracted from the image. Then, an image feature image histogram $F$ is generated according to the bag-of-features approach which is defined in Algorithm 1. The $Dist$ function calculates the Euclidean distance between two vectors. The smaller the distance, the greater similarity is between two vectors. Hence, a word that minimizes the distance from a descriptor is chosen as the best match, $bm$.

---

**Algorithm 1.** Feature generation using a bag-of-features approach

---

**for** $i = 1$ to $N$ **do**
  $bm \leftarrow \min_j Dist(D_i, CB_j)$
  $F_{bm} \leftarrow F_{bm} + 1$
**end for**

---

Our main research question in this work is straightforward: what new or superior properties we can achieve by replacing the c-means in the codebook generation with the self-organising map [16] and how these properties can be quantitatively measured? We claim that a proper evaluation procedure is to perform a complete experiment on visual object categorisation and then test the effect of replacing different parts in the system. In our work, we apply the

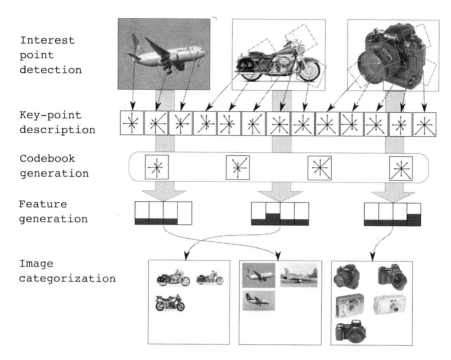

**Fig. 1.** General structure and information flow in the "bag-of-features" approach. Note that the codebook generation is performed only in the training phase for supervised methods.

simplest form of the BoF principle described, e.g., in [1]. We compare the two methods in supervised and unsupervised experiments with the same evaluation measures and data sets as in the recent state-of-the-art papers [1,4]. Moreover, we point out that their evaluation is lacking in some respect and claim that the evaluation should actually investigate the performance as a function of the number of categories. Only this asymptotic behaviour reveals information about generality and extensibility of a method.

## 4   Experiments

As discussed above, we do not support the idea that there would be a single evaluation criteria for the codebook selection, and therefore, we established the complete VOC framework and conducted experiments through the complete pipeline.

In the first experiment, we analyse BoF in its most typical structure, exactly the one depicted in Fig. 1, and supervised visual object categorisation. In the supervised VOC, we have a training set of labelled images (list of objects present in the images). In particular, we replicate the system and experiments in Dance et al. [1], except that we replace the support vector machine classifier with the

**Fig. 2.** Image set for the first test: CalTech 4 and side images of cars [17]. CalTech 4 contains of images of aeroplanes, cars (rear), faces and motorbikes.

**Table 1.** C-means vs. SOM generated codebooks in the VOC framework in Dance et al. [1] for the CalTech 4 + car side image set (optimal c-means codebook size is 100 and SOM 50)

| Category | c-means w/ 1-NN | SOM w/ 1-NN | (c-means w/ SVM) Dance et. al. [1] |
|---|---|---|---|
| Aeroplanes | 0.760 | 0.753 | 0.963 |
| Cars (rear) | 0.893 | 0.953 | 0.977 |
| Cars (side) | 0.980 | 0.953 | 0.996 |
| Faces | 0.787 | 0.833 | 0.940 |
| Motorbikes | 0.593 | 0.707 | 0.927 |
| **Average** | 0.803 | 0.840 | 0.961 |

simple 1-NN decision rule. In this experiment, Caltech 4 together with side images of cars were used. One example image from each category is shown in Fig. 2. The only tunable parameter for the SOM and c-means is the size of the codebook which was optimised for the best results to facilitate reliable comparison. The best results are given in Table 1 where it is evident that the basic SOM can easily match the performance of the c-means and, in this case, also outperform it. It should be noted that the results in [1] were achieved with tailored and heavily optimised support vector machine (SVM) classifier. However, the simple 1-NN classifier performed comparably with no special optimisation.

In the second experiment, we moved from the supervised VOC problem to the more recent challenge, unsupervised VOC. The unsupervised problem has been investigated hitherto only in a few papers, and we utilised the same data and the same performance measure as in Sivic et al. [4]. The performance of the system is defined in Eq. 2 as average performance of nodes. The node performance, $p_t$, is computed as

$$p_t = \max_i \frac{GT_i \cap P_t}{GT_i \cup P_t} \qquad (1)$$

where $GT_i$ is the number of ground truth images from the category $i$, $P_t$ is the number of images assigned to the node $t$. The average performance, $p$, is then

$$p = \frac{1}{N_c} \sum_{i=1}^{N_c} \max_t p_{(t,i)} \qquad (2)$$

where $N_c$ is the number of categories. In the equation, the highest performing node is chosen for each category and then adds performances together and

**Fig. 3.** Image set for the second test: MSRC v1 [18]

**Table 2.** C-means vs. SOM codebook generation & c-means vs. SOM unsupervised classification for the MSRC V1 image set

| **Category** | c-means w/ c-means | c-means w/ SOM | SOM w/ c-means | SOM w/ SOM |
|---|---|---|---|---|
| Aeroplanes | 0.248 | 0.263 | 0.430 | 0.485 |
| Bikes | 0.246 | 0.165 | 0.339 | 0.605 |
| Buildings | 0.258 | 0.165 | 0.149 | 0.251 |
| Cars | 0.123 | 0.187 | 0.211 | 0.271 |
| Cows | 0.217 | 0.261 | 0.159 | 0.263 |
| Grass | 0.203 | 0.356 | 0.174 | 0.461 |
| Faces | 0.252 | 0.178 | 0.250 | 0.245 |
| Sky | 0.196 | 0.206 | 0.170 | 0.209 |
| Trees | 0.265 | 0.233 | 0.346 | 0.450 |
| **Average** | 0.223 | 0.224 | 0.247 | 0.360 |

divided sum by the number of categories which give average categorisation accuracy over all categories.

In the unsupervised scheme, the 1-NN rule must be omitted and replaced with an unsupervised approach. As a simple approach, we fed the extracted codebook loadings (histograms) again to the clustering method, and assigned to each cluster the most representative category label afterwards. Knowledge about labels of the images is not used in learning phase, but they are needed for performance evaluation. Hence, data must have labels otherwise; it is not possible to measure performance. The data used in Sivic et al. consists of nine manually segmented object categories from the MSRC v1 image set [18]. Examples of the images are shown in Fig. 3. We also adopted their performance measure despite the fact that it is intended for measuring consistency of object hierarchies.

We tested all four combinations (c-means/SOM codebook generation & c-means/SOM "category clustering") and optimised the codebook sizes to report the best performances. The results are shown in Table 2 where it is evident

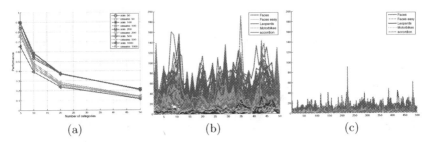

(a)                    (b)                    (c)

**Fig. 4.** Test results and feature histograms. (a) C-means vs. SOM codebook generation & 1-NN classification for the CalTech 101 database. Note that the SOM graphs are coded with red and c-means with green colour. (b) Feature histograms from five categories using 50 words and SOM. (c) Feature histograms from five categories using 500 words and SOM.

that the SOM-SOM combination provided distinctly better results than any other combination, and again the SOM generated codebooks outperformed the c-means. It should be noted that Sivic et al. reported the performance as high as 0.72, but it describes accuracy of object hierarchy which is not included to our method at all. Moreover, the actual level of supervision is not very clear from their report.

The previous two experiments demonstrated the superiority of the SOM in the two previously reported test cases. However, we claim that in those test cases the used performance measure and the amount of data were not adequate for a reliable evaluation of unsupervised VOC performance. The important factor is actually the asymptotic behaviour of the performance as a function of the number of categories. After all, it is more important to know how a method performs with hundreds and thousands of categories. Performance with some specific number of categories can tell only about performance of the system with a specific test set. When the performance of the system is tested with different number of categories, it can tell overall performance of the system more completely. To initiate a better practise, we performed the last experiment using a more proper performance measure and with the well-known Caltech 101 [17] database. Our evaluation procedure was adopted from Fei-Fei et al. [17], where 5 iterations were computed for 30 random images in the training set and another 20 random images in the testing set. We can observe two important results from this experiment (see Fig. 4(a)). At first, collapse of the performance occurs quite rapidly if more than 10 categories are used. Secondly, the SOM systematically outperforms the c-means algorithm. The best overall performances for the SOM were (codebook size 100) 0.898 accuracy for 5 categories, 0.589 for 10, 0.377 for 20 and 0.208 for 50 categories. The best performances for the c-means were 0.856, 0.471, 0.269 and 0.141 respectively. This experiment, we believe, is the strongest proof of superiority of the SOM algorithm in the codebook generation.

**Table 3.** Results in Fig. 4(a) listed in the table for different codebook sizes

| Num. of categer. | SOM | | | | | c-means | | | | |
|---|---|---|---|---|---|---|---|---|---|---|
| | 50 words | 100 words | 200 words | 500 words | 1000 words | 50 words | 100 words | 200 words | 500 words | 1000 words |
| 5 | 0.894 | **0.898** | **0.898** | 0.836 | 0.654 | 0.766 | 0.846 | 0.856 | 0.806 | 0.752 |
| 10 | 0.554 | **0.589** | 0.586 | 0.533 | 0.393 | 0.450 | 0.444 | 0.471 | 0.426 | 0.487 |
| 20 | 0.376 | **0.377** | 0.372 | 0.371 | 0.236 | 0.249 | 0.267 | 0.269 | 0.260 | 0.284 |
| 50 | 0.204 | 0.208 | **0.217** | 0.206 | 0.112 | 0.109 | 0.117 | 0.141 | 0.133 | 0.141 |

Figs. 4(b) and 4(c) shows feature histograms. These two figures discover the fact that when the size of the codebook increases, feature histograms gets less distinctive to each other and thus it is more difficult to separate different images and image categories.

## 5    Conclusions

In this work, we studied whether the self-organisation principle and especially the self-organising map algorithm could provide novel or superior properties in the codebook generation for the visual object categorisation problem. In all the performed experiments, it was shown how the SOM matches, and in the most of the cases, outperforms the c-means algorithm which is the standard in this task. Lower performance of the c-means is a result of poor clustering. C-means sets most of the cluster centre points near to density areas and thus centre points cover well only a fraction of the data. SOM assigns cluster centre points more evenly and thus they cover most of the data. It leads to better codebooks which increases the performance of VOC system. Quantization error could be decreased by increasing the size of the codebook, but it does not lead always to good performance of the system. When the size of the codebook increases, feature histograms get less distinctive and hence it is more difficult to separate images from each other. This affects to the performance in negative manner. This phenomenon is illustrated in Figures 4(b) and 4(c). The results motivate us in the future work to further study the self-organising principle as the predominant principle for realising visual object categorisation and especially unsupervised visual object categorisation.

## Acknowledgements

The authors would like to thank the Academy of Finland and partners of the VisiQ project (no. 123210) for support.

## References

1. Dance, C., Willamowski, J., Fan, L., Bray, C., Csurka, G.: Visual categorization with bags of keypoints. In: ECCV Workshop on Statistical Learning in Computer Vision (2004)

2. van Gemert, J., Geusebroek, J., Veenman, C., Smeulders, A.: Kernel codebooks for scene categorization. In: Proc. of the European Conf. on Computer Vision, pp. 696–709 (2008)
3. Marszałek, M., Schmid, C.: Constructing category hierarchies for visual recognition. In: Proc. of the European Conf. on Computer Vision (2008)
4. Sivic, J., Russell, B.C., Zisserman, A., Freeman, W.T., Efros, A.A.: Unsupervised discovery of visual object class hierarchies. In: Proc. of the Computer Vision and Pattern Recognition, pp. 1–8 (2008)
5. Bart, E., Porteous, I., Perona, P., Welling, M.: Unsupervised learning of visual taxonomies. In: Proc. of the Computer Vision and Pattern Recognition (2008)
6. Jurie, F., Triggs, B.: Creating efficient codebooks for visual recognition. In: Int. Conf. on Computer Vision, pp. 604–610 (October 2005)
7. Leibe, B., Ettlin, A., Schiele, B.: Learning semantic object parts for object categorization. Image and Vision Computing 26, 15–26 (2008)
8. Nowak, E., Jurie, F., Triggs, B.: Sampling strategies for bag-of-features image classification. In: Leonardis, A., Bischof, H., Pinz, A. (eds.) ECCV 2006. LNCS, vol. 3954, pp. 490–503. Springer, Heidelberg (2006)
9. Willamowski, J., Arregui, D., Csurka, G., Dance, C., Fan, L.: Categorizing nine visual classes using local appearance descriptor. In: ICPR Workshop Learning for Adaptable Visual Systems (2004)
10. Lazebnik, S., Schmid, C., Ponce, J.: Beyond bags of features: Spatial pyramid matching for recognizing natural scene categories. In: Conf. on Computer Vision and Pattern Recognition, pp. 2169–2178 (2006)
11. Lowe, D.: Distinctive image features from scale-invariant keypoints. Int. Journal of Computer Vision 20, 91–110 (2004)
12. Matas, J., Chum, O., Urban, M., Pajdla, T.: Robust wide-baseline stereo from maximally stable extremal regions. In: Proc. of the British Machine Vision Conf., pp. 384–393 (2002)
13. Kadir, T., Zisserman, A., Brady, M.: An affine invariant salient region detector. In: Pajdla, T., Matas, J(G.) (eds.) ECCV 2004. LNCS, vol. 3021, pp. 228–241. Springer, Heidelberg (2004)
14. Bay, H., Tuytelaars, T., Gool, L.V.: Surf: Speeded up robust features. In: Leonardis, A., Bischof, H., Pinz, A. (eds.) ECCV 2006. LNCS, vol. 3951, pp. 404–417. Springer, Heidelberg (2006)
15. Mikolajczyk, K., Tuytelaars, T., Schmid, C., Zisserman, A., Matas, J., Schaffalitzky, F., Kadir, T., Gool, L.V.: A comparison of affine region detectors. Int. Journal of Computer Vision 65(1/2), 43–72 (2005)
16. Kohonen, T.: The self-organizing map. Proc. of the IEEE 78(9), 1464–1480 (1990)
17. Fei-Fei, L., Fergus, R., Perona, P.: Learning generative visual models from few training examples: an incremental bayesian approach tested on 101 object categories. In: CVPR Workshop on Generative-Model Based Vision (2004)
18. Winn, J., Criminisi, A., Minka, T.: Object categorization by learned universal visual dictionary. In: Int. Conf. on Computer Vision, pp. 1800–1807 (2005)

# On the Quantization Error in SOM vs. VQ: A Critical and Systematic Study

Teuvo Kohonen, Ilari T. Nieminen, and Timo Honkela

Helsinki University of Technology, Centre of Adaptive Informatics
P.O. Box 5400, 02015 HUT, Finland
teuvo.kohonen@tkk.fi

**Abstract.** The self-organizing map (SOM) is related to the classical vector quantization (VQ). Like in the VQ, the SOM represents a distribution of input data vectors using a finite set of models. In both methods, the quantization error (QE) of an input vector can be expressed, e.g., as the Euclidean norm of the difference of the input vector and the best-matching model. Since the models are usually optimized in the VQ so that the sum of the squared QEs is minimized for the given set of training vectors, a common notion is that it will be impossible to find models that produce a smaller rms QE. Therefore it has come as a surprise that in some cases the rms QE of a SOM can be smaller than that of a VQ with the same number of models and the same input data. This effect may manifest itself if the number of training vectors per model is on the order of small integers and the testing is made with an independent set of test vectors. An explanation seems to ensue from statistics. Each model vector in the VQ is determined as the average of those training vectors that are mapped into the same Voronoi domain as the model vector. On the contrary, each model vector of the SOM is determined as a weighted average of all of those training vectors that are mapped into the "topological" neighborhood around the corresponding model. The number of training vectors mapped into the neighborhood of a SOM model is generally much larger than that mapped into a Voronoi domain around a model in the VQ. Since the SOM model vectors are then determined with a significantly higher statistical accuracy, the Voronoi domains of the SOM are significantly more regular, and the resulting rms QE may then be smaller than in the VQ. However, the effective dimensionality of the vectors must also be sufficiently high.

**Keywords:** self-organizing map, vector quantization, quantization error.

## 1 Introduction

Some attempts have been made to compare the quantization errors (QEs) in a *self-organizing map (SOM)* vs. the same errors in classical vector quantization (VQ), also named the *k-means (clustering) algorithm*. It is usually taken as self-evident that if the models or "codebook vectors" are optimized in the VQ so that the sum of the squared QEs is minimized for given training vectors, it will

J.C. Príncipe and R. Miikkulainen (Eds.): WSOM 2009, LNCS 5629, pp. 133–144, 2009.

be impossible to find any other set of models that produces a smaller *rms QE* (square root of the mean square of QE over independent test data). Thus the rms QE also in the SOM is supposed to be larger. Therefore it has come as a surprise that *the rms QE of the SOM may sometimes be smaller than that of the VQ* (cf., e.g., [1] and [13]).

In a careful and systematic examination we found out that *this effect depends most strongly on the ratio of the number of training vectors and the number of model vectors*. If this ratio is *small, on the order of small integers*, the rms QE of the SOM is usually smaller than that of the VQ. However, *the training vectors must also have a significant local variance in sufficiently many dimensions*, whereas the effect depends only weakly on the size of the SOM array.

## 2    A Possible Explanation of the Observed Ratio of Quantization Errors

In vector quantization, the *k-means algorithm* usually minimizes the root-mean-square quantization error. Then the point density of the VQ model vectors is approximately equal to $p(\mathbf{x})^{\alpha}$, where $p(\mathbf{x})$ is the probability density of the input vectors $\mathbf{x}$, $\alpha = n/(n+2)$ , and $n$ is the dimensionality of $\mathbf{x}$ [8], [11], [16]. The corresponding point density expression is not known for the SOM of arbitrary dimensionality, but counterexamples from the one-dimensional case [14], [4] prove that $\alpha$ can have a different value, whereupon the rms QE of the SOM is larger. At the borders of the SOM array the network of model vectors is "shrunk" with respect to the distribution of the input vectors. The strengths of these effects depend on the effective radius of the neighborhood function.

The situation, however, is different with a *small, finite set of training vectors*, when the evaluation of the rms QE is made using a *statistically independent set of test vectors*. Consider now that every model vector in the VQ coincides with the centroid of the training vectors mapped into the corresponding *Voronoi domain* around this model, and is defined by them [8]. On the other hand, it is generally known [9] that every model vector in the SOM coincides with the weighted average of those training vectors that are mapped into the "topological neighborhood" of this model vector. The weighting is made by the *neighborhood function*. The effective size of each neighborhood in the SOM, however, is a multiple of Voronoi domains. Since the sets of training vectors that determine the model vectors in the SOM are thus generally much larger than the corresponding sets that determine the codebook vectors in the VQ, *the models of the SOM are determined with a much higher statistical accuracy than the codebook vectors of the VQ*. Thus the spacings of the model vectors in the SOM are more regular than those in the VQ, whereupon the Voronoi domains of the SOM are also more regular than those of the VQ. We demonstrate first by a simple example how *the observed differences in the rms QEs may be explainable by this regularity/irregularity condition*.

Consider first two-dimensional input vectors $\mathbf{x}$. Assume tentatively that the centroid of the training vectors in a Voronoi domain coincides with that of the test vectors. Actually this is not exactly true with only few training vectors per model and a set of independent test vectors, but we shall revert to this restriction a bit later. If the set of test vectors is very large, and $p(\mathbf{x})$ is further selected as *locally constant*, the *mean-square QE over this Voronoi domain* is proportional to the *second (geometric) moment with respect to the centroid* (denoted here $M_2$) of this domain.

The two-dimensional Voronoi domains are polygons with varying numbers of sides. Consider the simplest polygon, *an arbitrary triangle,* where a perpendicular, drawn from some apex to the opposite side, divides this side into the parts $a$ and $b$, respectively. Let $h$ be the length of the perpendicular and $A = h(a+b)/2$ the area of the triangle, respectively. The $M_2$ of this triangle is known to be $A(h^2 + a^2 + b^2 + ab)$. Under the constraint that $A$ has a constant value, the minimal $M_2$ is obtained when the triangle is *equilateral.* A similar condition is derivable for an arbitrary polygon. Also it can be shown that the relative $M_2$ of a regular polygon is the smaller, the larger the number of its sides is.

In Fig. 1 we plot the variable $\sqrt{M_2}$ vs. $h/a$ for an *isosceles triangle* with the base $2a$ and the height $h$, respectively. The $\sqrt{M_2}$ is proportional to the rms QE with respect to the centroid of the triangle, when $A$ and $p(\mathbf{x})$ are constant.

Actually, with only few training vectors, the (weighted) centroids of the training vectors and those of the independent test vectors neither coincide fully in a Voronoi domain of the VQ, or in a "topological" neighborhood of the SOM. However, the difference of these centroids is generally smaller in the SOM than in the VQ, because the models of the SOM are determined with a much higher

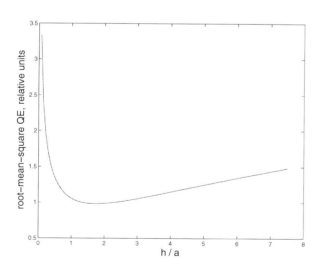

**Fig. 1.** The rms QE with respect to the centroid of a hypothetical Voronoi domain of the form of an *isosceles triangle,* when its area and $p(\mathbf{x})$ are constant. The triangle has the base $2a$ and the height $h$, respectively. The minimum is obtained at the argument value $h/a = \sqrt{3}$, i.e., when the triangle is equilateral.

statistical accuracy than those of the VQ, and thus the rms QE in the SOM, compared with that in the VQ, will be still smaller.

It may now be conjectured that *in the case of general dimensionality and form of the Voronoi domains, the relative rms QE will be the smaller, the closer to a hypersphere the Voronoi domains are on the average.* Any mathematical proof of this is beyond our capabilities for the time being.

Nonetheless, as we shall further specify in Subsections 4.1 and 4.4, *the training vectors must additionally have a significant (local) variance in sufficiently many dimensions* so that the training samples have enough *degrees of freedom into which they can be scattered randomly.* Only then are the Voronoi domains irregular enough in order to reflect significant differences between the rms QEs in the SOM and the VQ .

# 3   A Remark on the Convergence of the Batch SOM

The two main versions of the SOM [9] are the *stepwise corrective* algorithm and the *batchwise training* algorithm, respectively, which are not quite equivalent [6]. The former includes the learning rate as a parameter, while this parameter is lacking from the batch version. In both of these algorithms, the model vectors can be initialized either as random vectors, or as a regular two-dimensional sequence of vectors of the linear subspace spanned by those two eigenvectors of the input autocorrelation matrix that belong to the largest eigenvalues. The latter option is called *linear initialization*, and it results in a much faster convergence. In practice, the linear initialization, combined with the batch computation, also produces the most unique and stable SOMs.

We applied the batch training process with linear initialization to construct the SOMs. The software package that we used is called the *SOM Toolbox* [15], and it is downloadable from *www.cis.hut.fi/projects/somtoolbox.* The process usually consists of two phases: the coarse one, in which the "topological" ordering of the model vectors takes place, and the refining phase, in which the final values of the SOM model vectors are determined.

It may not be generally known, however, that the batch training version of the SOM algorithm may terminate *in a finite number of iterations,* i.e., the corrections will become exactly zero after a certain number of iterations, if the set of input vectors is finite and the neighborhood function does not change in time. This has happened at least in all of the cases we studied closer, when an approximate order had already been achieved in the coarse training phase. If, namely, the convergence process has proceeded to a state where the map is already "topologically" ordered and the model vectors are sufficiently close to their asymptotic values, there will be a positive probability for all of the training vectors being mapped into the same Voronoi sets as in the previous iteration. If this really happens, the centroids of the training vectors in each Voronoi set will be exactly the same as those in the previous iteration, and thus the next distribution of the training vectors into the Voronoi sets around the corresponding "winners" must be identical with the previous distribution. This kind of an exact termination of the learning process was particularly useful in the present numerical comparisons of the SOMs, for checking that at least a local optimum has been reached exactly.

## 4   Experiments

### 4.1   Artificially Generated Random Data

The first experiment was carried out with artificially generated random data. Using a random-number generator, the training and test vectors were drawn from high-dimensional normal distributions with zero mean and identity covariance.

The evaluation of the rms QEs was carried out for the dimensionalities of 10, 20, and 50 of the input vectors, and for two sizes of the SOM and the VQ (7x10 = 70 and 10x14 = 140 models, respectively.) The neighborhood function of the SOMs was a two-dimensional Gaussian.

As we wanted to point out that the effect studied in this work depends most strongly on the ratio of the number of input vectors and the number of model vectors, we evaluated the rms QEs for a series of this ratio. Let us call the selected values of the ratio the *argument values*.

The linear initialization of the SOM model vectors was used. This initialization was carried out separately for each of the above finite sets of training vectors. In order to increase the statistical accuracy, the initialization and training of the SOMs and the VQs were repeated 20 times for each argument value (training vectors per model), each time using a different (finite) set of training and test vectors.

During training, the effective radius (normalized central second moment) of the neighborhood function decreased linearly during the coarse training phase from the value 2 (for the 7x10 array) or 3 (for the 10x14 array) to the value 1 in 20 iterations, and during the refinement phase it was held at the value 1 until the iterations terminated automatically (cf. Sec. 3).

The model vectors of the VQs were computed by the *kmeans* algorithm of the Matlab Version 7, with special precautions to avoid "empty clusters." We used the model values computed by the SOM algorithm as initial values for the VQ, as often recommended.

Figs. 2 and 3 represent the results of this first experiment. With the input dimensionality 10 the rms QE of the VQ was always smaller. For the dimensionalities 20 and 50 and for the 7x10 SOM, the "break even" points (where the rms QEs of the SOM and the VQ are equal) occurred at the argument values 3.2 and 12.2, and for the 10x14 SOM, at the values 2.1 and 10.0, respectively. Below these points, the rms QEs were always smaller in the SOM. The lower limit of the input dimensionalities for this effect to occur is between 10 and 20. These results seem to verify the explanation suggested in Sec. 2, namely, that *the Voronoi domains are more irregular, when there are fewer training vectors per model, and there is significant variance in more dimensions. Accordingly, the rms QE in the SOM is then correspondingly smaller than in the VQ.* On the other hand, the size of the SOM array seems to have only a marginal effect on this effect.

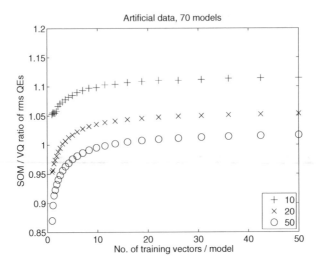

**Fig. 2.** Ratio of the rms QEs in the SOM and the VQ for the artificially generated random-data set, as a function of the number of training vectors per model, and for the dimensionalities 10, 20, and 50, respectively. The number of models was 70.

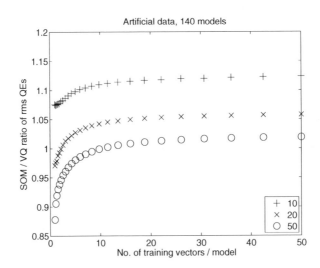

**Fig. 3.** Ratio of the rms QEs in the SOM and the VQ for the artificially generated random-data set, as a function of the number of training vectors per model, and for the dimensionalities 10, 20, and 50, respectively. The number of models was 140.

## 4.2   ISOLET Data

The input vectors in this experiment consisted of 617 acoustic features extracted from spoken letters of the English alphabet [2]. Our task was to compute the

SOM/VQ rms QE ratio for a series of argument values, viz. for different numbers of training vectors per model, like with the artificial data.

In the construction and testing of each SOM/VQ pair, the training and test vectors, as stated earlier, should be statistically independent of each other. Furthermore, in order to achieve a sufficient statistical accuracy, the construction and testing of each SOM/VQ pair should be repeated using independent data.

The problem in practice, however, is the limited availability of validated and verified high-dimensional experimental data. A generally used method to increase the statistical accuracy is the *repeated holdout validation*. In it, say, $M$ training samples are picked up at random from the available data set, while the rest of it is set aside for testing. This randomized division into separate training and test samples and performing of the experiment is repeated a wanted number of times, whereafter the results are averaged.

In the ISOLET experiment we had 7797 input vectors available. For instance, with the SOM array size 10x14 = 140 and the argument value 50, we selected $M$ = 140x50 = 7000 samples at random from the basic data set for the construction of one SOM/VQ pair, while the rest were set aside for testing. At lower argument values and with the 7x10 SOM array, less data are needed for the construction of a SOM/VQ pair, whereupon more data can be reserved for testing. This random selection of the training vectors was repeated 20 times for every argument value, and a new SOM/VQ pair was constructed every time. The averages over the repeated evaluations of the rms QEs were then formed for every argument value.

In the construction of the SOMs, linear initialization by the subset of the chosen training vectors was used. During training, the neighborhood function had the Gaussian form, and its effective radius decreased linearly with time during the coarse training phase from the value 2 (for the 7x10 array) or 3 (for the 10x14 array) to the value 1 in 20 iterations, and during the refinement phase, the value was held equal to 1 until the iterations terminated automatically (cf. Sec. 3). The Matlab *kmeans* function was used to construct the VQ models.

In Figs. 4 and 5 we display the ratio of the rms QEs in the SOMs and in the VQs as a function of the number of training vectors per model. It can be seen that the "break even" points (at which the rms QEs in the SOM and the VQ are equal) are about 3.2 and 3.7 for the two array sizes, respectively. While the dimensionality of the input data was 617 and thus much higher than that of the 20-dimensional artificial data, the effect is anyway of the same order of magnitude. A possible explanation of this result is the following. It is generally known that the local variances of natural experimental data are usually very different in different directions, whereupon the "average effective dimensionality" of the data is much lower than their true dimensionality (cf. Subsection 4.4).

## 4.3 Reuters Data

Our third experiment was based on the text corpus collected by the Reuters Corp. No original documents were available to us, but Lewis et al. [10] have prepared a test data set on the basis of this corpus for benchmarking purposes, preprocessing the textual data, removing the stop words, and reducing the words

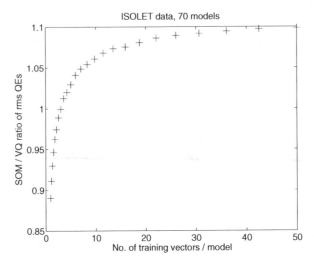

**Fig. 4.** Ratio of the rms QEs in the SOM and the VQ for the ISOLET data set, as a function of the number of training vectors per model and for 70 models

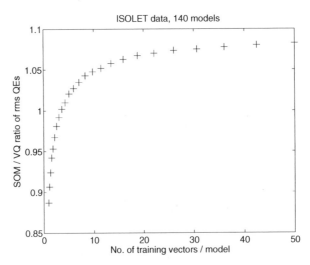

**Fig. 5.** Ratio of the rms QEs in the SOM and the VQ for the ISOLET data set, as a function of the number of training vectors per model and for 140 models

into their stems (called "term" here). In our experiments the term $i$ of document $j$ was weighted by the factor

$$w_{ij} = (1 + \log(TF_{ij}))\log(N/DF_i) \;, \tag{1}$$

where $TF_{ij}$ is the "term frequency" (frequency of term $i$ in document $j$), $DF_i$ is the "document frequency" (telling in how many documents term $i$ occurs), and $N$ is the number of documents [12].

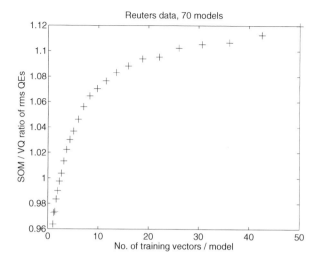

**Fig. 6.** Ratio of the rms QEs in the SOM and the VQ for the Reuters data set, as a function of the number of training vectors per model and for 70 models.

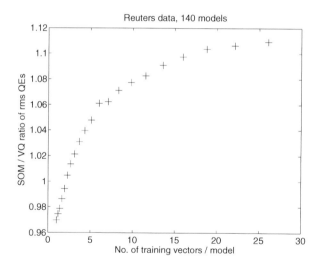

**Fig. 7.** Ratio of the rms QEs in the SOM and the VQ for the Reuters data set, as a function of the number of training vectors per model and for 140 models.

J. Salojärvi of our laboratory picked up those 233 terms that appeared at least 800 times in 4000 selected documents. Thus the dimensionality of the input vectors was 233.

The general arrangement of this experiment was similar to that with the ISO-LET data. This time we had only 4000 input samples available, which restricted the definition of the argument values for 140 models only to about 26.5. This does not present any problem, since the interesting effect occurs at much lower argument values.

The averaged results are shown in Figs. 6 and 7. The "break even" points with the array sizes 7x10 and 10x14 are now about 2.48 and 2.45, respectively. Again, a possible reason for these small values, in spite of the input dimensionality being 233, is the average effective dimensionality of the data manifold.

### 4.4 "Average Effective Dimensionality"

In this subsection we explicate the concept of the "average effective dimensionality" of the input data on which the ratio of the rms QEs in the SOM and the VQ was supposed to depend.

Our objective is to estimate *how many degrees of freedom* the training vectors have on the average. This would be given by the *fractal dimension* [5], which, however, is difficult to estimate accurately for high-dimensional data.

On the other hand, an early measure named the *intrinsic dimensionality* [7] related to the *local variances of the data manifold*. It was used mainly in optimal feature selection, and it was introduced before the development of effective VQ methods. Its computation is cumbersome, too. Therefore we suggest yet another solution.

The *linear extensions* of any subset of vectorial data are describable by the *standard deviations* $\sqrt{\lambda_i}$, where $\lambda_i$ is the $i$th eigenvalue of the autocorrelation matrix of the vectors in this subset. Consider the expression $D = \sum_i \sqrt{\lambda_i}/\sqrt{\lambda_m}$, where $\lambda_m$ is the maximum eigenvalue. If a subset of the $\lambda_i$ were equal to $\lambda_m$, while the rest of the eigenvalues were equal to zero, $D$ would define the number of degrees of freedom exactly. For general vectorial data $D$ is now supposed to give an estimate of the "effective dimensionality."

In order to find the *local dimensionalities*, we first divide the data manifold into *k-means clusters* and evaluate $D$ for each cluster. Then we average $D$ over the clusters. Since the average $D$ depends on $k$, we maximize it over $k$ for every data set. The average $D$, together with the optimal $k$, are given for each data set in the table below.

| Data | Dimensionality | Average $D$ | Optimal $k$ |
|---|---|---|---|
| Random | 10 | 8.6 | 3 |
| Random | 20 | 16.1 | 5 |
| Random | 50 | 33.8 | 6 |
| ISOLET | 617 | 34.8 | 15 |
| Reuters | 233 | 24.4 | 10 |

## 5    Discussion

It is easy to construct SOMs that *graphically* describe the metric relations and clustering tendencies of the input samples reasonably well and facilitate an overview of the data space. Contrary to that, if the SOM is intended for a tool in *numerical problems*, plenty of precautions must be taken into account in order to obtain reliable and accurate results.

The present study dealt with the *rms QE* in the SOM when compared with that in the VQ. Certain surprising results in some earlier studies gave rise to us to carry out an extensive critical comparison of the SOM vs. the VQ in this respect. It indeed transpired that the rms QE in the SOM *may* be smaller than that in the VQ, but only if the number of training vectors per model is small, on the order of small integers, and if the "average effective dimensionality" of the data vectors used for training exceeds a certain limit.

Although we have not provided a complete analysis, we believe that we have been able to narrow down the conditions in which this effect occurs. The main explanation seems to ensue from statistics. Each model vector in the VQ is determined as the centroid of those training vectors that belong to the same Voronoi domain as the model vector, whereas the SOM model vectors coincide with the weighted centroids of all those training vectors that belong to the much larger "topological" neighborhoods around the corresponding models. With a small number of training vectors per model, the statistical accuracy of the SOM model vectors is therefore significantly higher than that of the VQ model vectors, and their constellation in the data space is then significantly more *regular*, resulting in more regular forms of Voronoi domains and correspondingly smaller quantization errors on the average.

At any rate we can make the following conclusions. First, this phenomenon has occurred in all of our numerous experiments (of which only a few are reported here), when the input dimensionality has been high enough. The differences in the "break even" points are probably due to differences in the local dimensionalitites of the data manifolds and probably also in their forms. Second, the general nature of the experimental curves, in which this effect occurs, is almost similar, indicating that the phenomenon is some consistent function of the number of training vectors per model. Contrary to that, the number of models has only a marginal influence on the SOM/VQ rms QE ratio.

# References

1. Bação, O., Lobo, V., Painho, M.: Self-organizing maps as substitutes for k-means clustering. In: Sunderam, V.S., van Albada, G.D., Sloot, P.M.A., Dongarra, J. (eds.) ICCS 2005. LNCS, vol. 3516, pp. 476–483. Springer, Heidelberg (2005)
2. Cole, R.A., Muthusamy, Y., Fanty, M.A.: The ISOLET Spoken Letter Database, Technical Report 90-004, Computer Science Department, Oregon Graduate Institute (1994)
3. Cottrell, M., Fort, J.C., Pagès, G.: Theoretical aspects of the SOM algorithm. Neurocomputing 21(1), 119–138 (1998)
4. Dersch, D., Tavan, P.: Asymptotic level density in topological feature maps. IEEE Trans. Neural Networks 6(1), 230–236 (1995)
5. Falconer, K.: Fractal Geometry: Mathematical Foundations and Applications. Wiley, West Sussex (2003)

6. Fort, J.C., Cottrell, M., Letremy, P.: Stochastic on-line algorithm vs. batch algorithm for quantization and self-organizing maps. In: Neural Networks for Signal Processing XI: Proc. of the 2001 IEEE Signal Processing Society Workshop, pp. 43–52. IEEE, Piscataway (2001)
7. Fukunaga, K., Olsen, D.R.: An algorithm for finding intrinsic dimensionality of data. IEEE Trans. Computers C-20, 176–183 (1971)
8. Gersho, A.: On the structure of vector quantizers. IEEE Trans. Inform. Theory IT-25, 373–380 (1979)
9. Kohonen, T.: Self-Organizing Maps, 3rd edn. Springer, Heidelberg (2001)
10. Lewis, D.D., Yang, Y., Rose, T.G., Li, T.: RCV1: A new benchmark collection for text categorization research. J. Mach. Learn. Res. 5, 361–397 (2004)
11. Linde, Y., Buzo, A., Gray, R.M.: An algorithm for vector quantization. IEEE Trans. Communication COM-28, 84–95 (1980)
12. Manning, C.D., Schütze, H.: Foundations of Statistical Natural Language Processing. MIT Press, Cambridge (1999)
13. McAuliffe, J.D., Atlas, L.E., Rivera, C.: A comparison of the LBG algorithm and Kohonen neural network paradigm for image vector quantization. In: Proc. ICASSP-90, Acoustics, Speech and Signal Processing, vol. IV, pp. 2293–2296. IEEE Service Center, Piscataway (1990)
14. Ritter, H.: Asymptotic level density for a class of vector quantization processes. IEEE Trans. Neural Networks 2(1), 173–175 (1991)
15. Vesanto, J., Alhoniemi, E., Himberg, J., Kiviluoto, K., Parviainen, J.: Self-organizing map for data mining in Matlab: the SOM Toolbox. Simulation News Europe (25), 54 (1999)
16. Zador, P.L.: Asymptotic quantization error of continuous signals and the quantization dimension. IEEE Trans. Inform. Theory IT-28, 139–149 (1982)

# Approaching the Time Dependent Cocktail Party Problem with Online Sparse Coding Neural Gas

Kai Labusch, Erhardt Barth, and Thomas Martinetz

University of Lübeck, Institute for Neuro- and Bioinformatics
Ratzeburger Alle 160 23538 Lübeck, Germany

**Abstract.** We show how the "Online Sparse Coding Neural Gas" algorithm can be applied to a more realistic model of the "Cocktail Party Problem". We consider a setting where more sources than observations are given and additive noise is present. Furthermore, we make the model even more realistic, by allowing the mixing matrix to change slowly over time. We also process the data in an online pattern-by-pattern way where each observation is presented only once to the learning algorithm. The sources are estimated immediately from the observations. In order to evaluate the influence of the change rate of the time dependent mixing matrix and the signal-to-noise ratio on the reconstruction performance with respect to the underlying sources and the true mixing matrix, we use artificial data with known ground truth.

## 1 Introduction

The problem of following a party conversation by separating several voices from noise focusing on a single voice has been termed the "Cocktail Party Problem". A review is provided in [1]. This problem has been tackled by a number of researchers in the following mathematical setting:

We are given a sequence of observations $\mathbf{x}(1), \ldots, \mathbf{x}(t), \ldots$ with $\mathbf{x}(t) \in \mathbb{R}^N$ that are a linear mixture of a number of unknown sources $\mathbf{a}(1), \ldots, \mathbf{a}(t), \ldots$ with $\mathbf{a}(t) \in \mathbb{R}^M$:

$$\mathbf{x}(t) = C\mathbf{a}(t) \tag{1}$$

Here $C = (\mathbf{c}_1, \ldots, \mathbf{c}_M), \mathbf{c}_j \in \mathbb{R}^N$ denotes the mixing matrix. We may consider the observations $\mathbf{x}(t)$ to be what we hear and the sources $\mathbf{a}(t)$ to be the voices of $M$ persons at time $t$. The sequence $\mathbf{s}_j = a(1)_j, \ldots, a(t)_j, \ldots$ consists of all statements of person $j$. Is it possible to estimate the sources $\mathbf{s}_j$ only from the mixtures $\mathbf{x}(t)$ without knowing the mixing matrix $C$? In the past, a number of methods have been proposed that can be used to estimate the statements $\mathbf{s}_j$ and $C$ when only the mixtures $\mathbf{x}(t)$ are known and one can assume that $M = N$ [2], [3]. Moreover, some methods assume statistical independence of the sources [3]. Unfortunately the number of sources is not always equal to the number of observations, i.e., often $M > N$ holds. Humans have two ears but a large number of persons may be present at a party. The problem of having an overcomplete

J.C. Príncipe and R. Miikkulainen (Eds.): WSOM 2009, LNCS 5629, pp. 145–153, 2009.

set of sources has been studied more recently [4,5,6,7]. Due to the presence of a certain amount of additional background noise the problem may become even more difficult. A model that introduces a certain amount of additive noise,

$$\mathbf{x}(t) = C\mathbf{a}(t) + \epsilon(t) \quad \|\epsilon(t)\| \leq \delta, \tag{2}$$

has also been studied in the past [8]. The "Sparse Coding Neural Gas" (SCNG) algorithm [9,10] can be used to perform overcomplete blind source separation under the presence of noise as shown in [11]. Here we want to consider an even more realistic setting by allowing the mixing matrix to be time dependent:

$$\mathbf{x}(t) = C(t)\mathbf{a}(t) + \epsilon(t) \quad \|\epsilon(t)\| \leq \delta. \tag{3}$$

For the time dependent mixing matrix $C(t) = (\mathbf{c}_1(t), \ldots, \mathbf{c}_M(t)), \mathbf{c}_j(t) \in \mathbb{R}^N$, we require $\|\mathbf{c}_j(t)\| = 1$ without loss of generality. For instance, in the case of the cocktail party setting, this corresponds to party guests who change their position during the conversation. We want to process the observations in an online pattern-by-pattern mode, i.e., each observation is presented only once to the learning algorithm and the sources are estimated immediately. We do not make assumptions regarding the type of noise but our method requires that the underlying sources $\mathbf{s}_j$ are sufficiently sparse, in particular, it requires that the $\mathbf{a}(t)$ are sparse, i.e., only a few persons talk at the same time. The noise level $\delta$ and the number of sources $M$ have to be known.

## 1.1   Source Separation and Orthogonal Matching Pursuit

We here briefly discuss an important property of the orthogonal matching pursuit algorithm (OMP) [12] with respect to the obtained performance on the representation level that has been shown recently [13]. It provides the theoretical foundation that allows us to apply OMP to the problem of source separation.

Our method does not require that the sources $\mathbf{s}_j$ are independent but it requires that only few sources contribute to each mixture $\mathbf{x}(t)$, i.e., that the $\mathbf{a}(t)$ are sparse. However, an important observation is that if the underlying sources $\mathbf{s}_j$ are sparse and independent, for a given mixture $\mathbf{x}(t)$ the vector $\mathbf{a}(t)$ will be sparse, too.

Let us assume that we know the mixing matrix $C(t)$ at time $t$. Let us further assume that we know the noise level $\delta$. Let $\mathbf{a}(t)$ be the vector containing a small number $k$ of non-zero entries such that equation (3) holds for a given observation $\mathbf{x}(t)$. OMP provides an estimation $\mathbf{a}(t)^{\mathrm{OMP}}$ of $\mathbf{a}(t)$ by iteratively constructing $\mathbf{x}(t)$ out of the columns of $C(t)$. Let $C(t)\mathbf{a}(t)^{\mathrm{OMP}}$ denote the current approximation of $\mathbf{x}(t)$ in OMP and $\epsilon(t)$ the residual that still has to be constructed. Let $U$ denote the set of indices of those columns of $C(t)$ that already have been used during OMP. The number of elements in $U$, i.e., $|U|$, equals the number of OMP iterations that have been performed so far. The columns of $C(t)$ that are indexed by $U$ are denoted by $C(t)^U$. Initially, $\mathbf{a}(t)^{\mathrm{OMP}} = 0$, $\epsilon(t) = \mathbf{x}(t)$ and $U = \emptyset$. OMP works as follows:

1. Select $\mathbf{c}_{l_{\text{win}}}(t)$ by $\mathbf{c}_{l_{\text{win}}}(t) = \arg\max_{\mathbf{c}_l(t),l\notin U}(\mathbf{c}_l(t)^T\boldsymbol{\epsilon}(t))$
2. Set $U = U \cup l_{\text{win}}$
3. Solve the optimization problem $\mathbf{a}(t)^{\text{OMP}} = \arg\min_{\mathbf{a}} \|\mathbf{x}(t) - C(t)^U\mathbf{a}\|_2^2$
4. Obtain current residual $\boldsymbol{\epsilon}(t) = \mathbf{x}(t) - C(t)\mathbf{a}(t)^{\text{OMP}}$
5. Continue with step 1 until $\|\boldsymbol{\epsilon}(t)\| \leq \delta$

It can be shown that

$$\|\mathbf{a}(t)^{\text{OMP}} - \mathbf{a}(t)\| \leq \Lambda_{\text{OMP}}\,\delta \tag{4}$$

holds if the smallest entry in $\mathbf{a}(t)$ is sufficiently large and the number of non-zero entries in $\mathbf{a}(t)$ is sufficiently small. Let

$$H(C(t)) = \max_{1\leq i,j\leq M,i\neq j} |\mathbf{c}_i(t)^T\mathbf{c}_j(t)| \tag{5}$$

be the mutual coherence of the mixing matrix $C(t)$. The smaller $H(C(t))$, $N/M$ and $k$ are, the smaller $\Lambda_{\text{OMP}}$ becomes and the smaller $\min(\mathbf{a}(t))$ is allowed to be [13]. Since (4) only holds if the smallest entry in $\mathbf{a}(t)$ is sufficiently large, OMP has the property of local stability with respect to (4) [13]. Furthermore it can be shown that under the same conditions $\mathbf{a}(t)^{\text{OMP}}$ contains only non-zeros that also appear in $\mathbf{a}(t)$ [13]. An even globally stable approximation of $\mathbf{a}(t)$ can be obtained by methods such as basis pursuit [13,14].

## 1.2 Optimized Orthogonal Matching Pursuit (OOMP)

The "Sparse Coding Neural Gas" algorithm is based on "Optimised Orthogonal Matching Pursuit" (OOMP) which is an improved variant of OMP[15]. In general, the columns of $C(t)$ are not pairwise orthogonal. Hence, the criterion of OMP that selects the column $\mathbf{c}_{l_{\text{win}}}(t)$, $l_{\text{win}} \notin U$ of $C(t)$ that is added to $U$ is not optimal with respect to the minimization of the residual that is obtained after the column $\mathbf{c}_{l_{\text{win}}}(t)$ has been added. Hence OOMP runs through all columns of $C(t)$ that have not been used so far and selects the one that yields the smallest residual:

1. Select $\mathbf{c}_{l_{\text{win}}}(t)$ such that $\mathbf{c}_{l_{\text{win}}}(t) = \arg\min_{\mathbf{c}_l(t),l\notin U}\min_{\mathbf{a}} \|\mathbf{x}(t) - C(t)^{U\cup l}\mathbf{a}\|$
2. Set $U = U \cup l_{\text{win}}$
3. Solve the optimization problem $\mathbf{a}(t)^{\text{OOMP}} = \arg\min_{\mathbf{a}} \|\mathbf{x}(t) - C(t)^U\mathbf{a}\|_2^2$
4. Obtain current residual $\boldsymbol{\epsilon}(t) = \mathbf{x}(t) - C(t)\mathbf{a}(t)^{\text{OOMP}}$
5. Continue with step 1 until $\|\boldsymbol{\epsilon}(t)\| \leq \delta$

Step (1) involves $M - |U|$ minimization problems. In order to reduce the computational complexity of this step, we employ a temporary matrix $R$ that has been orthogonalized with respect to $C(t)^U$. $R$ is obtained by removing the projection of the columns of $C(t)$ onto the subspace spanned by $C(t)^U$ from $C(t)$ and setting the norm of the residuals $\mathbf{r}_l$ to one. The residual $\boldsymbol{\epsilon}(t)^U$ is obtained by removing the projection of $\mathbf{x}(t)$ to the subspace spanned by $C(t)^U$ from $\mathbf{x}(t)$.

Initially, $R = (\mathbf{r}_1, \ldots, \mathbf{r}_l, \ldots, \mathbf{r}_M) = C(t)$ and $\epsilon(t)^U = \mathbf{x}(t)$. In each iteration, the algorithm determines the column $\mathbf{r}_l$ of $R$ with $l \notin U$ that has maximum overlap with respect to the current residual $\epsilon(t)^U$:

$$l_{\text{win}} = \arg\max_{l, l \notin U} (\mathbf{r}_l^T \epsilon(t)^U)^2 . \tag{6}$$

Then, in the construction step, the orthogonal projection with respect to $\mathbf{r}_{l_{\text{win}}}$ is removed from the columns of $R$ and $\epsilon(t)^U$:

$$\mathbf{r}_l = \mathbf{r}_l - (\mathbf{r}_{l_{\text{win}}}^T \mathbf{r}_l)\mathbf{r}_{l_{\text{win}}}, \tag{7}$$

$$\epsilon(t)^U = \epsilon(t)^U - (\mathbf{r}_{l_{\text{win}}}^T \epsilon(t)^U)\mathbf{r}_{l_{\text{win}}} . \tag{8}$$

After the projection has been removed, $l_{\text{win}}$ is added to $U$, i.e., $U = U \cup l_{\text{win}}$. The columns $\mathbf{r}_l$ with $l \notin U$ may be selected in the subsequent iterations of the algorithm. The norm of these columns is set to unit length. If the stopping criterion $\|\epsilon(t)^U\| \le \delta$ has been reached, the final entries of $\mathbf{a}(t)^{\text{OOMP}}$ can be obtained by recursively collecting the contribution of each column of $C(t)$ during the construction process, taking into account the normalization of the columns of $R$ in each iteration. The selection criterion (6) ensures that the norm of the residual $\epsilon(t)^U$ obtained by (8) is minimal. Hence, the OOMP algorithm can provide an approximation of $\mathbf{a}(t)$ containing even less non-zeros than the approximation provided by OMP.

## 2   Learning the Mixing Matrix

We want to estimate the mixing matrix $C(t) = (\mathbf{c}_1(t), \ldots, \mathbf{c}_M(t))$ from the mixtures $\mathbf{x}(t)$ given the noise level $\delta$ and the number of underlying sources $M$. As a consequence of the sparseness of the underlying sources $\mathbf{a}(t)$, we are looking for a mixing matrix $C(t)$ that minimizes the number of non-zero entries of $\mathbf{a}(t)^{\text{OOMP}}$, i.e., the number of iteration steps required by the OOMP algorithm to approximate $\mathbf{a}(t)$ up to the noise level $\delta$. Furthermore, let us assume that the mixing matrix changes slowly over time such that $C(t)$ is approximately constant for some time interval $[t - T, t]$. Hence, we look for the mixing matrix which minimizes

$$\min_{C(t)} \sum_{t'=t-T}^{t} \|\mathbf{a}(t')^{\text{OOMP}}\|_0 \quad \text{subject to} \quad \|\mathbf{x}(t') - C(t)\mathbf{a}(t')^{\text{OOMP}}\| \le \delta . \tag{9}$$

Here $\|\mathbf{a}(t')^{\text{OOMP}}\|_0$ denotes the number of non-zero entries in $\mathbf{a}(t')^{\text{OOMP}}$. The smaller the norm of the current residual $\epsilon(t')^U$ is, the fewer OOMP iterations have to be performed until the stopping criterion $\|\epsilon(t')^U\| \le \delta$ has been reached and the smaller $\|\mathbf{a}(t')^{\text{OOMP}}\|_0$ becomes. In order to minimize the norm of the residuals and thereby the expression (9), we have to maximize $(\mathbf{r}_{l_{\text{win}}} \epsilon(t')^U)^2$. Therefore, we consider the following optimization problem

$$\max_{\mathbf{r}_1, \ldots, \mathbf{r}_M} \sum_{t'=t-T}^{t} \max_{l, l \notin U} (\mathbf{r}_l^T \epsilon(t')^U)^2 \quad \text{subject to} \quad \|\mathbf{r}_l\| = 1 . \tag{10}$$

We maximize (10) by updating $R$ and $C(t)$ prior to the construction step (7) and (8). The update step in each iteration of the OOMP algorithm is a combination of Oja's learning rule [16] and the Neural Gas [17,18]. As introduced in [9] one obtains what we called "Sparse Coding Neural Gas" (SCNG) learning rule

$$\Delta\mathbf{r}_{l_k} = \Delta\mathbf{c}_{l_k}(t) = \alpha(t)e^{-k/\lambda(t)}y\left(\boldsymbol{\epsilon}(t)^U - y\,\mathbf{r}_{l_k}\right) \qquad (11)$$

with learning rate

$$\alpha(t) = \alpha_0\left(\alpha_{\text{final}}/\alpha_0\right)^{t/t_{\max}}, \qquad (12)$$

and neighbourhood-size

$$\lambda(t) = \lambda_0\left(\lambda_{\text{final}}/\lambda_0\right)^{t/t_{\max}} \qquad (13)$$

where

$$-\left(\mathbf{r}_{l_0}^T\boldsymbol{\epsilon}(t)^U\right)^2 \leq \ldots \leq -\left(\mathbf{r}_{l_k}^T\boldsymbol{\epsilon}(t)^U\right)^2 \leq \ldots \leq -\left(\mathbf{r}_{l_{M-|U|}}^T\boldsymbol{\epsilon}(t)^U\right)^2, l_k \notin U \quad (14)$$

and $y = \mathbf{r}_{l_k}^T\boldsymbol{\epsilon}(t)^U$. We have shown [10] that (11) implements a gradient descent with respect to

$$\max_{\mathbf{r}_1,\ldots,\mathbf{r}_M}\sum_{t'=t-T}^{t}\sum_{l=1}^{M}h_{\lambda_t}(k(\mathbf{r}_l,\boldsymbol{\epsilon}(t')^U))(\mathbf{r}_l^T\boldsymbol{\epsilon}(t')^U)^2 \quad \text{subject to} \quad \|\mathbf{r}_l\| = 1, \quad (15)$$

with $h_{\lambda_t}(v) = e^{-v/\lambda_t}$. $k(\mathbf{r}_l,\boldsymbol{\epsilon}(t')^U)$ denotes the number of $\mathbf{r}_j$ with $(\mathbf{r}_j^T\boldsymbol{\epsilon}(t')^U)^2 < (\mathbf{r}_l^T\boldsymbol{\epsilon}(t')^U)^2$, i.e., (15) is equivalent to (10) for $\lambda(t) \to 0$. Due to (11) the updates of all OOMP iterations are accumulated in the learned mixing matrix $C(t)$. Due to the orthogonal projection (7) and (8) performed in each iteration, these updates are pairwise orthogonal. The columns of the original matrix emerge in random order in the learned mixing matrix. The sign of the columns of the mixing matrix $\mathbf{c}_l(t)$ cannot be determined because multiplying $\mathbf{c}_l(t)$ by $-1$ corresponds to multiplying $\mathbf{r}_l$ by $-1$ which does not change (15).

What happens for $t > t_{\max}$? Assuming that after $t_{\max}$ learning steps have been performed the current learned mixing matrix is close to the true mixing matrix, we track the slowly changing true mixing matrix by setting $\alpha(t) = \alpha_{\text{final}}$ and $\lambda(t) = \lambda_{\text{final}}$.

## 3   Experiments

We performed a number of experiments on artificial data in order to study whether the underlying sources can be reconstructed from the mixtures. We consider sequences

$$\mathbf{x}(t) = C(t)\mathbf{a}(t) + \boldsymbol{\epsilon}(t), \; t = 1,\ldots,L, \qquad (16)$$

where $\|\boldsymbol{\epsilon}(t)\| \leq \delta$, $\mathbf{x}(t) \in \mathbb{R}^N$, $\mathbf{a}(t) \in \mathbb{R}^M$. The true mixing matrix $C(t)$ slowly changes from state $C_{i-1}$ to state $C_i$ in $P$ time steps. We randomly chose a sequence of true mixing matrices $C_i$, $i = 1, \ldots, \lceil L/P \rceil$ with entries taken from a uniform distribution. The columns of these mixing matrices were set to unit norm. At time $t$ with $(i-1)P \leq t \leq iP$ the true mixing matrix $C(t)$ is chosen according to

$$C(t) = \left(1 - \frac{(t - (i-1)P)}{P}\right) C_{i-1} + \frac{(t - (i-1)P)}{P} C_i. \qquad (17)$$

The norm of the columns of each true mixing matrix $C(t)$ is then set to unit norm. The sources $\mathbf{a}(t)$ were obtained by setting up to $k$ entries of the $\mathbf{a}(t)$ to uniformly distributed random values in $[-1, 1]$. For each $\mathbf{a}(t)$ the number of non-zero entries was obtained from a uniform distribution in $[0, k]$. Uniformly distributed noise $\mathbf{e}(t) \in \mathbb{R}^M$ in $[-1, 1]$ was added such that

$$\mathbf{x}(t) = C(t)(\mathbf{a}(t) + \mathbf{e}(t)) = C(t)\mathbf{a}(t) + \boldsymbol{\epsilon}(t) . \qquad (18)$$

We want to asses the error that is obtained with respect to the recontruction of the sources. Hence, we evaluate the difference between the sources $\mathbf{a}(t)$ and the estimation $\mathbf{a}(t)^{\mathrm{OOMP}}$ that is obtained from the OOMP algorithm on the basis of the mixing matrix $C^{learn}(t)$ that is provided by the SCNG algorithm:

$$\|\mathbf{a}(t) - \mathbf{a}(t)^{\mathrm{OOMP}}\|_2. \qquad (19)$$

With $(\mathbf{s}_1^{\mathrm{OOMP}}, \ldots, \mathbf{s}_M^{\mathrm{OOMP}})^T = (\mathbf{a}(1)^{\mathrm{OOMP}}, \ldots, \mathbf{a}(L)^{\mathrm{OOMP}})$ we denote the estimated underlying sources obtained from the OOMP algorithm. In order to evaluate (19) we have to assign the entries in $\mathbf{a}(t)^{\mathrm{OOMP}}$ to the entries in $\mathbf{a}(t)$ which is equivalent to assigning the true sources $\mathbf{s}_j$ to the estimated sources $\mathbf{s}_j^{\mathrm{OOMP}}$. This problem arises due to the random order in which the columns of the true mixing matrix appear in the learned mixing matrix. Due to the time dependent mixing matrix the assignment may change over time. In order to obtain an assignment at time $t$, we consider a window of size $s_w$:

$$(\mathbf{w}_1(t)^{\mathrm{OOMP}}, \ldots, \mathbf{w}_M(t)^{\mathrm{OOMP}})^T = (\mathbf{a}(t - s_w/2)^{\mathrm{OOMP}}, \ldots, \mathbf{a}(t + s_w/2)^{\mathrm{OOMP}}) \qquad (20)$$

and

$$(\mathbf{w}_1(t), \ldots, \mathbf{w}_M(t))^T = (\mathbf{a}(t - s_w/2), \ldots, \mathbf{a}(t + s_w/2)). \qquad (21)$$

We obtain the assignment by performing the following procedure:

1. Set $I_{\mathrm{true}} : \{1, \ldots, M\}$ and $I_{\mathrm{learned}} : \{1, \ldots, M\}$.
2. Find and assign $\mathbf{w}_i(t)$ and $\mathbf{w}_j(t)^{\mathrm{OOMP}}$ with $i \in I_{\mathrm{true}}, j \in I_{\mathrm{learned}}$ such that

$$\frac{|\mathbf{w}_j(t)^{\mathrm{OOMP}} \mathbf{w}_i(t)^T|}{\|\mathbf{w}_i(t)\| \|\mathbf{w}_j(t)^{\mathrm{OOMP}}\|} \text{ is maximal.}$$

3. Remove $i$ from $I_{\mathrm{true}}$ and $j$ from $I_{\mathrm{learned}}$.
4. If $\mathbf{w}_j(t)^{\mathrm{OOMP}} \mathbf{w}_i(t)^T < 0$ set $\mathbf{w}_j(t)^{\mathrm{OOMP}} = -\mathbf{w}_j(t)^{\mathrm{OOMP}}$.
5. Proceed with (2) until $I_{\mathrm{true}} = I_{\mathrm{learned}} = \emptyset$.

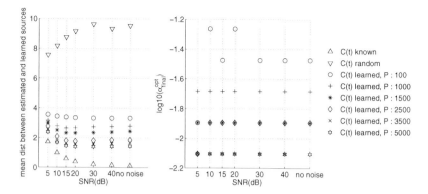

**Fig. 1.** Left: Mean distance between $\mathbf{a}(t)$ and $\mathbf{a}(t)^{\text{OOMP}}$ for different SNR and $P$. Right: Best performing final learning rate for each $P$ and SNR.

For all experiments, we used $L = 20000$ and $\alpha_0 = 1$ for the learing rate as well as $\lambda_0 = M/2$ and $\lambda_{\text{final}} = 10^{-10}$ for the neighbourhood size and $t_{\max} = 5000$. We repeated all experiments 20 times and report the mean result over the 20 runs. For the evaluation of the reconstruction error the window size $s_w$ was set to 30 and the reconstruction error was only evaluated in the time interval $t_{\max} < t < L$. In all experiments an overcomplete setting was used consisting of $M = 30$ underlying sources and $N = 15$ available oberservations. Up to $k = 3$ underlying sources were active at the same time.

In the first experiment, we varied the parameter $P$ which controls the change rate of the true mixing matrix as well as the SNR. The final learning rate $\alpha_{\text{final}}$ was varied for each combination of $P$ and SNR such that the minimal reconstruction error was obtained. For comparison purposes, we also computed the reconstruction error that is obtained by using the true mixing matrix as well as the error that is obtained by using a random matrix. The results of the first experiment are shown in figure 1. On the left side the mean distance between $\mathbf{a}(t)$ and $\mathbf{a}(t)^{\text{OOMP}}$ is shown for different SNR and $P$. It can be seen that the larger the change rate of the true mixing matrix (the smaller $P$) and the stronger the noise, the more the reconstruction performance degrades. But even for strong noise and a fast changing true mixing matrix, the estimation provided by SCNG clearly outperforms a random matrix. Of course, the best reconstruction performance is obtained by using the true mixing matrix. On the right side of the figure the best performing final learning rate for each $P$ and SNR is shown. It can be seen that the optimal final learning rate depends on the change rate of the true mixing matrix but not on the strength of the noise. In order to assess how good the true mixing matrix is learned, we perform an experiment that is similar to an experiment that has been used to asses the performance of the K-SVD algorithm with respect to the learing of the true mixing matrix [19]. Note, that the K-SVD algorithm cannot be applied to the setting that is described in the following. We compare the learned mixing matrix to the true mixinig matrix

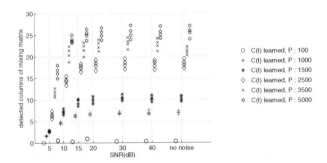

**Fig. 2.** We sorted the 20 trials according to the number of successfully learned columns of the mixing matrix and order them in groups of five experiments. The figure shows the mean number of successfully detected columns of the mixing matrix for each of the five groups.

using the maximum overlap between each column of the true mixing matrix and each column of the learned mixing matrix, i.e, whenever

$$\max_{j} \left(1 - |\mathbf{c}_i(t)\mathbf{c}_j^{\text{learn}}(t)|\right) \tag{22}$$

is smaller than 0.05, we count this as a success. We repeat the experiment 20 times with a varying SNR as well as zero noise. For each SNR, we sort the 20 trials according to the number of successfully learned columns of the mixing matrix and order them in groups of five experiments. Figure 2 shows the mean number of successfully detected columns of the mixing matrix for each of the five groups for each SNR and $P$. The smaller the SNR and the smaller the change rate of the true mixing matrix is, the more columns are learned correctly. If the true mixing matrix changes very fast ($P = 100$) almost no column can be learned with the required accuracy.

## 4    Conclusion

We showed that the "Sparse Coding Neural Gas" algorithm can be applied to a more realistic model of the "Cocktail Party Problem" that allows for more sources than observations, additive noise and a mixing matrix that is time dependent, which corresponds to persons that change their position during the conversation. The proposed algorithm works online, the estimation of the underlying sources is provided immediately. The method requires that the sources are sparse enough, that the mixing matrix does not change too quickly and that the additive noise is not too strong. In order to apply this algorithm to real-world data, future work is required. The problem of (i) choosing the number of sources $M$ based solely on the observations, (ii) determining the noise level $\delta$ based solely on the observations and (iii) obtaining the temporal assignment of the sources based solely on the estimated sources, i.e., thereby not using the sliding window procedure described in the experiments section have to be investigated.

# References

1. Haykin, S., Chen, Z.: The cocktail party problem. Neural Comput. 17(9), 1875–1902 (2005)
2. Bell, A.J., Sejnowski, T.J.: An information-maximization approach to blind separation and blind deconvolution. Neural Computation 7(6), 1129–1159 (1995)
3. Hyvärinen, A.: Fast and robust fixed-point algorithms for independent component analysis. IEEE Transactions on Neural Networks 10(3), 626–634 (1999)
4. Hyvärinen, A., Cristescu, R., Oja, E.: A fast algorithm for estimating overcomplete ica bases for image windows. In: Proceedings of the International Joint Conference on Neural Networks, IJCNN 1999, vol. 2, pp. 894–899 (1999)
5. Lee, T.W., Lewicki, M., Girolami, M., Sejnowski, T.: Blind source separation of more sources than mixtures using overcomplete representations. IEEE Signal Processing Letters 6(4), 87–90 (1999)
6. Lewicki, M.S., Sejnowski, T.J.: Learning Overcomplete Representations. Neural Computation 12(2), 337–365 (2000)
7. Theis, F., Lang, E., Puntonet, C.: A geometric algorithm for overcomplete linear ica. Neurocomputing 56, 381–398 (2004)
8. Hyvärinen, A.: Gaussian moments for noisy independent component analysis. IEEE Signal Processing Letters 6(6), 145–147 (1999)
9. Labusch, K., Barth, E., Martinetz, T.: Learning Data Representations with Sparse Coding Neural Gas. In: Verleysen, M. (ed.) Proceedings of the 16th European Symposium on Artificial Neural Networks, pp. 233–238. D-Side Publishers (2008)
10. Labusch, K., Barth, E., Martinetz, T.: Sparse Coding Neural Gas: Learning of Overcomplete Data Representations. Neurocomputing 72(7-9), 1547–1555 (2009)
11. Labusch, K., Barth, E., Martinetz, T.: Sparse Coding Neural Gas for the Separation of Noisy Overcomplete Sources. In: Koutník, J., Krurková, V., Neruda, R. (eds.) ICANN 2008, Part I. LNCS, vol. 5163, pp. 788–797. Springer, Heidelberg (2008)
12. Pati, Y., Rezaiifar, R., Krishnaprasad, P.: Orthogonal matching pursuit: Recursive function approximation with applications to wavelet decomposition. In: Proceedings of the 27 th Annual Asilomar Conference on Signals, Systems (November 1993)
13. Donoho, D.L., Elad, M., Temlyakov, V.N.: Stable recovery of sparse overcomplete representations in the presence of noise. IEEE Transactions on Information Theory 52(1), 6–18 (2006)
14. Chen, S.S., Donoho, D.L., Saunders, M.A.: Atomic decomposition by basis pursuit. SIAM Journal on Scientific Computing 20(1), 33–61 (1998)
15. Rebollo-Neira, L., Lowe, D.: Optimized orthogonal matching pursuit approach. IEEE Signal Processing Letters 9(4), 137–140 (2002)
16. Oja, E.: A simplified neuron model as a principal component analyzer. J. Math. Biol. 15, 267–273 (1982)
17. Martinetz, T., Schulten, K.: A "Neural-Gas Network" Learns Topologies. Artificial Neural Networks I, 397–402 (1991)
18. Martinetz, T., Berkovich, S., Schulten, K.: "Neural-gas" Network for Vector Quantization and its Application to Time-Series Prediction. IEEE-Transactions on Neural Networks 4(4), 558–569 (1993)
19. Aharon, M., Elad, M., Bruckstein, A.: K-SVD: An Algorithm for Designing Overcomplete Dictionaries for Sparse Representation. IEEE Transactions on Signal Processing 54(11), 4311–4322 (2006); see also IEEE Transactions on Acoustics, Speech, and Signal Processing

# Career-Path Analysis Using Optimal Matching and Self-Organizing Maps

Sébastian Massoni[1], Madalina Olteanu[2], and Patrick Rousset[3]

[1] CES, Université Paris 1
112 Bd de l'Hopital, Paris, France
[2] SAMOS - CES, Université Paris 1
90 Rue de Tolbiac, Paris, France
[3] CEREQ
10 Place de la Joliette, Marseille, France

**Abstract.** This paper is devoted to the analysis of career paths and employability. The state-of-the-art on this topic is rather poor in methodologies. Some authors propose distances well adapted to the data, but are limiting their analysis to hierarchical clustering. Other authors apply sophisticated methods, but only after paying the price of transforming the categorical data into continuous, via a factorial analysis. The latter approach has an important drawback since it makes a linear assumption on the data. We propose a new methodology, inspired from biology and adapted to career paths, combining optimal matching and self-organizing maps. A complete study on real-life data will illustrate our proposal.

## 1   Introduction

The question of analyzing school-to-work transitions is a challenging topic for the economists working on the labor market. In the current economic context of the world, characterized by a significant unemployment rate of young people (in France, 19.1% of the young people under 25 were unemployed during the second semester of 2008), it is interesting to study the insertion of graduates and the evolution of their career paths. The aim of this paper is to identify and analyze career-paths typologies.

Let us recall that a career path is defined as a sequence of labor-market statuses, recorded monthly. The number and the labels associated to the statuses are defined by experts. The experts take into account the fact that the more the number of labels is detailed, the more the analysis of the career paths will be accurate. We shall also remark that the labelling of the statuses is not neutral. Indeed, the choice of different criteria for identifying the statuses introduces an *a priori* separation between "good" and "bad" statuses. Generally, the number and the labelling of the statuses are quite similar in the literature, with different labels corresponding to employment and unemployment situations. The data "Generation 98" used for this study are labelled according to nine possible statuses, five employment statuses (permanent-labor contract, fixed-term contract,

J.C. Príncipe and R. Miikkulainen (Eds.): WSOM 2009, LNCS 5629, pp. 154–162, 2009.

apprenticeship contract, public temporary-labor contract, on-call contract) and four unemployment statuses (unemployed, inactive, military service, education).

In the state-of-the-art, two approaches seem to be currently used for clustering career paths. The first approach consists in transforming categorical variables into continuous variables by a factorial analysis and then apply usual clustering algorithms for Euclidean data ([9],[8]). The second approach consists in computing adapted distances for the data ([12], [4]) and then cluster using an hierarchical tree and a proximity criterion based on the distance matrix only. Both approaches have some drawbacks: in the first case, the use of factorial methods implies quite strong hypothesis concerning the linearity of the data; in the second case, hierarchical clustering is not suited for large data sets and does not provide any tools for displaying and visualizing the results. In order to address these drawbacks, we propose to cluster career paths using a two-step methodology. The two steps of the algorithm are independent and quite general. First, we compute a dissimilarity matrix between the career paths. Second, a self-organizing map for dissimilarity matrices is trained. Besides identifying the main typologies of career paths, we are also looking for a graphical output representing the proximities and the evolutions in the different career paths.

The rest of the document is organized as follows: Section 2 is devoted to a description of the methodology and a short state-of-the-art on the subject. Section 3 contains the results on the data "Generation 98" (CEREQ, France). The conclusion and some perspectives are presented in the last section.

## 2    Methodology

From a statistical point of view, several problems arise when analyzing school-to-work transitions. The data sets are usually containing categorical variables, often in high dimension and have an important sample size. In order to handle these data, we made the choice of splitting the analysis into two steps. The first step consists in defining a distance or a dissimilarity measure well-suited to the data. In the second step, a clustering method is used to build and define typologies. The clustering method has to be general enough to "forget" the initial structure of the data and determine classes on the unique basis of the dissimilarity matrix computed in the previous step. This approach has the advantage of allowing a wide choice for the dissimilarity measure in the first step.

### 2.1    Step 1 (Optimal Matching) – Choosing a Good Distance

The first step of the analysis consists in choosing a dissimilarity measure between the career paths. Previous studies suggest the use of a multiple correspondence analysis ([9]) and a transformation of the categorical variables into continuous. This way, usual clustering algorithms based on the Euclidean distance can be trained on the factorial components. This approach has an important drawback, since it makes the assumption that data are linear. As there is no reason for this assumption to hold in our case, we prefer to use a distance which avoids this hypothesis.

Optimal matching, also known as "edit distance" or "Levenshtein distance", was first introduced in biology by [13] and used for aligning and comparing sequences. In social sciences, the first applications are due to [1]. Let us consider two sequences of different lengths $a = (a_1, ..., a_{n_1})$ and $b = (b_1, ..., b_{n_2})$, $a_i, b_j \in S$ where $S$ is a finite state-space. How may the two sequences be aligned and what is the distance between them? The underlying idea of optimal matching is to transform the sequence $a$ into the sequence $b$ using three possible operations: insertion, deletion and substitution. A cost is associated to each of the three operations. The distance between $a$ and $b$ is computed as the cost associated to the smallest number of operations which allow to transform $a$ into $b$. The method seems simple and relatively intuitive, but the choice of the costs is a delicate operation in social sciences. This topic is subject to lively debates in the literature ([3], [14]) mostly because of the difficulties in establishing an explicit and sound theoretical frame. The interested reader may refer to [7] for a state-of-the-art.

In the data set "Generation 98", all career paths have the same length, the status of the graduate students being observed during 94 months. Thus, we may suppose that there are no insertions or deletions and that the only cost to be computed is the substitution cost. The latter was computed using the transition matrix between the statuses as proposed in [12]: the less transitions between two statuses are observed, the more the statuses are different and the substitution cost is high. The cost $w$ for transforming $a_i$ into $a_j$ is computed as a function of the observed longitudinal transitions:

$$w(a_i, a_j) = 2 - P(a_i|a_j) - P(a_j|a_i).$$

## 2.2   Step 2 – Self-Organizing Maps for Categorical Data

The methodology "optimal matching - clustering" has already been used ([4], [12]) during the past few years for the analysis of career or life paths. The general approach consists in computing a dissimilarity matrix using optimal matching, build an hierarchical tree and make a description of the resulting typologies. However, hierarchical clustering is limited in terms of displaying and visualizing the results. Instead, we suggest a clustering method which provides, besides clusters, a graphical representation preserving the proximity between paths.

Self-organizing maps (Kohonen algorithm, [11]) are, at the same time, a clustering algorithm and a nonlinear projection method. The input data are projected on a grid, generally rectangular or hexagonal. The grid has the important property of topology preservation: close inputs will be projected in the same class or in neighbor classes. The algorithm was initially developed for continuous data with a Euclidean distance. Since its first application, new versions, suited for particular data structures, are regularly proposed. Let us mention [10], who are using for the first time self-organizing maps and optimal matching for the analysis of biological sequences.

For career-paths data, we used a general self-organizing map algorithm, proposed by [5]. Their method does not take into account the initial structure of the data and uses only a dissimilarity matrix as input. The size and the structure of the grid must also be provided as input. At the first step of the algorithm, the prototypes are chosen at random in the input data. The rest of the algorithm consists in repeating the following steps, until the partition is stable:

- Allocating step: each input is assigned to the class of the closest prototype by minimizing a criterion of intra-class variance, extended to the neighbors. During this step, the prototypes are fixed.
- Representing step : once the new partition is determined, the new prototypes are computed by minimizing the same criterion.

The price of the generality of the method is the poverty of the space where the inputs lie: the new prototypes have to be computed and chosen between the inputs.

This algorithm is a generalization of *K-means* by introducing with a neighborhood relation defined between classes. Since the algorithm is BATCH, it is quite sensitive to the initializations.

## 3   Real-Life Data – "Generation 98" Survey

For illustrating the proposed methodology, we used the data in the survey "Generation 98" from CEREQ, France (http://www.cereq.fr/). The data set contains information on 16040 young people having graduated in 1998 and monitored during 94 months after having left school. The labor-market statuses have nine categories, labelled as follows: "permanent-labor contract", "fixed-term contract", "apprenticeship contract", "public temporary-labor contract", "on-call contract", "unemployed", "inactive", "military service", "education". The following stylized facts are highlighted by a first descriptive analysis of the data (Fig. 1):

- permanent-labor contract are representing more than 20% of all statuses after one year and their ratio continues to increase until 50% after three years and almost 75% after seven years;
- the ratio of fixed-terms contracts is more than 20% after one year on the labor market, but it is decreasing to 15% after three years and then seems to converge to 8%;
- almost 30% of the young graduates are unemployed after one year. This ratio is decreasing and becomes constant, 10%, after the fourth year.

Considering the important ratio of permanent contracts obtained relatively fast and its absorbing character, we decided to focus our analysis on the career paths which don't enter the "permanent contract" status in less than two years. Thus, the data set is reduced to 11777 inputs, which represent almost 3/4 of the initial data. This decision is also justified by some numerical problems such as the storage of the dissimilarity matrix and the computation time.

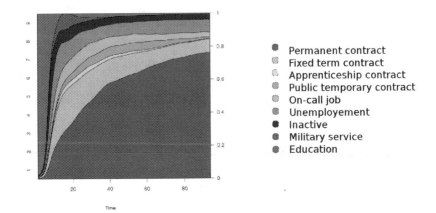

**Fig. 1.** Labor market structure

The inputs excluded from the analysis are grouped into a class corresponding to a "fast access to stable employment". The class contains 2032 inputs having obtained a permanent contract in less than a year and 4263 inputs having obtained a permanent contract in less than two years.

After having preprocessed the data, the analysis is conducted in three steps:

1. the transition matrix and the associated substitution-cost matrix are computed;
2. the dissimilarity matrix is computed by optimal matching with the substitution-cost matrix in step 1;
3. career paths are clustered with the self-organizing map algorithm, according to the dissimilarity matrix in step 2.

For the first two steps, we used the R-package TraMineR, available in [6]. The third step was implemented in R and is available on demand.

The cost matrix computed on the 11777 input data is the following:

$$C = \begin{pmatrix} 0 & 1.968 & 1.976 & 1.989 & 1.977 & 1.973 & 1.975 & 1.985 & 1.987 \\ 1.968 & 0 & 1.991 & 1.994 & 1.978 & 1.927 & 1.957 & 1.979 & 1.976 \\ 1.976 & 1.991 & 0 & 1.999 & 1.994 & 1.980 & 1.989 & 1.998 & 1.997 \\ 1.989 & 1.994 & 1.999 & 0 & 1.998 & 1.984 & 1.993 & 1.998 & 1.997 \\ 1.977 & 1.978 & 1.994 & 1.998 & 0 & 1.951 & 1.973 & 1.979 & 1.988 \\ 1.973 & 1.927 & 1.980 & 1.984 & 1.951 & 0 & 1.954 & 1.971 & 1.966 \\ 1.975 & 1.957 & 1.989 & 1.993 & 1.973 & 1.954 & 0 & 1.977 & 1.947 \\ 1.985 & 1.979 & 1.998 & 1.998 & 1.979 & 1.971 & 1.977 & 0 & 1.996 \\ 1.987 & 1.976 & 1.997 & 1.997 & 1.988 & 1.966 & 1.947 & 1.996 & 0 \end{pmatrix}$$

Let us remark that the values in the cost matrix are very similar. Different approaches for improving cost computation should be investigated, some perspectives are given in the conclusion.

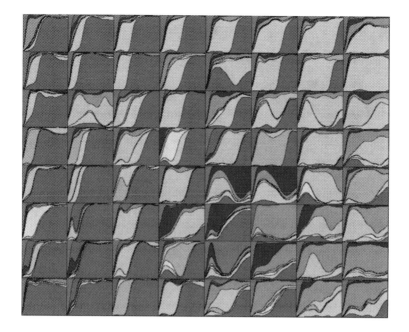

**Fig. 2.** SOM 8x8

When trying to compute the dissimilarity matrix, we were confronted to a numerical problem: the impossibility of storing a matrix of size 11777. Actually, this numerical problem constitutes the main drawback of the proposed methodology: the sample size must be "reasonable". The size of the data has to be reduced before training the self-organizing map algorithm. In order to summarize the career-path data, we used a *K-modes* algorithm, which may be trained directly on the initial data set. Thus, the 11777 career-paths were summarized by 1000 representative paths and these 1000 paths were clustered with self-organizing maps.

The dissimilarity matrix between the 1000 paths was computed with the optimal matching distance and the substitution-cost matrix $C$. Then, we trained a self-organizing map on a rectangular grid $8 \times 8$. The resulting map is plotted in Fig.2.

A lecture of the map summarizes the information on the career paths. Thus, we can stress out the proximities between different paths and the evolution of the career paths.

The most striking opposition appears between the career paths leading to a stable-employment situation (permanent contract and/or fixed-term contract) and the career-paths more "chaotic" (unemployment, on-call contracts, apprenticeship contracts). On the map in Fig.2, the "stable" situations are mainly situated in the *west* region of the map.

Thus, the first three columns contain essentially the classes where a permanent contract was rapidly obtained. However, the *north* and the *south* regions of these columns are quite different: in the *north* region, the access to a permanent contract is achieved after a first contact with the labor-market through a fixed-term contract, while the *south* classes are only subject to transitions through a military service or education periods. At the halfway between *north* and *south* regions, we may find several apprenticeship contracts or public temporary contracts.

The stability of the career-paths noticed in the *west* region of the map is getting worse as we move to the *east*. In the *north* region, the initial fixed-term contract is getting longer until becoming a poor employment situation in the *north-east* corner. Thus, all the *east* region of the map is revealing for difficult school-to-work transitions. Let us remark the on-call contracts situations which may end by a stable contract or by unemployment. At the opposite, the career-paths starting with an apprenticeship contract are most of the time ending with a permanent contract. Finally, the *south-east* corner is characterized by exclusion career-paths: inactive and unemployment. The inactivity may appear immediately after the education period or after a first failure on the labor market.

The clustering with SOM provided interesting results by highlighting proximities or oppositions between the career paths. In order to determine a small number of typologies, we compute an hierarchical classification tree on the 64 prototypes of the map. Ward criterion was used for aggregating classes. The clustering is represented in Fig.3. The final configuration with nine classes allows to describe 8 career-path typologies:

1. relatively fast access to a stable situation (*class 1*)
2. transition through a fixed-term contract before obtaining a permanent contract(*class 2*)
3. fixed-term contracts (*class 3*)

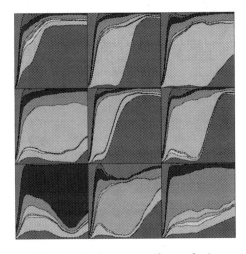

**Fig. 3.** Final career-path typologies

4. on-call contracts with periods of unemploymentnt (*class 4*)
5. shorter or longer on-call contracts and finally a permanent contract(*classes 5 and 6*)
6. inactivity after loosing a first job or immediately after school (*class 7*)
7. apprenticeship contract ending with a fixed-term or a permanent-term contract(*class 8*)
8. long unemploymentnt period with a gradual return to employment (*class 9*)

We may also add the first class which was excluded from the analysis:

9. fast access to stabl employment after leaving school.

The relative importance of every class is given in Table 1.

**Table 1.** Importance of final classes

| Class | Size | Weight in the sample | Weight in the whole sample |
|-------|------|----------------------|----------------------------|
| 1. | 5475 | 46.5% | 34.1% |
| 2. | 968 | 8.2% | 6.0% |
| 3. | 1082 | 9.2% | 6.8% |
| 4. | 514 | 4.4% | 3.2% |
| 5. | 1122 | 9.5% | 7.0% |
| 6. | 328 | 2.8% | 2.0% |
| 7. | 1002 | 8.5 | 6.3% |
| 8. | 1286 | 10.9 | 8.0% |
| 9. | 4263 | NA | 26.6% |

## 4    Conclusion and Perspectives

Several typologies for career paths were highlighted by our analysis. The self-organizing map allowed a detailed characterization of the proximities, oppositions and transitions between the different career paths. The proposed methodology is thus well suited for this kind of data.

However, several aspects should be improved in this approach. Concerning the labelling of the statuses, we remarked that the "military service" is not representative in a employment-unemployment discrimination. The solution is either to change the label of this status, either to delete this period. In the latter case, we would no longer have equal-sized sequences and the question of computing an insertion/deletion cost would arise.

The second remark concerns the computation of the substitution-cost matrix. In this paper, it was computed using the observed transitions, but without taking into account that these transitions change in time. Indeed, in a more realistic frame we should consider the non-homogeneity of the transitions and probably use non-homogeneous Markov chains in order to estimate the costs.

# References

1. Abbott, A., Forrest, J.: Optimal matching methods for historical sequences. Journal of Interdisciplinary History 16, 471–494 (1986)
2. Abbott, A., Hrycak, A.: Measuring resemblance in sequence data: An optimal matching analaysis of musicians carrers. American Journal of Sociolgy 96(1), 144–185 (1990)
3. Abbott, A., Tsay, A.: Sequence analysis and optimal matching methods in sociology. Review and prospect. Sociological Methods and Research 29(1), 333 (2000)
4. Brzinsky-Fay, C.: Lost in Transition? Labour Market Entry Sequences of School Leavers in Europe. European Sociological Review 23(4), 409–422 (2007)
5. Conan-Guez, B., Rossi, F., El Golli, A.: Fast algorithm and implementation of dissimilarity Self-Organizing Maps. Neural Networks 19(6-7), 855–863 (2006)
6. Gabadinho, A., Ritschard, G., Studer, M., Müller, N.S.: Mining Sequence Data in R with TraMineR: a user's guide (2008), http://mephisto.unige.ch/traminer
7. Gauthier, J.A., Widmer, E.D., Bucher, P., Notredame, C.: How much does it cost? Optimization of costs in sequence analysis in social science data. Sociological Methods and Research (in press) (2008)
8. Giret, J.F., Rousset, P.: Classifying qualitative time series with SOM: the typology of career paths in France. In: Computational and ambient intelligence, IWANN 2007 Proceedings, pp. 757–764. Springer, Berlin (2007)
9. Grelet, Y., Fenelon, J.-P., Houzel, Y.: The sequence of steps in the analysis of youth trajectories. European Journal of Economic and Social Systems 14(1) (2000)
10. Kohonen, T., Somervuo, P.: Self-organizing maps for symbol strings. Neurocomputing 21, 19–30 (1998)
11. Kohonen, T.: Self Organizing Maps. Springer, Berlin (1995)
12. Müller, N.S., Ritschard, G., Studer, M., Gabadinho, A.: Extracting knowledge from life courses: Clustering and visualization. In: Song, I.-Y., Eder, J., Nguyen, T.M. (eds.) DaWaK 2008. LNCS, vol. 5182, pp. 176–185. Springer, Heidelberg (2008)
13. Needleman, S., Wunsch, C.: A general method applicable to the search for similarities in the amino acid sequence of two proteins. Journal of Molecular Biology 48(3), 443–453 (1970)
14. Wu, L.: Some comments on Sequence analysis and optimal matching methods in sociology, review and prospect. Sociological methods and research 29(1), 41–64 (2000)

# Network-Structured Particle Swarm Optimizer with Various Topology and Its Behaviors

Haruna Matsushita and Yoshifumi Nishio

Tokushima University, 2-1 Minami-Josanjima, 770-8506 Tokushima, Japan
{haruna,nishio}@ee.tokushima-u.ac.jp

**Abstract.** This study proposes Network-Structured Particle Swarm Optimizer (NS-PSO) with various neighborhood topology. The proposed PSO has the various network topology as rectangular, hexagonal, cylinder and toroidal. We apply NS-PSO with various topology to optimization problems. We investigate their behaviors and evaluate what kind of topology would be the most appropriate for each function.

**Keywords:** Particle Swarm Optimization (PSO), network structure, Self-Organizing Map (SOM).

## 1 Introduction

Particle Swarm Optimization (PSO) [1] is an evolutionary algorithm to simulate the movement of flocks of birds. Due to the simple concept, easy implementation and quick convergence, PSO has attracted much attention and is used to wide applications in different fields in recent years. In PSO algorithm, there are no special relationships between particles. Each particle position is updated according to its personal best position and the best particle position among the all particles, and their weights are determined at random in every generation.

On the other hand, the Self-Organizing Map (SOM) [2] is an unsupervised learning and is a simplified model of the self-organizing process of the brain. The map consists of neurons located on a hexagonal or rectangular grid. The neurons self-organize statistical features of the input data according to the neighborhood relationship of the map structure.

Various topological neighborhoods for PSO have been considered by researches [3]–[7]. Each particle shares its best position among neighboring particles on the network. However, the information of each particle is not updated according to the neighborhood distance on the network.

In our past study, we have applied the concept of SOM to PSO and have proposed a new PSO algorithm with topological neighborhoods; Network-Structured Particle Swarm Optimizer considering neighborhood relationships (NS-PSO) [8]. All particles of NS-PSO are connected to adjacent particles by a neighborhood relation, which dictates the topology of the 2-dimensional network. The connected particles, namely neighboring particles on the network, share the information of

J.C. Príncipe and R. Miikkulainen (Eds.): WSOM 2009, LNCS 5629, pp. 163–171, 2009.

their own best position. In every generation, we find a winner particle, whose function value is the best among all particles, as SOM algorithm, and each particle is updated depending on the neighborhood distance between it and the winner on the network. However, the relevance between the efficiently of optimization and the shape of network topology of NS-PSO was not completely clear.

In this study, we propose NS-PSO with various neighborhood topology. We apply NS-PSO to the various network topology as rectangular, hexagonal, cylinder and toroidal. NS-PSO with various topology are applied to eight test functions which are unimodal and multimodal. We investigate their behaviors and evaluate what kind of topology would be the most appropriate for each function. From results, we find that the circular-topology is effective for the simple unimodal functions, because this topology easily transmits the information of each best position to the whole particles. We also confirm that the hexagonal-topology is appropriate for the complex multimodal functions, because this topology contains various kinds of particles and this effect averts the premature convergence.

## 2   Network-Structured Particle Swarm Optimizer Considering Neighborhood Relationships (NS-PSO)

In the algorithm of the standard PSO, multiple solutions called "particles" coexist. At each time step, the particle flies toward its own past best position and the best position among all particles. Each particle has two informations; position and velocity. The position vector of each particle $i$ and its velocity vector are represented by $\boldsymbol{X}_i = (x_{i1}, \cdots, x_{id}, \cdots, x_{iD})$ and $\boldsymbol{V}_i = (v_{i1}, \cdots, v_{id}, \cdots, v_{iD})$, respectively, where $(d = 1, 2, \cdots, D)$, $(i = 1, 2, \cdots, M)$ and $x_{id} \in [x_{\min}, x_{\max}]$.

The algorithm of NS-PSO is based on both two structures; the standard PSO and SOM. NS-PSO has following three key features.

**1.** All particles are connected to adjacent particles by a neighborhood relation which dictates the topology of the network. In this study, we use various topology networks shown in Fig. 1 and investigate their behaviors. The rectangular-topology and the hexagonal-topology as Figs. 1(a)–(b) are the sheet shapes, and the cylinder-topology and toroidal-topology as Figs. 1(c)–(d) are circular map.

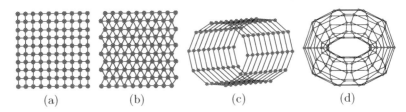

(a)              (b)              (c)              (d)

**Fig. 1.** Different map shapes with $10 \times 10$ particles used in this study. (a) Rectangular-topology. (b) Hexagonal-topology. (c) Cylinder-topology. (d) Toroidal-topology.

**2.** The particles share the local best position between the neighborhood particles directly connected.

**3.** In every generation, we find a winner particle with best function value among all particle as SOM learning.

By these features, each particle of NS-PSO is updated depending on its own best position, the position of the winner and the neighborhood distance between it and the winner on the network.

**(NS-PSO1)** (Initialization) Let a generation step $t = 0$. Randomly initialize the particle position $\boldsymbol{X}_i$, initialize its velocity $\boldsymbol{V}_i$ for each particle $i$ to zero, and initialize $\boldsymbol{P}_i = (p_{i1}, p_{i2}, \cdots, p_{iD})$ with a copy of $\boldsymbol{X}_i$. Evaluate the objective function $f(\boldsymbol{X}_i)$ for each particle $i$ and find $\boldsymbol{P}_g$ with the best function value among all the particles. Define $g$ as the winner $c$. Find $\boldsymbol{L}_i = (l_{i1}, l_{i2}, \cdots, l_{iD})$ with the best function value among the directly connected particles, namely own neighbors.

**(NS-PSO2)** Evaluate the fitness $f(\boldsymbol{X}_i)$ and find a winner particle $c$ with the best fitness among the all particles at current time $t$;

$$c = \arg\min_i \{f(\boldsymbol{X}_i(t))\}. \tag{1}$$

For each particle $i$, if $f(\boldsymbol{X}_i) < f(\boldsymbol{P}_i)$, the personal best position (called *pbest*) $\boldsymbol{P}_i = \boldsymbol{X}_i$. Let $\boldsymbol{P}_g$ represents the best position with the best fitness among all particles so far (called *gbest*). If $f(\boldsymbol{X}_c) < f(\boldsymbol{P}_g)$, update *gbest* $\boldsymbol{P}_g = \boldsymbol{X}_c$, where $\boldsymbol{X}_c = (x_{c1}, x_{c2}, \cdots, x_{cD})$ is the position of the winner $c$.

**(NS-PSO3)** Find each local best position (called *lbest*) $\boldsymbol{L}_i$ among the particle $i$ and its neighborhoods, which are directly connected with $i$ on the network, so far. For each particle $i$, update *lbest* $\boldsymbol{L}_i$, if needed.

**(NS-PSO4)** Update $\boldsymbol{V}_i$ and $\boldsymbol{X}_i$ of each particle $i$ depending on its *lbest*, position of the winner $\boldsymbol{X}_c$ and the distance on the network between $i$ and the winner $c$, according to

$$
\begin{aligned}
v_{id}(t+1) &= wv_{id}(t) + c_1 \mathrm{rand}(\cdot)\left(l_{id} - x_{id}(t)\right) + c_2 h_{c,i}\left(x_{cd} - x_{id}(t)\right), \\
x_{id}(t+1) &= x_{id}(t) + v_{id}(t+1),
\end{aligned} \tag{2}
$$

where $w$ is the inertia weight determining how much of the previous velocity of the particle is preserved. $c_1$ and $c_2$ are two positive acceleration coefficients, generally $c_1 = c_2$, $\mathrm{rand}(\cdot)$ is an uniform random number sample from $U(0,1)$. $h_{c,i}$ is the fixed neighborhood function defined by

$$h_{c,i} = \exp\left(-\frac{\|\boldsymbol{r}_i - \boldsymbol{r}_c\|^2}{2\sigma^2}\right), \tag{3}$$

where $\|\boldsymbol{r}_i - \boldsymbol{r}_c\|$ is the distance between network nodes $c$ and $i$ on the network, and the fixed parameter $\sigma$ corresponds to the width of the neighborhood function.

Therefore, large $\sigma$ strengthens particles' spreading force to the whole space, and small $\sigma$ strengthens their convergent force toward the winner.

**(NS-PSO5)** Let $t = t + 1$ and go back to (NS-PSO2).

## 3    Experimental Results

In order to evaluate the performance of NS-PSO with various topology, we use eight benchmark optimization problems summarized in Table 1. $f_1$, $f_2$, $f_3$ and $f_4$ are unimodal functions, and $f_5$, $f_6$, $f_7$ and $f_8$ are multimodal functions with numerous local minima. All the functions have $D$ variables, and the symmetric landscape maps of Sphere, Rosenbrock, Rastrigin and Ackley functions with two variables are shown in Fig. 2. Table 2 lists the dimensionality $D$, the optimum solution $x^*$, the optimum function value $f(x^*)$ and the initialization ranges. In order to investigate the behaviors in various initialization spaces, we use the symmetric and the asymmetric initialization spaces. The population size $M$ is set to 36 in PSO, and the network size is $6 \times 6$ in NS-PSO with each topology. For PSO and NS-PSO, the parameters are set as $w = 0.7$ and $c_1 = c_2 = 1.6$. The neighborhood radius $\sigma$ of all NS-PSOs are 1.5. We carry out the simulations repeated 30 times for all the optimization functions with 3000 generations.

**Table 1.** Eight Test Functions

| Function name | Test Function |
|---|---|
| Sphere function; | $f_1(x) = \displaystyle\sum_{d=1}^{D-1} x_d^2$ |
| Rosenbrock's function; | $f_2(x) = \displaystyle\sum_{d=1}^{D-1} \left( 100 \left( x_d^2 - x_{d+1} \right)^2 + (1 - x_d)^2 \right)$ |
| $3^{\text{rd}}$ De Jong's function; | $f_3(x) = \displaystyle\sum_{d=1}^{D} |x_d|$ |
| $4^{\text{th}}$ De Jong's function; | $f_4(x) = \displaystyle\sum_{d=1}^{D} d x_d^4$ |
| Rastrigin's function; | $f_5(x) = \displaystyle\sum_{d=1}^{D} \left( x_d^2 - 10 \cos\left(2\pi x_d\right) + 10 \right)$ |
| Ackley's function; | $f_6(x) = \displaystyle\sum_{d=1}^{D-1} \left( 20 + e - 20 e^{-0.2\sqrt{0.5(x_d^2 + x_{d+1}^2)}} \right.$ $\left. - e^{0.5(\cos(2\pi x_d) + \cos(2\pi x_{d+1}))} \right)$ |
| Stretched V sine wave; | $f_7(x) = \displaystyle\sum_{d=1}^{D-1} (x_d^2 + x_{d+1}^2)^{0.25} \left(1 + \sin^2(50(x_d^2 + x_{d+1}^2)^{0.1})\right)$ |
| Griewank's function; | $f_8(x) = \displaystyle\sum_{d=1}^{D} \dfrac{x_d^2}{4000} - \displaystyle\prod_{d=1}^{D} \cos\left(\dfrac{x_d}{\sqrt{d}}\right) + 1$ |

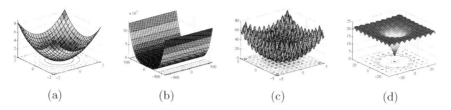

| (a) | (b) | (c) | (d) |

**Fig. 2.** Symmetric landscape of four test functions with two variables. First and second variables are on the x-axis and y-axis, respectively, and z-axis shows its function value. (a) Sphere. (b) Rosenbrock. (c) Rastrigin. (d) Ackley.

**Table 2.** Parameters for test functions

| $f$ | $D$ | $x^*$ | $f(x^*)$ | Initialization Space | |
| --- | --- | --- | --- | --- | --- |
| | | | | Symmetric | Asymmetric |
| $f_1$ | 50 | $[0, 0, \ldots, 0]$ | 0 | $[-5.12, 5.12]^D$ | $[-2.56, 5.12]^D$ |
| $f_2$ | 50 | $[1, 1, \ldots, 1]$ | 0 | $[-2.048, 2.048]^D$ | $[-1.024, 2.048]^D$ |
| $f_3$ | 50 | $[0, 0, \ldots, 0]$ | 0 | $[-2.048, 2.048]^D$ | $[-1.024, 2.048]^D$ |
| $f_4$ | 50 | $[0, 0, \ldots, 0]$ | 0 | $[-1.28, 1.28]^D$ | $[-0.64, 1.28]^D$ |
| $f_5$ | 50 | $[0, 0, \ldots, 0]$ | 0 | $[-5.12, 5.12]^D$ | $[-2.56, 5.12]^D$ |
| $f_6$ | 50 | $[0, 0, \ldots, 0]$ | 0 | $[-10, 10]^D$ | $[-5, 10]^D$ |
| $f_7$ | 50 | $[0, 0, \ldots, 0]$ | 0 | $[-30, 30]^D$ | $[-15, 30]^D$ |
| $f_8$ | 50 | $[0, 0, \ldots, 0]$ | 0 | $[-600, 600]^D$ | $[-300, 600]^D$ |

## 3.1 Symmetric and Asymmetric Functions

The performances with the minimum and mean function values over 30 independent runs on eight functions with the symmetric initialization are listed in Table 3. The best results of the mean values among all the algorithms are shown in bold. All NS-PSOs with various topology evidently surpasses the standard PSO on all the eight functions. In fact, the standard PSO has not obtained better results than any other algorithms which consider the network-structure. From these results, we can say that PSO, which has the specific network-structure, is more effective than the standard PSO, which has no neighborhood relationship, for the symmetric functions.

Table 4 shows the best result among the five algorithms and the difference between the best result and the result of each algorithm. NS-PSO with rectangular-topology, with hexagonal-topology, with cylinder-topology and with toroidal-topology achieve the best values 0, 3, 2 and 3 times, respectively. For the unimodal functions as $f_1$, $f_2$, $f_3$ and $f_4$, NS-PSO with toroidal-topology has obtained the best results most frequently, and the cylinder-topology delivers a very small difference from the best results. However, for the multimodal functions as $f_5$, $f_6$, $f_7$ and $f_8$, the differences between the results of toroidal-topology and the best results are bigger than other three NS-PSOs although it is the best topology for the unimodal functions. Meanwhile, NS-PSO with hexagonal-topology

**Table 3.** Comparison results of PSO and NS-PSO with symmetric initialization on 8 test functions with $D = 50$

| $f$ | | PSO | NS-PSO | | | |
|---|---|---|---|---|---|---|
| | | | Rectangular | Hexagon | Cylinder | Toroidal |
| $f_1$ | Mean | 2.29e-20 | 8.22e-25 | 1.50e-23 | 1.62e-25 | **1.58e-25** |
| | Minimum | 4.09e-27 | 1.51e-29 | 1.29e-22 | 9.59e-32 | 1.78e-31 |
| $f_2$ | Mean | 55.24 | 43.61 | 42.56 | **38.80** | 40.82 |
| | Minimum | 36.74 | 38.48 | 35.56 | 31.04 | 30.33 |
| $f_3$ | Mean | 7.49e-06 | 1.23e-07 | **7.37e-09** | 3.93e-08 | 4.50e-08 |
| | Minimum | 9.41e-11 | 3.15e-12 | 5.81e-13 | 1.85e-12 | 3.19e-11 |
| $f_4$ | Mean | 1.58e-35 | 1.51e-41 | 1.32e-41 | 2.90e-42 | **3.53e-44** |
| | Minimum | 9.86e-42 | 7.96e-47 | 2.84e-46 | 3.90e-49 | 1.32e-50 |
| $f_5$ | Mean | 148.31 | 92.80 | 104.44 | **88.32** | 115.45 |
| | Minimum | 94.52 | 52.73 | 60.69 | 45.77 | 29.85 |
| $f_6$ | Mean | 249.67 | 159.62 | **157.28** | 193.50 | 205.75 |
| | Minimum | 97.84 | 67.60 | 41.13 | 64.90 | 66.46 |
| $f_7$ | Mean | 65.62 | 41.35 | **33.46** | 41.06 | 43.04 |
| | Minimum | 39.36 | 21.95 | 17.68 | 18.90 | 21.78 |
| $f_8$ | Mean | 0.2440 | 0.0853 | 0.0448 | 0.0924 | **0.0350** |
| | Minimum | 0 | 0 | 1.11e-16 | 1.11e-16 | 0 |

**Table 4.** Difference from the best result with symmetric initialization

| $f$ | Best Mean Result | Difference from the best mean result | | | | |
|---|---|---|---|---|---|---|
| | | PSO | NS-PSO | | | |
| | | | Rectangular | Hexagon | Cylinder | Toroidal |
| $f_1$ | 1.58e-25 | 2.29e-20 | 6.64e-25 | 1.49e-23 | 4.72e-27 | 0 |
| $f_2$ | 38.80 | 16.44 | 4.82 | 3.77 | 0 | 2.02 |
| $f_3$ | 7.37e-09 | 7.48e-06 | 1.15e-07 | 0 | 3.19e-08 | 3.76e-08 |
| $f_4$ | 3.53e-44 | 1.58e-35 | 1.50e-41 | 1.32e-41 | 1.32e-41 | 0 |
| $f_5$ | 88.319 | 60.00 | 4.48 | 16.12 | 0 | 27.13 |
| $f_6$ | 157.28 | 92.39 | 2.34 | 0 | 36.22 | 48.47 |
| $f_7$ | 33.4615 | 32.16 | 7.89 | 0 | 7.60 | 9.58 |
| $f_8$ | 0.0350 | 0.2090 | 0.0503 | 0.0098 | 0.0574 | 0 |

obtains the best results on $f_6$ and $f_7$, and it can obtain stable good results, which are small differences from the best results, for other two multimodal functions. NS-PSO with rectangular-topology achieves the stable good results for both the unimodal and multimodal functions even if it can not obtain the best results among NS-PSOs for any benchmarks.

The performances over 30 independent runs on asymmetric functions are listed in Table 5. Since the standard PSO can not obtain the best results among all five PSOs for any benchmarks, PSO with some networks is more suitable for the optimization problems than the standard PSO.

**Table 5.** Comparison results of PSO and NS-PSO with asymmetric initialization on 8 test functions with $D = 50$

| $f$ | | PSO | NS-PSO | | | |
|---|---|---|---|---|---|---|
| | | | Rectangular | Hexagon | Cylinder | Toroidal |
| $f_1$ | Mean | 2.31e-21 | 2.03e-24 | 1.13e-23 | 4.95e-22 | **8.15e-26** |
| | Minimum | 3.58e-26 | 2.96e-29 | 8.27e-28 | 3.99e-30 | 8.90e-31 |
| $f_2$ | Mean | 55.96 | 64.80 | 50.02 | 65.96 | **40.23** |
| | Minimum | 6.22 | 15.47 | 0.1998 | 13.19 | 0.2033 |
| $f_3$ | Mean | 2.06e-05 | 2.56e-08 | **8.32e-09** | 7.60e-08 | 1.02e-07 |
| | Minimum | 1.65e-10 | 5.86e-12 | 3.29e-12 | 2.38e-11 | 1.15e-11 |
| $f_4$ | Mean | 4.82e-36 | 4.72e-43 | 3.60e-39 | 3.44e-43 | **4.02e-44** |
| | Minimum | 8.51e-41 | 2.74e-47 | 1.79e-46 | 1.41e-48 | 8.03e-51 |
| $f_5$ | Mean | 150.70 | 96.54 | 89.65 | **87.16** | 153.61 |
| | Minimum | 104.47 | 57.71 | 49.75 | 53.73 | 34.63 |
| $f_6$ | Mean | 190.24 | 177.03 | **142.66** | 207.75 | 217.90 |
| | Minimum | 69.48 | 69.13 | 37.87 | 81.76 | 35.02 |
| $f_7$ | Mean | 61.90 | 42.19 | **34.93** | 37.18 | 41.14 |
| | Minimum | 37.86 | 21.84 | 19.00 | 21.87 | 18.33 |
| $f_8$ | Mean | 0.0521 | **0.0240** | 0.1199 | 0.0249 | 0.1576 |
| | Minimum | 0 | 0 | 0 | 0 | 1.11e-16 |

**Table 6.** Difference from the best result with asymmetric initialization

| $f$ | Best Mean Result | Difference from the best mean result | | | | |
|---|---|---|---|---|---|---|
| | | PSO | NS-PSO | | | |
| | | | Rectangular | Hexagon | Cylinder | Toroidal |
| $f_1$ | 8.15e-26 | 2.31e-21 | 1.95e-24 | 1.13e-23 | 4.95e-22 | 0 |
| $f_2$ | 40.23 | 15.73 | 24.57 | 9.79 | 25.73 | 0 |
| $f_3$ | 8.32e-09 | 2.06e-05 | 1.73e-08 | 0 | 6.77e-08 | 9.38e-08 |
| $f_4$ | 4.02e-44 | 4.82e-36 | 4.32e-43 | 3.60e-39 | 3.04e-43 | 0 |
| $f_5$ | 87.16 | 63.54 | 9.39 | 2.49 | 0 | 66.45 |
| $f_6$ | 142.66 | 47.58 | 34.37 | 0 | 65.09 | 75.24 |
| $f_7$ | 34.93 | 26.96 | 7.26 | 0 | 2.25 | 6.21 |
| $f_8$ | 0.0240 | 0.0281 | 0 | 0.0959 | 9.13e-04 | 0.1336 |

Table 6 shows the the difference between the best result and the result of each algorithm. For the unimodal functions, NS-PSO with toroidal-topology can obtain the best results on $f_1$, $f_2$ and $f_4$, and also on $f_3$, it is a very small difference from the best results. Therefore, we can say that toroidal-topology is the most effective for the asymmetric unimodal functions as same as the symmetric unimodal functions. However, for the multimodal functions, NS-PSO with toroidal-topology obtain the worst results 3 times among five algorithms including the standard PSO. On the other hand, NS-PSO with hexagonal-topology obtains the best results 2 times, in particular, it evidently surpasses other four algorithms on $f_6$.

From these results, on both symmetric and asymmetric spaces, the circular-shaped NS-PSO as toroidal-topology is more suitable for the unimodal functions, and the sheet-shaped NS-PSO as hexagonal-topology is more effective for the multimodal functions. In particular, we found that the toroidal-topology is not suitable on the asymmetric multimodal functions.

## 3.2   Behaviors of NS-PSO with Various Topology

The convergence rate of NS-PSO is almost same or slower than the standard PSO. In the standard PSO, the particles move toward *gbest* or toward *pbest*, however, the direction, which more particles move toward, is decided at random on every generation. On the other hand, the neighborhood gaussian function is used in NS-PSO, then, the particles move according to the neighborhood distance between the winner and them. The winner's neighborhood particles move toward the winner, so that they spread to whole space. For the particles which are not 1-neighbors of the winner but are connected near the winner, the gravitation toward the winner is strong. The other particles fly toward their *lbest*. In other words, the roles of the NS-PSO particles are determined by the connection relationship, and they produce the diversity of the particles. These effects avert the premature convergence, and the particles of NS-PSO can easily escape from the local optima.

**Discussion about evaluation of each topology:** Let us consider the network topology and its behavior in terms of average node-to-node distance $L$, which is also known as average shortest path length, and the average number of particles in local neighbor $N_l$. On $6 \times 6$ map, the average shortest path length $L$ of respective topology; rectangular, hexagonal, cylinder and toroidal, are 6.6, 5.34, 5.8 and 5.0, respectively. The average number of particles $N_l$ in local neighbor including itself of respective topology; rectangular, hexagonal, cylinder and toroidal, are 4.6, 6.18, 4.8 and 5.0. Because the cylinder and toroidal are the circular topology, the individuality of each particle is almost same. In other words, on toroidal-topology, $L$ and $N_l$ is completely same for any of the particles. Furthermore, $L$ of toroidal-topology is the smallest in four NS-PSOs. From these effects, it is easy to transmit the information of *lbest* to the whole particles, therefore, the circular topology is effective for the unimodal function which is simple. However, the premature communication produces the premature convergence, then, toroidal-topology easily goes into local optima in the multimodal functions. On the other hand, NS-PSO with hexagonal-topology contains various kinds of particles which has different shortest path length and different size of local neighbors, although $L$ is small and $N_l$ is big. Because these effects produce the diversity of the particles and avert the premature convergence, the particles of NS-PSO with hexagonal-topology can easily escape from the local optima.

## 4   Conclusions

In this study, we have proposed Network-Structured Particle Swarm Optimizer (NS-PSO) with various neighborhood topology which is a collaboration between

Self-Organizing Map (SOM) and PSO. All particles of NS-PSO are connected to adjacent particles by a neighborhood relation, and their information are updated by the neighborhood topology. We have applied NS-PSO with various topology to optimization problems. and have confirmed that PSO, which has the specific network-structure, is more effective than the standard PSO, which has no neighborhood relationship. Furthermore, we have found that the toroidal-topology and the hexagonal-topology are suitable for the unimodal and for the multimodal function, respectively.

# References

1. Kennedy, J., Eberhart, R.C.: Particle swarm optimization. In: Proc. of IEEE. Int. Conf. on Neural Netw., pp. 1942–1948 (1995)
2. Kohonen, T.: Self-organizing Maps. Springer, Berlin (1995)
3. Kennedy, J.: Small worlds and mega-minds: effects of neighborhood topology on particle swarm performance. In: Proc. of Cong. on Evolut. Comput., pp. 1931–1938 (1999)
4. Kennedy, J., Mendes, R.: Population structure and particle swarm performance. In: Proc. of Cong. on Evolut. Comput., pp. 1671–1676 (2002)
5. Mendes, R., Kennedy, J., Neves, J.: The Fully Informed Particle Swarm: Simpler, Maybe Better. IEEE Trans. Evolut. Comput. 8(3), 204–210 (2004)
6. Lane, J., Engelbrecht, A., Gain, J.: Particle Swarm Optimization with Spatially Meaningful Neighbours. In: Proc. of IEEE Swarm Intelligence Symposium, pp. 1–8 (2008)
7. Akat, S.B., Gazi, V.: Particle Swarm Optimization with Dynamic Neighborhood Topology: Three Neighborhood Strategies and Preliminary Results. In: Proc. of IEEE Swarm Intelligence Symposium, pp. 1–8 (2008)
8. Matsushita, H., Nishio, Y.: Network-Structured Particle Swarm Optimizer Considering Neighborhood Relationships. In: Proc. of IEEE. Int. Jont Conf. on Neural Netw. (accepted, 2009)

# Representing Semantic Graphs in a Self-Organizing Map

Marshall R. Mayberry and Risto Miikkulainen

University of California, Merced
University of Texas, Austin
mmayberry@ucmerced.edu, risto@cs.utexas.edu

**Abstract.** A long-standing problem in the field of connectionist language processing has been how to represent detailed linguistic structure. Approaches have ranged from the encoding of syntactic trees in RAAM to the use of a mechanism to query meanings in a "gestalt layer". In this article, a technique called semantic self-organization is presented that allows for the optimal allocation and explicit representation of semantic dependency graphs on a SOM-based grid. This technique has been successfully used in a connectionist natural language processing architecture called INSOMNET to scale up the subsymbolic approach to represent sentences in the LINGO Redwoods HPSG Treebank drawn from the VerbMobil Project and annotated with rich semantic information. INSOMNET was also shown to retain the cognitively plausible behavior detailed in psycholinguistics research. Consequently, semantic self-organization holds considerable promise as a basis for real-world natural language understanding systems that mimic human linguistic performance.

**Keywords:** Connectionist Natural Language Processing, Cognitive Architecture, Self-Organizing Maps, Semantic Dependency Graphs.

## 1 Introduction

Historically research in statistical natural language processing has focused on parsing sentences into syntactic representations as exemplified in the Penn Treebank [1]. While the proper syntactic representation of a sentence does remain controversial, the situation is even less clear with respect to semantics. This lack of an annotation standard has left semantic corpora lagging far behind syntactic corpora, although recent work has progressed on representing the semantics of sentences in such projects as the LINGO Redwoods HPSG Treebank [2] and FRAMENET [3]. Yet, the accurate representation of sentence or discourse semantics is absolutely imperative if true natural language systems are to be realized.

The technique of explicitly representing semantics introduced in this article – callled *semantic self-organization* – has been developed not only to fill the need for the representation of semantics, but also to do so in a manner that is cognitively plausible. That is, the semantic representation of a sentence should develop incrementally as a sentence is processed; sentence processing should be

J.C. Príncipe and R. Miikkulainen (Eds.): WSOM 2009, LNCS 5629, pp. 172–181, 2009.

robust to dysfluencies in the input; ambiguities should be resolved based on context when possible; meaning should be graded; semantic priming should follow from expectations about how a sentence is likely to continue; and sentence meaning should be revisable if those expectations are contradicted by later sentence input. These cognitive properties emerge automatically from subsymbolic systems, and the use of self-organization is a way to develop a map-like layout so that subsymbolic processes can operate on it.

The INSOMNET sentence-processing system, for which the technique of semantic self-organization was developed, has been described elsewhere [4,5,6]. This article will focus on the primary innovation of INSOMNET: the self-organization of compressed semantic frame representations.

Section 2 will briefly describe Minimal Recursion Semantics (MRS; [7]), the semantic formalism used in INSOMNET. Section 3 will then describe how that formalism is rendered into a set of representations that are used to train and test the network so that it processes sentences in a cognitively plausible manner. Section 4 describes the results of a series of experiments designed to demonstrate the feasibility of the approach. Section 5 provides a discussion of the system's capabilities and directions for future research.

## 2   Minimal Recursion Semantics

Minimal Recursion Semantics uses flat semantics in contrast to the nested structure of Predicate Logic traditionally used to represent sentence semantics. The flat semantics of MRS is thus amenable to representation on a grid with dependencies represented by pointers. Figure 1 illustrates the MRS representation for the sentence *the boy hit the girl with the ball*, that features prepositional phrase attachment ambiguity. Abbreviations are used for semantic annotations. Each node in the graph has a label called a handle and a named predicate. The subcategorization argument roles of the predicate are represented by the arcs emanating from the node, labelled with abbreviations for the roles. Because the MRS representation is a DAG, some nodes are leaves. These leaves are called *feature nodes* that characterize the predicates that reference them. Their roles are filled with nominal or verbal features such as gender or tense, respectively.

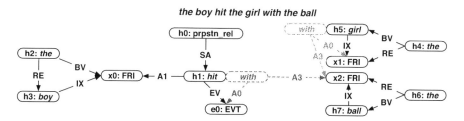

**Fig. 1. An MRS Dependency Graph.** Sentence meaning is represented in a pointer-based directed acyclic semantic graph.

In order to understand the MRS example completely, the semantic annotation used in Figure 1 needs to be briefly explained. The sentence is a declarative proposition, as indicated by the semantic relation **prpstn_rel** in the top node labeled by the handle **h0**. Sentence types subcategorize for a single role, *state-of-affairs*, as indicated by the arc labeled **SA**. The filler for this role is the handle **h1**, which is the label of the node for the main predication of the sentence, the word *hit*. Transitive verbs such as *hit* have *event*, *arg1*, and *arg3* arguments, that denote the thematic roles of **Event**, **Agent** and **Patient**, respectively. These three roles are represented in Figure 1 by the arcs labeled **EV**, **A1**, and **A3**.

The semantics of the noun phrases are represented by three sets of nodes. Each noun phrase contains the following arguments: a noun that subcategorizes for an *instance* (**IX**) argument (**FRI**), that describes the noun's features; a determiner that has both a *bound variable* (**BV**) argument, filled with the noun's instance handle and a *restriction* (**RE**) argument, filled with the noun's handle. In Figure 1, the agent *boy* and patient *girl* are labelled by the handles **x0** and **x1**, respectively. The handle **x2** denotes the noun *ball* and fills the **A3** role of the preposition *with*. The preposition can either modify the verb for the instrumental sense of the sentence, or the noun for the modifier sense. In MRS, modification is represented by *conjunction* of predicates; thus, *big red ball* would be denoted by $\bigwedge[big(x),\ red(x),\ ball(x)]$. The $n$-ary connective $\bigwedge$ is replaced by a handle, which is distributed so that each predicate shares the same handle. For verb-attachment, the verb *hit* and the preposition *with* both share the handle **h1**, and the preposition's **A0** role is filled with the verb's event structure handle **e0**. For noun-attachment, *with* has the same handle **h5** as *girl*, and its **A0** role points to the index **x1** of *girl*. In this way, MRS is a symbolic linguistic representation that can be encoded with neural networks, as described next.

# 3    Learning and Representing Semantic Graphs

In this section, the technique of semantic self-organization used to develop representations suitable for MRS semantic dependency graphs is described. The INSOMNET architecture, for which the technique was originally developed, is also briefly presented to make clear how those representations are utilized.

## 3.1    Semantic Graph Representation

The MRS dependency graph is transformed into a *frameset* of representations for each node and arc that serve as inputs and targets for INSOMNET system, where each frame shown in Figure 2 has the format

| **Handle Word Semantic-Relation Subcategorization <Arguments>** |

For example, the node **h1**: *hit* in the MRS graph in is represented by the frame

|  **h1**  *hit*  **_arg13_rel**  **A0A1A3DMEV**  _   x0   x1 _   e0 _  |

The first element, **h1**, is the **Handle** (label) of the frame. The **Handle** serves as a *pointer* that fills roles of other nodes in the semantic graph. Here, **h1** fills

**the boy hit the girl with the ball**

| Handle | Word | Relation | Subcat | | | | | | |
|---|---|---|---|---|---|---|---|---|---|
| h0 | __ | prpstn_rel | SA | h1 | __ | __ | __ | __ | __ |
| h1 | hit | arg13_rel | A0A1A3DMEV | ___ | x0 | x1 | __ | x0 | ___ |
| h1 | with | miscprep_ind_rel | A0A3DMEV | e0 | x2 | __ | __ | __ | __ |
| e0 | __ | EVT | DVASMOTN | bool | -asp | ind | past | __ | __ |
| h2 | the | def_explicit_rel | BVDMRESC | x0 | __ | h3 | __ | __ | __ |
| h3 | boy | diadic_nom_rel | A3IX | __ | x0 | __ | __ | __ | __ |
| x0 | __ | FRI | DVGNPNPT | __ | masc | 3sg | prn | __ | __ |
| h4 | the | def_explicit_rel | BVDMRESC | x1 | __ | h5 | __ | __ | __ |
| h5 | girl | diadic_nom_rel | A3IX | __ | x1 | __ | __ | __ | __ |
| h5 | with | miscprep_ind_rel | A0A3DMEV | x1 | x2 | __ | __ | __ | __ |
| x1 | __ | FRI | DVGNPNPT | __ | fem | 3sg | prn | __ | __ |
| h6 | the | def_explicit_rel | BVDMRESC | x2 | __ | h7 | __ | __ | __ |
| h7 | ball | reg_nom_rel | IX | x2 | __ | __ | __ | __ | __ |
| x2 | __ | FRI | DVGNPNPT | __ | neu | 3sg | prn | __ | __ |

**Fig. 2. The MRS Frameset.** The MRS Dependency Graph in Figure 1 is converted into a set of frames that can be represented in a neural network.

the *state-of-affairs* (**SA**) slot in the topmost node, **h0: prpstn_rel**. The second element, *hit*, gives the **Word** for this frame. The third element is the **Semantic-Relation** for the frame. The fourth element, **A0A1A3DMEV**, represents the **Subcategorization** and is shorthand for the sequence **A0 A1 A3 DM EV**: the argument roles that the semantic relation subcategorizes for. Recall that *hit* is a transitive verb with three **Arguments** (the agent (**A1**), the patient (**A3**), and the event (**EV** roles), the fillers for which are listed in the rest of the frame as <__ x0 x1 __ e0> (the **A0** and **DM** arguments are not filled). These handles label those frames that represent the arguments' role fillers.

There are two properties of MRS **Handle**s that have influenced the design of INSOMNET. First, a given **Handle** may denote more than one frame due to the the need to represent predicate conjunction, as described in Section 2. Second, in the symbolic MRS formalism, **Handle**s are arbitrary designators. However, in a neural network, **Handle**s have to be encoded as patterns of activation in a vector space and that can be learned. In INSOMNET, the **Handle**s are designed to be dynamically associated with patterns that represent core semantic features, such as predicate argument structure and role fillers. In this way, INSOMNET can learn to generalize these semantic features to process novel sentences. These **Handle** representations are derived from compression of the frames they label via auto-association of the frame components. How these **Handle** representations (which will be simply referred to as HANDLEs to distinguish them from the field **Handle**) are developed and self-organized is a central aspect of the model that will be described in detail in Section 3.3 following a brief description of the INSOMNET sentence processing architecture in the next section to illustrate how a flat semantic representation such as MRS can be incorporated into a connectionist model.

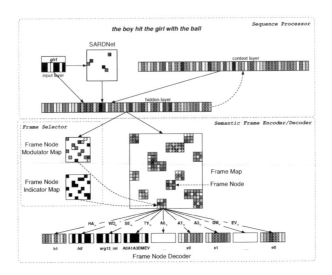

**Fig. 3. The INSOMNET Architecture.** The INSOMNET model consists of three operational modules based on how they function together. The Sequence Processor reads the input sentence incrementally and activates both the Frame Selector and the Semantic Frame Encoder/Decoder. The Semantic Frame Encoder/Decoder encodes the MRS dependency graph representing the semantic interpretation of the sentence. The Frame Selector select frames in the Frame Map in a graded manner.

## 3.2   The INSOMNET Architecture

The INSOMNET model, shown in Figure 3, will be used to motivate the technique of *semantic self-organization* that is used to encode, decode, and cluster the HANDLES described above. The model consists of three operational components: the `Sequence Processor`, the `Semantic Frame Encoder/Decoder`, and the `Frame Selector`. Each is described in turn below.

The `Sequence Processor` is based on the SRN and processes a sentence one word at a time. A SARDNET MAP [8], a type of sequential SOM, retains an exponentially decaying activation of the input sequence.

The self-organized Frame Map of the `Semantic Frame Encoder/Decoder` is the main innovation of INSOMNET. In the current model, each Frame Node in the map itself consists of 100 units. The Frame Map itself is a $12 \times 12$ assembly of these nodes. As a result of processing the input sequence, the `Frame Selector` activates a number of these nodes to different degrees; that is, a particular pattern of activation appears over the units of these nodes. Through the weights in the Frame Node Decoder, these patterns are decoded into the corresponding MRS case-role frame representations. The same weights are used for the same role in each Frame Node in the map. This weight-sharing enforces generalization among common elements across all frames in the training and test sets.

The Frame Map is self-organized on the basis of HANDLEs. This process serves to identify which nodes in the Frame Map correspond to which case-role frames in the MRS structure. Because the HANDLEs are distributed representations of case-role frames, similar frames will cluster together on the map according to common semantic features. Accordingly, each node becomes tuned to particular kinds of frames, but no particular Frame Node becomes dedicated to any particular frame. Rather, the nodes are flexible enough to represent different frames from different sentences on the basis of semantic similarity.

During training, each Frame Node serves as a second hidden layer and its corresponding case-role frame as the output layer. The appropriate frames are presented as targets for the decoded Frame Node, and the resulting error signals are backpropagated through the Frame Node Decoder weights to the Frame Node and the weights connecting it to the SRN.

While the Frame Map represents the contents of the target semantic frames, The Frame Selector is used to indicate the degree of activation (ranging from 0.0 to 1.0) of those frames that comprise the current interpretation of a sentence. For this reason, a threshold based on precision and recall is used to optimize target frame activation against that of false positives and negatives. At the same time, the Frame Selector also plays a crucial role in identifying the Frame Nodes that correspond to the HANDLEs for different frames. This process – the linchpin of semantic self-organization – is described next.

## 3.3   Semantic Self-Organization

As previously mentioned, The HANDLEs are developed by compressing frames through auto-association. Frames with similar components develop similar representations. In the limit, there would be as many distinct handles as there are distinct frames, with the typical problem of information loss through repeated compression using RAAM. To avoid this problem, semantic self-organization relies on *pointer quantization*. Pointer quantization instead uses the prototypical codebook vectors of the Frame Node Indicator Map closest to the HANDLEs to represent the HANDLEs themselves, as shown in Figure 4. Self-organization of

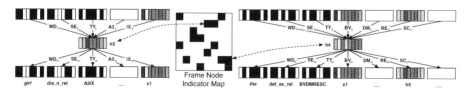

**Fig. 4.** HANDLEs **as the Basis of Semantic Self-organization.** Compressed representations for two frames with **Handles h4** and **h5** are developed using auto-association. A process called *pointer quantization* facilitates learning by restricting the representations of HANDLEs to the small set of codebook vectors developed in the Frame Node Indicator Map, which, as a result of semantic similarities among frames, becomes self-organized, and, for the same reason, the Frame Map that encodes the frames.

the Frame Node Indicator Map (which is in one-to-one correspondence with the Frame Map), occurs simultaneously with frame compression.

Pointer Quantization has two primary advantages. First, it constrains the number of distinct pointer representations to the number of units in the Frame Node Indicator Map, thus facilitating training. Second, whereas two similar frames can map onto the same unit in a regular SOM, the base architecture for the Frame Node Indicator Map is SARDNET (see Section 3.2). Because winning units are removed from further competition, identical (or very similar) frame representations are assigned to distinct nodes with distinct codebook vectors. In this way, the quantization of the HANDLE representations function as *content-addressable* pointers to the Frame Node that in turn encodes the complete frame.

Figure 4 shows how the frame | *girl* diadic_nom_rel **A3IX** _ x1 | (without the **Handle** field) is auto-associated through input and output weights for each frame component, resulting in a HANDLE for the frame **h5** in the auto-association network. The codebook vector to this HANDLE in the Frame Node Indicator Map is assigned **h5**, the label for this frame. The codebook vector is then used as the HANDLE for fillers of other frames that reference **h5**, such as the frame **h4**: | *the* _def_rel **BVDMRESC** x1 _ **h5** _ |. These components are auto-associated in turn, the closest codebook vector substituted as the HANDLE for **h4**, and the process repeated over the entire DAG. Because the graph is acyclic, a compressed representation for each frame in the MRS dependency graph of a sentence can be developed. The leaves of the graph serve as stable starting points, and compressed representations of the frame corresponding to each frame are developed recursively. While **Handles** such as **h4** and **h5** are arbitrarily generated during training, the patterns they designate develop non-arbitrary semantic content. Indeed, the network is able to predict which nodes frames are likely to map to in order to generalize to novel sentences.

# 4   Simulations

Two simulations using INSOMNET will be briefly described. Details are available in the forementioned articles on INSOMNET [4,5,6].

## 4.1   The Redwoods Treebank

To demonstrate that INSOMNET could be scaled up to a corpus of realistic natural language sentences, the model was evaluated on 4817 sentences from the Redwoods Treebank for which at least one analysis was selected as correct by the treebank annotator. The dataset contains the full MRS representation, resulting in 18,225 unique frames overall.

INSOMNET was evaluated with respect to both its *parsing* performance; (i.e., precision and recall of targeted frames), as well as with respect to its *comprehension* performance; (i.e., were the targeted frames in fact correctly decoded?). Ten-fold cross-validation was carried out, with an average parsing performance

F-measure of 0.81 at a threshold of 0.7 on the test sets. Average comprehension performance on the test sets was 85%. A comparison of INSOMNET with a state-of-the-art grammar-based conditional log-linear parsing model [9] was performed, with both models trained using ten-fold cross-validation on a significantly simpler version of the dataset described above. INSOMNET achieved an overall parsing performance F-measure of 0.75 at a threshold of 0.7 and comprehension performance of 85%, compared with an accuracy of 74.9% for Toutanova's model (Toutanova, personal communication).

A further simulation was carried out to evaluate the robustness of the model when dysfluencies from the original VerbMobil transcriptions were included in the test sentences. The parsing F-measure was 0.77 at a threshold of 0.7, and comprehension was 84%, demonstrating that the network showed human-like tolerance to typical speech errors. Also, the model was lesioned with increasing levels of Gaussian noise added to the network's weights, and its performance degraded gracefully, as expected of subsymbolic models.

These simulations demonstrated that a connectionist system could be scaled up to processing realistic natural language in a robust manner.

## 4.2   Cognitive Plausibility

INSOMNET was further evaluated on a corpus originally created by McClelland and Kawamoto in [10] and since used as a benchmark to demonstrate that a connectionist system could learn to map syntactic constituents to thematic roles. The corpus – expanded to 1475 sentences over 30 words for which semantic features (e.g., animacy) were prespecified – was used [11]. Chief among the thematic roles of interest were instrument and modifier, with some sentences being globally ambiguous. In the general case, however, which of the two roles was correct was determined by sentence format, frequency, and the particular words in the sentence. For example, the word *ball* in the sentence *the boy hit the girl with the ball* was interpreted as an instrument, whereas the word *doll* in the sentence *the boy hit the girl with the doll* was interpreted as a modifier. Other roles such as the three elicited by *the boy broke the window, the ball broke the window,* and *the window broke,* were also part of the cognitive phenomena tested.

INSOMNET demonstrated nearly perfect performance on the unambiguous sentences, correctly identifying the targeted thematic roles and their fillers. Furthermore, the model showed sensitivity to frequency effects, both with respect to individual words and their correlations with thematic roles, as well as to sentence prefixes. Besides co-activating multiple senses in the face of ambiguity, the model correctly disambiguated temporarily ambiguous sentences. INSOMNET exhibited expectations of likely sentence continuations, and semantic priming of related words (such as *hammer* and *hatchet*). Finally, the model was able to revise its interpretation when later input overrode a given thematic role assignment.

This simulation showed that INSOMNET not only scales up, but does so in a cognitively valid manner.

## 5    Discussion and Future Work

The technique of semantic self-organization described in this article has been instrumental in allowing a connectionist model to scale up to real-world language
processing, while at the same time retaining the cognitively plausible behavior
that have made connectionist models the subject of considerable psycholinguistic
research. The resulting semantic representations are not only *explicit*, but also
exhibit the *gradience* that has long been recognized as a hallmark of natural language [12]. Concepts tend to defy easy categorization at all levels of description,
a problem commonly recognized as the *grain problem* [13].

While connectionist models have become the systems of choice for researchers
interested in various aspects of human language performance, they have been
limited in their scope to well-defined, often "toy" problems. Due to memory limitations, they have proven exceedingly difficult to scale up to processing corpora
that more traditional grammar-based statistical models routinely handle.

The use of a flat semantic representation such as MRS laid out on a grid of cells
representing compressed frames and self-organized allows the circumvention of
the limitations of earlier models that used RAAM to represent semantics based on
Predicate Logic. Such systems were very limited in the semantic detail that could
be captured due to the problem of repeated compression. Moreover, sentence
semantics that are best expressed in a graph structure were forced to conform to
tree structures for use with RAAM, although clever ways were devised to allow
the representation of more generalized structure [14].

Future work will focus on improving memory and reusing nodes in the Frame
Map whose activation has fallen below threshold.

## 6    Conclusion

INSOMNET is a subsymbolic sentence processing system that produces explicit
and graded semantic graph representations. The novel technique of semantic self-
organization allows the network to learn typical semantic dependencies between
arguments and their fillers that generalizes to novel sentences. The technique
makes it possible to assign case roles flexibly, while retaining the cognitively plausible behavior that characterizes connectionist modeling. INSOMNET was shown
to scale up to sentences of realistic complexity, including those with dysfluencies
in the input and damage in the network. The network also exhibits the crucial
cognitive properties of incremental processing, expectations, semantic priming,
and nonmonotonic revision of an interpretation during sentence processing. Semantic self-organization therefore constitutes a significant step towards building
a cognitive parser that works with the everyday language that people use.

## Acknowledgments

This research was supported in part by NIH under grant R21-DC009446.

# References

1. Marcus, M.P., Santorini, B., Marcinkiewicz, M.A.: Building a large annotated corpus of English. Computational Linguistics 19, 313–330 (1993)
2. Oepen, S., Flickinger, D., Toutanova, K., Manning, C.: LinGO redwoods. In: Proceedings of the First Workshop on Treebanks and Linguistic Theories (TLT 2002), Sozopol, Bulgaria, TLT (2002)
3. Ruppenhofer, J., Ellsworth, M., Petruck, M.R.L., Johnson, C.R., Scheffczyk, J.: FrameNet II: Extended theory and practice. Technical report, ICSI (2005)
4. Mayberry, M.R., Miikkulainen, R.: Incremental nonmonotonic sentence processing through semantic self-organization. Cognitive Science (submitted)
5. Mayberry, M.R., Miikkulainen, R.: Incremental nonmonotonic parsing through semantic self-organization. In: Proceedings of the 25th Annual Conference of the Cognitive Science Society, pp. 367–372. Erlbaum, Mahwah (2003)
6. Mayberry, M.R.: Incremental Nonmonotonic Parsing through Semantic Self-Organization. PhD thesis, Department of Computer Sciences, The University of Texas at Austin, Austin, TX (2003)
7. Copestake, A., Flickinger, D., Sag, I.A., Pollard, C.: Minimal recursion semantics. Research on Language and Computation 3(2-3), 281–332 (2005)
8. Mayberry, M.R., Miikkulainen, R.: Using a sequential SOM to parse long-term dependencies. In: Proceedings of the 21st Annual Conference of the Cognitive Science Society, Hillsdale, NJ, pp. 367–372. Erlbaum, Mahwah (1999)
9. Toutanova, K., Manning, C.D., Shieber, S.M., Flickinger, D., Oepen, S.: Parse disambiguation for a rich HPSG grammar. In: First Workshop on Treebanks and Linguistic Theories (TLT2002), Sozopol, Bulgaria, pp. 253–263 (2002)
10. McClelland, J.L., Kawamoto, A.H.: Mechanisms of sentence processing. In: McClelland, J.L., Rumelhart, D.E. (eds.) Parallel Distributed Processing: Explorations in the Microstructure of Cognition. Psychological and Biological Models, vol. 2, pp. 272–325. MIT Press, Cambridge (1986)
11. Miikkulainen, R.: Natural language processing with subsymbolic neural networks. In: Browne, A. (ed.) Neural Network Perspectives on Cognition and Adaptive Robotics, pp. 120–139. Institute of Physics Publishing, Bristol (1997)
12. Crocker, M.W., Keller, F.: Probabilistic grammars as models of gradience in language processing. In: Fanselow, G., Féry, C., Vogel, R., Schlesewsky, M. (eds.) Gradience in Grammar: Generative Perspectives, pp. 227–245. Oxford University Press, Oxford (2006)
13. Mitchell, D.C., Cuetos, F., Corley, M., Brysbaert, M.: Exposure-based models of human parsing. Journal of Psycholinguistic Research 24, 469–488 (1995)
14. Sperduti, A.: Encoding of Labeled Graphs by Labeling RAAM. In: Cowan, J.D., Tesauro, G., Alspector, J. (eds.) Advances in Neural Information Processing Systems, vol. 7, pp. 1125–1132. MIT Press, Cambridge (1995)

# Analytic Comparison of Self-Organising Maps

Rudolf Mayer[1], Robert Neumayer[2], Doris Baum[3], and Andreas Rauber[1]

[1] Vienna University of Technology, Austria
[2] Norwegian University of Science and Technology, Norway
[3] Fraunhofer Institute for Intelligent Analysis and Information Systems, Germany

**Abstract.** SOMs have proven to be a very powerful tool for data analysis. However, comparing multiple SOMs trained on the same data set using different parameters or initialisations is still a difficult task. In most cases it is performed only via visual inspection or by utilising one of a range of quality measures to compare vector quantisation or topology preservation characteristics of the maps. Yet, comparing SOMs systematically is both necessary as well as a powerful tool to further analyse data: necessary, because it may help to pick the most suitable SOM out of different training runs; a powerful tool because it allows analysing mapping stabilities across a range of parameter settings. In this paper we present an analytic approach to compare multiple SOMs trained on the same data set. Analysis of output space mapping, supported by a set of visualisations, reveals data co-locations and shifts on pairs of SOMs, considering both different neighbourhood sizes at source and target maps. A similar concept of mutual distances and relationships can be analysed at a cluster level. Finally, Comparisons aggregated automatically across several SOMs are strong indicators for strength and stability of mappings.

## 1 Introduction

Self-Organising Maps (SOMs) enjoy high popularity in various data analysis applications. Experimenting with SOMs of different sizes, initialisations or different values for other parameters, is an essential part of this analysis process. In many cases, users want to detect the influence of certain parameters or generally want more details about the relations and differences between input data and resultant clusters across these varying maps. In this paper we thus propose a method to compare two or more SOMs, indicating the differences in how the data was mapped on either of the SOMs. We introduce three quality measures with supporting visualisations for comparing multiple SOMs. Its remainder is structured as follows. Section 2 describes related work in the field of SOM quality measures and comparisons. Section 3 then describes three types of analysis, which are illustrated along with experimental results in Section 4. In Section 5 we draw conclusions and give an outlook on future work.

J.C. Príncipe and R. Miikkulainen (Eds.): WSOM 2009, LNCS 5629, pp. 182–190, 2009.
© Springer-Verlag Berlin Heidelberg 2009

## 2    Quality Measures for and Comparison of SOMs

A range of measures have been described for assessing the quality of either a SOM's quantisation, projection, or both; an overview is given in [7]. The probably best known *quantisation* measure is the *Quantisation Error*, which sums the distances between the input vectors and their best matching unit (BMU). Among the measures assessing the *projection* quality, the *Topographic Error* increases an error value if the BMU and the second BMU of an input vector are not adjacent to each other on the map. The normalised sum over all local errors is used as a global error value of a given map. The *Topographic Product* [1] measures for each unit whether its $k$ nearest neighbour units coincide, regardless of their order, by assessing the distances of the model vectors in the input and output space. Its result indicates whether the dimensionality of the output space is too large or too small. The *Neighbourhood Preservation* [8] measure is similar to the Topographic Product, but operates on the input data. Additionally [8] introduces *Trustworthiness*, measuring whether the k-nearest neighbours of data vectors in the output space are also close to each other in the input space. It thus gives and indication of the expressiveness and reliability of a given mapping.

Only limited research has been reported for comparing two or more SOMs with each other. An analysis of different distance measures for a supervised version of the SOM and it's application to the classification of rail defects, for example, is studied in [2]. Quality measures for the evaluation of data distribution across maps trained on multi-modal data are explored in [5], where the effect of multiple modalities is shown by the example of song lyrics and acoustic features for audio files. Both types of features are used for the same collection and the resultant map is compared according to spreading features. These help to identify musical genres with respect to their homogeneity in both dimensions. Analysis of different map sizes or other parameter variations are not considered. *Aligned Self-organising Maps* [6] are composed of several SOMs which are trained on the same data with differently weighted features, with the aim of exploring the impact of these differences on the resultant mappings. The maps are aligned as layers in a stack, and a distance measure is defined between stacks for comparison of units across layers. This measure is then used analogous to the distance between units on one layer to preserve the topology across the stack. The Aligned SOMs changes the SOM training algorithm so that each data vector is mapped onto a similar position also in the vertically stacked SOMs. However, this method can not be applied to maps with different sizes.

## 3    Analysing Data Shifts and Co-locations

The following methods allow comparisons of two or more SOMs trained on the same data set. The parameters for the SOM training such as the size of the map, the neighbourhood function, or the learning rate, can differ. Herein lies the strength of these visualisations, namely to compare differences in these parameters or of SOMs trained with identical parameters but different (random) initialisations of the model vectors, with respect to distributions in the output space. All the methods proposed below rely on one *source map* and one or more *target maps* the source is compared to. The resulting description may either be visualised by colour-coding the units, on the *source map*, or actually display-ing the detailed components of the resulting measurement using the source and target maps. In order to compare SOMs of radically different sizes, all methods make use of a neighbourhood definition in both the source and target maps.

## 3.1    Data Shifts Analysis

This method analyses and displays changes in the position of co-located data across multiple maps. For a given vector, it shows the position of the other vectors mapped onto the same unit (or within a given source neighbourhood) on a target map. This can be used to find out how stable the mapping is, and how steadily a data vector is put into a neighbourhood of other vectors on different SOMs. Put more abstractly, it measures how much of the data topology on the map really is caused by attributes of the data, and how much of it is simply an effect of different SOM parameters or initialisations, i.e. is caused by differences in parameter settings and training process.

An introductory example for the Data Shifts Visualisation is given in Fig-ure 1(a). The figure shows positions of data vectors between two maps in terms of data and cluster shifts. The SOMs in Figure 1(a) are visualised by the two rectangular grids (each square represents a unit of the SOM and the numbers indicate the number of instances mapped to the respective unit). The arrows show the movement of the four vectors lying on the lower left unit of the left map. Three out of four vectors move to the unit of the right map pointed to by the thick arrow.

The data shifts and their types can be formalised as follows: Let $r_1$ and $r_2$ be the radii of the source and target neighbourhoods, and let $d_1$ and $d_2$ be the dis-tance functions in the output space of the two SOMs. Let $c_s$ be the stable count threshold and $c_o$ be the outlier count threshold, which can be adjusted to ignore shifts concerning only "few" vectors, and to define what "few" means. With $x_i$

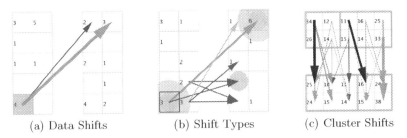

(a) Data Shifts          (b) Shift Types          (c) Cluster Shifts

**Fig. 1.** Positioning of data vectors across different SOMs 1(a). 1(b) shows all types of shifts, and neighbourhood radii. The movement of clusters is shown in 1(c). The arrows denote the movement of data vectors/clusters, respectively.

denoting the data vector in question, its source and target neighbourhoods $U_{1i}$ and $U_{2i}$ contain other data vectors as follows:

$$U_{1i} = \{x_j | d_1(x_j, x_i) \leq r_1\}, \qquad U_{2i} = \{x_j | d_2(x_j, x_i) \leq r_2\} \qquad (1)$$

The set of neighbours that are in both neighbourhoods, $S_i$ can easily be found, as well as the set of vectors that are neighbours in the first SOM but not in the second, $O_i$:

$$S_i = U_{1i} \cap U_{2i}, \qquad O_i = U_{1i} \setminus U_{2i} \qquad (2)$$

The input vector $x_i$'s data shift is stable for a given absolute threshold if $|S_i| \geq c_s$, or if $\frac{|S_i|}{|U_{1i}|} \geq c_s$ in the case of a relative threshold.

If the data shift is not a stable shift, it is an adjacent shift if there is another data vector $x_s$ whose data shift is stable and it lies within the neighbourhood radii.

$$d_1(x_i, x_s) \leq r_1 \wedge d_2(x_i, x_s) \leq r_2. \qquad (3)$$

Finally, if the shift is neither stable nor adjacent, it is an outlier shift if $|O_i| \geq c_o$ in case of absolute, and $\frac{|O_i|}{|U_{1i}|} \geq c_o$ for relative count threshold values.

Figure 1(b) illustrates all types of shifts, i.e. stable, adjacent and outlier shifts, by green, cyan and red arrows, respectively. The circles indicate the neighbourhood for determining the neighbour count (green) and the adjacent shifts (cyan).

## 3.2   Cluster Shifts Analysis

The Cluster Shifts Analysis is conceptionally similar to the Data Shifts Analysis but compares SOMs on a more aggregate level, by comparing clusters in the SOM instead of singular units or neighbourhoods. Thus, we first employ Ward's linkage clustering [4] on the SOM units, to compute the same (user-adjustable) number of clusters for both SOMs. The clusters found in both SOMs are linked to each other, determined by the highest matching number of data points for pairs of clusters on both maps – the more data vectors from cluster $A_i$ in the first SOM are mapped into cluster $B_j$ in the second SOM, the higher the confidence $p_{ij}$ that the two clusters correspond to each other. This can be formalised as follows: let the set $M_{ij}$ contain all data vectors $x$ which are mapped onto the units in $A_i$ and in $B_j$. To compute the confidence $p_{ij}$ that $A_i$ should be assigned to $B_j$, the cardinality of $M_{ij}$ is divided by the cardinality of $A_i$.

$$M_{ij} = \{x | x \in A_i \wedge x \in B_j\}, \qquad p_{ij} = \frac{|M_{ij}|}{|A_i|} \qquad (4)$$

We then compute all pairwise confidence values between all clusters $C_i$ in the maps. Finally, they are sorted and we repeatedly select the match with the highest values, until all clusters have been assigned exactly once. When the matching is determined, the visualisation can easily be created, analogously to the Visualisation of the Data Shifts. An example is depicted in Figure 1(c), which shows a map trained on synthetic data of two slightly overlapping Gaussian clusters. The number of clusters to find was set to two. The cluster mappings are indicated by blue arrows, whose thickness corresponds to the confidence of the match. Data vectors which move from a cluster in the first SOM to the matched cluster in the second SOM are considered 'stable' shifts', and indicated with green arrows; the red arrows represent 'outlier' shifts into other clusters.

### 3.3    Multi-SOM Comparison Analysis

While the previous two methods focus on a pair-wise comparison, the Multi-SOM Comparison Analysis can be used to compare multiple SOMs trained on the same data set. Its main focus is one specific SOM, the 'source SOM', to be compared against a number of other maps. More precisely, the visualisation colours each unit in the main SOM according to the average pairwise distance between the unit's mapped data vectors in the other $s$ SOMs. To this end, we find all $k$ possible pairs of the data vectors on $u$, and compute the distances $d_{ij}$ of the pair's positions in the other SOMs. These distances are then summed and averaged over the number of pairs and the number of compared SOMs, respectively. The mean pairwise distance $v_u$ of unit $u$ is thus calculated as follows:

$$v_u = \frac{\sum_{j=1}^{s} \frac{\sum_{i=1}^{k} d_{ji}}{k}}{s} \tag{5}$$

Similarly, the computation of the variance $w_u$ is defined as:

$$w_u = \frac{\sum_{j=1}^{s} \frac{\sum_{i=1}^{k} d_{ji}^2}{k}}{s} - v_u^2 \tag{6}$$

where $d_{ji}$ denotes the distance between the vectors of pair $i$ in the output space of SOM $j$.

When applied to the cluster based evaluation, we use the single linkage distance between the respective clusters $r$ and $s$ and their cluster members $x_{ri}$ and $x_{sj}$ as follows:

$$d^{SL}(r, s) = min(d(x_{ri}, x_{sj})) \tag{7}$$

Herein, the distance between two clusters is defined as the minimum distance between any of their respective members. In our case, we use unit coordinates of clusters in the SOMs as the features describing them. As a result of the computations described in this section, we obtain quality measures for single units with respect to the mapping of their data vectors on other SOMs.

## 4    Experiments

We present two sets of experiments, first with an artificial data set tailored to specific challenges in data mining, and then with the Iris benchmark data set.

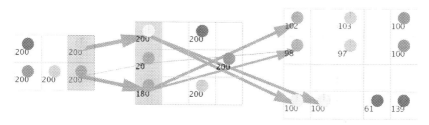

**Fig. 2.** Data shifts of the multi-challenge data set across three SOMs of different sizes trained on the same data

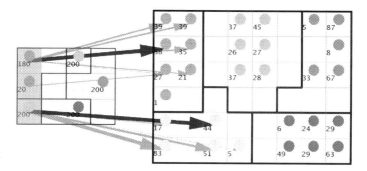

**Fig. 3.** Cluster shifts of the multi-challenge data set for two SOMs of varying sizes

## 4.1  Artificial 'Multi-challenge' Data Set

We created a 10-dimensional synthetic data set, which is used to demonstrate how a data analysis method deals with clusters of different densities and shapes when these different characteristics are present in the same data set [1]. It consists of five sub-sets, four of which live in a three-dimensional space. The subsets themselves are composed of several clusters, thus in total we have 14 distinguishable patterns of data. The first subset consists of one Gaussian cluster, and another cluster formed of three Gaussians, all of which are well separated. The second subset consists of two overlapping, three-dimensional Gaussians, while the third set is similar, but of ten dimensions. The fourth subset is the well-known *chainlink* problem of two intertwined rings. Finally, the fifth subset is sampled along a curve that consists of four lines that are patched together at their endpoints.

Figure 2 illustrates three different map sizes trained on this data set, and shows how the clusters slowly separate into their sub-clusters they are composed of, when the map size increases. In the middle illustration, even with doubling the number of units, only one cluster splits into two sub-clusters; finally, in the right image, all clusters have split on two different units. Figure 3 shows the cluster shifts for three selected clusters from a smaller map with twelve units

---

[1] The data set is available at http://www.ifs.tuwien.ac.at/dm/

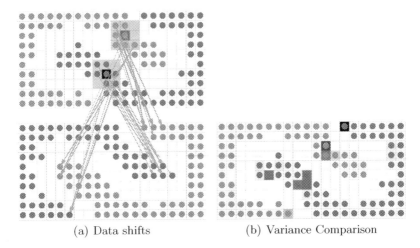

(a) Data shifts                    (b) Variance Comparison

**Fig. 4.** Data Shifts and Variance Analysis on the Chainlink data set

to a bigger map with 48 units. The clusters are identical on both maps, thus with a confidence of 100% each. It is, however, interesting to note that for the cluster arranged in the top-left corner of both maps, the initial separation on the smaller map does not prevail any more on the bigger map. Thus, the initial assumption that could be drawn from the smaller map, namely that the items found on the two units are clearly separable, could be refuted.

Figure 4 illustrates one specific subset, the *Chainlink* problem, for which it is known that it cannot be projected to a two-dimensional space without severely breaching the topology. The two rings are indicated by red and blue colour, respectively. It can be well observed from the visualisation of the Data Shifts in 4(a) that even though the projection looks very similar in both cases, the breaking points in the two rings are actually different in the two maps. Further, the illustration also depicts the mean values of the Multi-SOM comparison, evaluated across eight target SOMs trained with different initialisation and iteration parameters, with two nodes having high pairwise distances, and thus colour black. Figure 4(b) shows the distance variance of the same map. It can be noted that with this measure, we find a higher number of possible breaching points than we were able to detect with the mean pairwise distance only. The intensity of the grey-shade used denotes a higher variance of the distance in the different SOMs, and thus indicates dislocations of vectors, which in this case reveal the topology breach, with the black-filled units marking the points with the highest probability.

## 4.2   Iris Data Set

Finally, we performed experiments on the benchmark data set Iris [3]. Two maps were trained, with 25 and 100 units, respectively. The three different classes in the data set are marked with yellow (setosa), dark blue (versicolor) and light

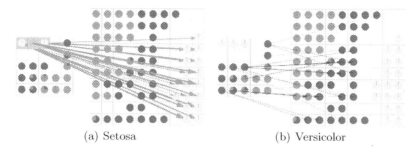

(a) Setosa                              (b) Versicolor

**Fig. 5.** Data Shifts: stable shifts from *Setosa* (a), outlier shifts from *Versicolor* (b)

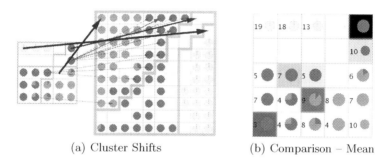

(a) Cluster Shifts                        (b) Comparison – Mean

**Fig. 6.** Iris: Cluster Shifts with three clusters and outliers (a), mean comparison (b)

blue (virginica). In Figure 5(a), we can see the data shifts from the setosa class. The topology-preservation ability of the SOM can be easily observed: the vectors from the rightmost setosa unit in the left map are mapped onto the top of the elongated setosa area in the large map, the vectors from the middle unit onto the middle of the elongated area, and the vectors on the leftmost unit are mapped onto the bottom of the elongated area. The borders of the virginica area and the versicolor area, however, are not as cohesive and spread over a wider area than the border between the other two classes. Figure 5(b) shows only the outlier shifts for the data shifts visualisation of the two SOMs. Most of the outlier shifts emerge from units in the versicolor area or the border of the virginica area.

Figure 6(a) shows a Cluster Shifts Visualisation based on three clusters. The setosa cluster is clearly separated from the others, and its mapping has 100% confidence. The other two clusters each represent one of the other two classes in the small map. In the large map the virginica cluster gets assigned quite a few versicolor samples as well. These show up as the outlier shifts drawn in in red. The virginica cluster match confidence is 100%, the versicolor clusters' confidence is only 69%.

Finally, a Multi-SOM comparison was used to find the units in the smaller SOM where the projection onto the two-dimensional SOM-grid is unstable, which is visualised in Figure 6(b). The minimum pairwise distance threshold was set to 2.5, to reduce the impact of the bigger size of the larger SOM – the data vectors spread over more units in the larger SOM, thus the vectors that lie on one unit in the small SOM will spread over a couple of neighbouring units in the large SOM. This would distort the conclusions we wish to draw from the visualisation, therefore the threshold is used to compensate for the difference in size. The units with the high mean pairwise distances (marked in shades of grey) all are either on the border between the versicolor and virginica classes or within the versicolor class. This points to the relative instability of the projection of the versicolor class onto the SOM-grid: data vectors from the versicolor class are projected differently in both SOMs. Yet again, these results suggest that the setosa class and to some extend the core of the virginica class are well-defined and distinct, while the border between virginica and versicolor and versicolor class itself are a relatively unstable area in a SOM projection. Thus, the results from the three visualisations support and reinforce each other.

## 5    Conclusions and Future Work

In this paper, we presented methods to analytically compare two or more SOMs with each other, and showed the feasibility of the approach on two data sets. Due to space limitations, we could not present experiments on further data sets and had to limit the level of detail in our experiment discussion; more details are availbale at http://www.ifs.tuwien.ac.at/dm/. Future work includes more extensive experiments to provide evidence for certain types of shifts and violations, to eventually automate the process of SOM interpretation, as well as for automatically setting useful threshold and analysis neighbourhood parameters.

## References

1. Bauer, H.-U., Pawelzik, K.R.: Quantifying the neighborhood preservation of self-organizing feature maps. Trans. Neural Networks 3(4), 570–579 (1992)
2. Fessant, F., Aknin, P., Oukhellou, L., Midenet, S.: Comparison of supervised self-organizing maps using euclidian or mahalanobis distance in classification context. In: Mira, J., Prieto, A.G. (eds.) IWANN 2001. LNCS, vol. 2084, pp. 637–644. Springer, Heidelberg (2001)
3. Fisher, R.A.: The use of multiple measurements in taxonomic problems. In: Annual Eugenics, Part II, vol. 7, pp. 179–188 (1936)
4. Ward Jr., J.H.: Hierarchical grouping to optimize an objective function. Journal of the American Statistical Association 58(301), 236–244 (1963)
5. Neumayer, R., Rauber, A.: Multi-modal music information retrieval - visualisation and evaluation of clusterings by both audio and lyrics. In: Proc. 8th Conf. Recherche d'Information Assistée par Ordinateur (RIAO 2007) (2007)
6. Pampalk, E.: Aligned self-organizing maps. In: Proc. 4th Workshop on Self-Organizing Maps (WSOM 2003), pp. 185–190 (2003)
7. Pölzlbauer, G.: Survey and comparison of quality measures for self-organizing maps. In: Proc. 5th Workshop on Data Analysis (WDA 2004), pp. 67–82 (2004)
8. Venna, J., Kaski, S.: Neighborhood preservation in nonlinear projection methods: An experimental study. In: Dorffner, G., Bischof, H., Hornik, K. (eds.) ICANN 2001. LNCS, vol. 2130, pp. 485–491. Springer, Heidelberg (2001)

# Modeling the Bilingual Lexicon
# of an Individual Subject

Risto Miikkulainen[1] and Swathi Kiran[2]

[1] The University of Texas at Austin, Austin, TX 78712, USA
[2] Boston University, Boston, MA 02215, USA

**Abstract.** Lexicon is a central component in any language processing system, whether human or artificial. Recent empirical evidence suggests that a multilingual lexicon consists of a single component representing word meanings, and separate component for the symbols in each language. These components can be modeled as self-organizing maps, with associative connections between them implementing comprehension and production. Computational experiments in this paper show that such a model can trained to match the proficiency and age of acquisition of particular bilingual individuals. In the future, it may be possible to use such models to predict the effect of rehabilitation of bilingual aphasia, resulting in more effective treatments.

**Keywords:** Lexicon, semantics, speech, language acquisition.

## 1 Introduction

The mental lexicon, i.e. the storage of word forms and their associated meanings, is a major component of language processing. It is also one that is perhaps the best understood in terms of computational modeling. Partly the reason is that abundant data exists about how the lexicon develops, how it is organized, how it functions, and how it breaks down in dyslexia and aphasia; partly the reason is that lexical processes are rather modular and therefore tend to be amenable to computational theories.

Although the physiological implementation and even the location of the lexicon in the brain is still open to some debate, there is evidence from MRI and electrophysiology that the lexicon may be laid out as a map, or multiple maps [16]. As a result, self-organizing map (SOM) models are a natural way to model the lexicon. SOM models have been developed to understand e.g. how ambiguity is processed by the lexicon, and how it breaks down in dyslexia and aphasia [12,13], and how the lexicon is acquired during development [11]. The SOM-based lexicon has also turned out useful as a component in larger systems of natural language processing [12].

Given that the majority of the world's population is bilingual (or multilingual) [17], an important extension of the SOM-based lexicon models, as well as lexical research in general, is to account for lexica with multiple languages. A theoretically based approach is developed in this paper, with the eventual aim of using the model to help rehabilitate bilingual patients with aphasia.

More specifically, a bilingual lexicon model is built for individuals with different proficiencies in the two languages, and different age at which the two languages were

J.C. Príncipe and R. Miikkulainen (Eds.): WSOM 2009, LNCS 5629, pp. 191–199, 2009.

acquired. Based on recent theoretical models, a single semantic map is used, with separate maps for the words in the two languages, and six separate sets of connections between the maps. Frequency of exposure, as opposed to age, is shown to primarily determine the proficiency, with age affecting the way the maps are organized. In the future, the model can be used to study how the multilingual lexicon breaks down with damage and how it can be rehabilitated. By matching the proficiency and age of acquisition with those of patients suffering from aphasia, it may be possible to use the model to derive most effective treatment regimes for them individually.

## 2   Bilingual Lexical Processing

Current theoretical models of the bilingual lexicon generally agree that bilingual individuals have a shared semantic (or conceptual) system and that there are separate lexical representations of the two languages. However, the models differ on how the lexica interact with the semantic system and with each other.

The concept-mediation model [15] (Fig. 1a), proposes that both the first (L1) and the second-language lexica directly access concepts. In contrast, the word-association model assumes that second-language words (L2) gain access to concepts only through first-language mediation (Fig. 1b). Empirical evidence [8] suggests that the word association model is appropriate for low-proficiency bilinguals and concept mediation model for high-proficiency bilinguals. As an explanation, De Groot [5] proposed the mixed model (Fig. 1c), where the lexica of a bilingual individual are directly connected to each other as well as indirectly (by way of a shared semantic representation). This model was further revised with asymmetry by Kroll & Stewart [9] (Fig. 1d). The associations from L2 to L1 are assumed to be stronger than those from L1 to L2, and the links between the semantic system and L1 are assumed to be stronger than those between the semantic system and L2.

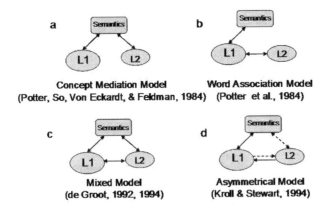

**Fig. 1. Theoretical models of the bilingual lexicon.** All four theories posit a common semantic system with language specific representations in L1 and L2. The most recent theory (d) includes connections between all maps, with connections of the most dominant language (L1 in this figure) stronger than the others (solid lines indicate strong connections and dashed lines weak connections). This theory is used as the starting point for the computational model.

A second important issue is whether activation of the semantic system spreads to both lexica or only within that of the language being used. The prevailing theory suggests that lexical access is target-language nonspecific [2], although, this view is controversial [3]. A third issue is the extent to which proficiency in the two languages and the age at which they are acquired (AoA) affect lexical access. There is evidence that language proficiency, and not AoA, primarily determines the nature of semantic processing [6]. For instance, Li and Farkas [10] showed that novice bilinguals have a fuzzier representation of semantics and phonology than proficient bilinguals.

Adopting the asymmetric mixed model of Fig. 1d, this paper will systematically examine the extent to which language proficiency and AoA influence activation of targets in the lexicon. The work will form a foundation for studying damage and rehabilitation of aphasia in bilingual patients in the future.

## 3   Computational Models of the Lexical System

Artificial neural networks have been used to model various aspects of the lexical system for over two decades. Most of the models aim to explain lexical processing with low-level mechanisms, focusing on the timing of the process as well as on certain types of performance errors and deficits. They are primarily process models, detached from the physical structures, and designed as controlled demonstrations of how disambiguation and production could be carried out in the lexical system [4,7,14].

One exception is the DISLEX model by Miikkulainen [12,13], which was further developed as DEVLEX by Li and colleagues [11]. Its organization is modeled after the cortical maps that underlie many perceptual processes and may also be the substrate for the lexical system in the brain [16]. DISLEX consists of two self-organizing maps, one for lexical symbols and the other for word meanings, as well as associative connections between them (Fig. 2). The lexical map is a layout of the orthographic or phonetic symbols in the language (orthography is used in the examples in this paper). It is a two-dimensional array of computational units, or neurons, trained to represent the symbols using the self-organizing map method. The symbols are vectors of gray-scale values [0..1], representing the orthographic features of the word. Each word is represented by a "blurry bitmap", i.e. a coarse visual image of the word as a series of letters.

During training, such vectors are presented to the map one at a time, and each unit computes the Euclidean distance $d$ between its weight vector $w$ and the symbol representation $v$:

$$d = \sqrt{\sum_k (w_k - v_k)^2}. \tag{1}$$

The unit with the smallest distance (unit $(r, s)$) is then found, and the weights of that unit and those in its neighborhood (units $(i, j)$) are adapted towards the input vector:

$$w'_{k,ij} = w_{k,ij} + \alpha(v_k - w_{k,ij})h_{rs,ij}, \tag{2}$$

where $h_{rs,ij}$ is a Gaussian function defining and $\alpha$ is the learning rate. This process has two effects: the weight vectors become representations of the symbol vectors, and the

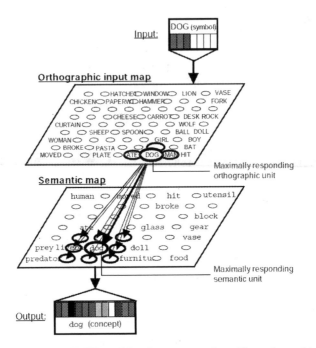

**Fig. 2. A single-language DISLEX model and representations.** The orthographic input symbol DOG is translated into the semantic concept dog in this example. The representations are vectors of gray-scale values between 0 and 1, stored in the weights of the map units. The size of the unit on the map indicates how strongly it responds. Only a few strongest associative connections of the orthographic input unit DOG (and only that unit) are shown.

neighboring weight vectors become similar. Over several presentations of each lexical symbol, the array of units then learns to represent the space of symbols in the language.

The semantic map is organized in a similar manner. The input vectors to this map represent semantic meanings of words. Each component of the vector represents a semantic microfeature (learned automatically based on word cooccurrence [12], and the vectors as a whole represent similarities between word meanings. The self-organizing map therefore learns the layout of the semantic space, i.e. the possible meanings of the words in the language.

Associations between the two maps are learned at the same time as the two maps are organized. A lexical symbol and its meaning are presented at the same time, resulting in activations on both maps. Associative connections between the maps are then adapted based on Hebbian learning, i.e. by strengthening those connections that link active units, and normalizing all connections of each unit:

$$a'_{ij,mn} = \frac{a_{ij,mn} + \alpha\eta_{ij}\eta_{mn}}{\sum_{uv}(a_{ij,uv} + \alpha\eta_{ij}\eta_{mn})}, \tag{3}$$

where $a_{ij,mn}$ is the weight on the associative connection from unit $(i,j)$ in one map to unit $(m,n)$ in the other map and $\eta_{ij}$ is the activation of the unit. As a result of this learning process, when a word is presented to the lexical map, its associated meaning is activated in the semantic map, and vice versa. DISLEX therefore models both comprehension and production in the lexicon.

The DISLEX model was evaluated in prior work in three ways: First, it was shown to function well as a practical lexicon component in a large language processing system called DISCERN [12], for processing script-based stories. Second, it was tested as a cognitive model of human lexical processing. The spatial organization of the lexicon is motivated by the maps in the brain, such as those suggested to underlie the lexical system [16]. Because of this structure, local lesions to the system result in category-specific impairments, similar to those documented in aphasic patients [18]. Noisy propagation between maps gives a possible computational explanation to many dyslexic phenomena [1], such as lexical errors ("ball" → "doll"), semantic errors ("lion" → "tiger") and combined errors ("sympathy" → "orchestra"). Third, the model was extended to model lexical development [11,10,19]. The extended model, called DEVLEX, is trained with gradually more words. It accounts for a range of phenomena in lexical acquisition, including effects of lexical categories such as representation of nouns/verbs, word frequency, word length and word density. AoA is considered to be an important factor in the developing bilingual lexicon as well. For instance, the timing of L2 acquisition impacts the structural representation of L1 and L2 maps. When L2 is acquired later than L1, it becomes dependent upon the L1 semantic map in a "parasitic" way and induces higher rates of errors [19].

DISLEX therefore forms a solid foundation for modeling bilingual lexical processing and its breakdown and recovery as well. The first step is to extend it to two languages with different proficiency and AoA, as described below.

## 4    The Bilingual DISLEX Model

The original DISLEX model was extended to include lexica for two languages, L1 and L2. By varying the amount and timing of training in the two languages, the model can represent an individual with a given proficiency and AoA of L1 and L2.

The bilingual lexicon model was constructed to match typical subjects in the empirical studies: L1 is Spanish, L2 is English, and the AoA of English varies. Furthermore, because the patients are typically more proficient in English than Spanish, English is the dominant language (L2d) and Spanish is the weaker language (L1w). The L1 and L2 lexical symbols are each laid out in a different map in the model (Fig. 3). To construct such a model, 30 corresponding words, represented orthographically, were used in each language, together with 18 distinct semantic meanings; the mapping from words to meanings was thus many-to-many. As an example, the model was organized to simulate both early (Fig. 3a) and late acquisition of L2 (Fig. 3b) where L2 was the dominant language.

In the early L2 acquisition model, each Spanish symbol - English symbol - meaning triple was presented 45 times, and during the same time period, each English symbol - meaning pair was presented an additional 15 times, modifying the input and associative weights each time as described above. Such different amount of training in English and Spanish corresponds to the different exposure to the two languages during acquisition. In the late L2 acquisition model, the Spanish symbol - meaning pairs were first presented 25 times; then, the English - Spanish - meaning triples were shown 20 times,

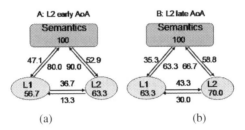

**Fig. 3. Bilingual DISLEX models of two typical (hypothetical) individuals.** Following the general architecture of Kroll & Stewart [9] (Fig. 1d), the bilingual DISLEX model consists of a map of lexical symbols for each of the two languages, a common map of word meanings, and associative connections between them. The solid arrows indicate strong and dashed weak associative connections; the numbers stand for percentage proficiency of each component. The model was trained in two different ways to match two different hypothetical patients, each with a different L2 AoA but high L2 proficiency. In this manner, the level of AOA, proficiency and impairment can each be altered to understand the individual contribution of each variable.

interleaved with 40 presentations of the English symbol - meaning pairs. In both cases, the lexicon self-organized to represent the two languages and the mapping between them (Fig. 4). Each lexical map is organized according to the similarity of the word shapes (mostly word length and matching letters, which are the most prominent characteristics of the "blurry bitmap" representation used). The semantic map is organized according to the word meaning (i.e. animate words are clustered together, as are verbs and objects).

When a meaning representation is given an input to the semantic map, the corresponding units in the English or Spanish map are activated through associative connections. The main difference between early and late L2 maps is that while the early L2 map is organized relatively smoothly, the late L2 map is irregular and uneven. To quantify the behavior of these models, the 18 meaning representations were each presented to the semantic map in turn and propagated to the English map and to the Spanish map, modeling naming in the two languages; similarly, the 30 English words and 30 Spanish words were each presented as input to the appropriate map and propagated to the semantic map, modeling word comprehension. In each case, the unit with the maximally responding unit was found in the input map and in the associated map. If its weight vector was closest to the correct representation, the output for the word in the lexicon was correct.

A typical bilingual performance was observed in this process in both early and late L2 models (Fig. 3). With early English AoA, English (L2d) dominated Spanish (L1w) in production (53 vs. 47% accuracy) as well as in comprehension (90 vs. 80%). Importantly, the same was true of late English AoA both in production (58 vs. 35%) and in comprehension (67 vs. 63%). (Note that only the relative proficiency is important; the absolute proficiency can be adjusted by changing the amount and rate of learning in the model). Most interestingly, an important asymmetry emerged between the L1 and L2 maps: An L1 word activates the corresponding L2 word more strongly than the other way around (37 vs. 13% early, 43 vs. 30% late), showing more proficiency in English

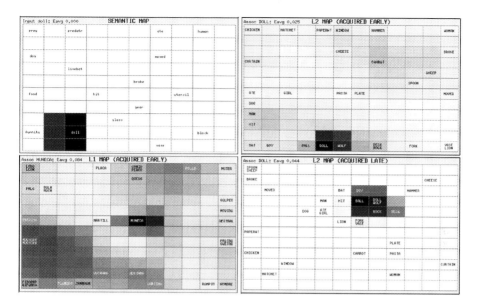

**Fig. 4. A DISLEX model of normal bilingual behavior.** The L1 and L2 maps are organized according to the perceptual similarities (mostly word length), and the semantic map according to semantic meaning (e.g. animals, objects, verbs). A word presented to one of the maps (in this case semantic meaning `doll`) activates that map, and the activity propagates through associative connections to the other maps (in this case L1 and L2, activating the units representing DOLL and MUNECA, as well as a few neighboring units). The colors white → yellow → red → black indicate increasing response. The connections to L1 are less specific (activating a wider area) and therefore result in more errors than those to L2. Also, although both the early AoA L2 (bottom right) and late AoA L2 (top right) result in roughly equal performance, the early map is better organized than the late map (which consists of several small discontinuous clusters). In this manner, DISLEX models bilingual naming and comprehension with different proficiencies and different age of acquisition in the two languages.

(L2d) than in Spanish (L1w), and demonstrated dominance of English over Spanish in transfer between the languages. This asymmetry is consistent with behavioral data (Section 2) showing that lexical activation in the non-dominant language results in activation of corresponding representations of the dominant language.

In a subsequent experiment, the role of the languages was reversed, resulting in a model with Spanish as the dominant (L1d) and English as the weaker (L2w) language. In a third experiment, the model was exposed equally frequently to the two languages (L1 = L2), and neither language dominated the other in the resulting model. These results were obtained both with early and with late L2 training, with consistent differences in the L2 map organization (Fig. 4). Together these experiments suggest that by training the model with different starting times and frequencies, the relative proficiency and organization in the two languages can be tuned, making it possible to fit the model to a particular individual's performance and learning history.

## 5  Discussion and Future Work

The model is consistent with behavioral data in two important ways. First, relative language proficiency is modulated by the amount of exposure to each specific language. Second, various levels of L2 language performance can be achieved with both early and late L2 AoA. Further, the late L2 AoA results in less well organized L2 map even when L2 achieves eventual high proficiency.

In the future, it may be possible to use the model to fit the pre-stroke performance and learning history of an individual patient, and then used to derive an optimal treatment for that patient. Damage to the lexical system can be modeled in DISLEX in two ways: (a) Units or connections can be deleted from the model, and (b) noise can be added to the connections. By controlling the type and extent of damage, it will be possible to fit the model to the profile of an individual patient. In rehabilitation training, then, the model will be presented with selected word-meaning pairs in the two languages and it will continue self-organizing using the same mechanisms as during initial training. By varying the types of words (such as concrete vs. abstract words, rare vs. frequent words, short vs. long words, and word categories) and numbers of training pairs in the two languages systematically, it should be possible to determine a training recipe that leads to fastest and most complete recovery. Such an ability could greatly improve our ability to treat bilingual aphasia in the future.

## 6  Conclusion

A bilingual lexicon model consisting of a common map of word meanings and a separate maps for the words in the two languages is consistent with current theory of bilingual processing in the mental lexicon. By varying the frequencies with which the maps are trained, and the timing of the training in the two languages, models of individual proficiency and order in the two languages can be developed. The approach can be used to develop models of individuals, and in the future, may form a foundation for discovering individually optimized recipes for treatment of aphasia.

## Acknowledgments

This research was supported in part by NIH under grant R21-DC009446.

## References

1. Coltheart, M., Patterson, K., Marshall, J.C. (eds.): Deep Dyslexia, 2nd edn. Routledge and Kegan Paul, London, New York (1988)
2. Costa, A., Heij, W., Navarette, E.: The dynamics of bilingual lexical access. Bilingualism: Language and Cognition 9, 137–151 (2006)
3. Costa, A., Miozzo, M., Caramazza, A.: Lexical selection in bilinguals: Do words in the bilingual's two lexicons compete for selection? Journal of Memory and Language 43, 365–397 (1999)

4. Cottrell, G.W., Small, S.L.: A connectionist scheme for modelling word sense disambigua-tion. Cognition and Brain Theory 6, 89–120 (1983)
5. de Groot, A.: Determinants of word translation. Journal of Experimental Psychology: Learn-ing, Memory and Cognition 18, 1001–1018 (1992)
6. Edmonds, L., Kiran, S.: Lexical selection in bilinguals: Do words in the bilingual's two lexicons compete for selection. Aphasiology 18, 567–579 (2004)
7. Hinton, G.E., Shallice, T.: Lesioning an attractor network: Investigations of acquired dyslexia. Psychological Review 98, 74–95 (1991)
8. Kroll, J., Curley, J.: Lexical memory in novice bilinguals: The role of concepts in retrieving second language words. In: Practical Aspects of Memory, vol. 2. Wiley, New York (1988)
9. Kroll, J.F., Stewart, E.: Category interference in translation and picture naming: Evidence for asymmetric connections between bilingual memory representations. Journal of Memory and Language 33, 149–174 (1994)
10. Li, P., Farkas, I.: A self-organizing connectionist model of bilingual processing. In: Heredia, R., Altarriba, J. (eds.) Bilingual Sentence Processing. Elsevier, Amsterdam (in press) (2002)
11. Li, P., Zhao, X., MacWhinney, B.: Dynamic self-organization and early lexical development in children. Cognitive Science 31, 581–612 (2007)
12. Miikkulainen, R.: Subsymbolic Natural Language Processing: An Integrated Model of Scripts, Lexicon, and Memory. MIT Press, Cambridge (1993)
13. Miikkulainen, R.: Dyslexic and category-specific impairments in a self-organizing feature map model of the lexicon. Brain and Language 59, 334–366 (1997)
14. Plaut, D.C.: A connectionist approach to word reading and acquired dyslexia: Extension to sequential processing. Cognitive Science 23, 543–568 (1999)
15. Potter, M., So, K., von Eckardt, B., Feldman, L.: Lexical and conceptual representation in beginning and proficient bilinguals. Journal of Verbal Learning and Verbal Behavior 23, 23–38 (1984)
16. Spitzer, M., Kischka, U., Gückel, F., Bellemann, M.E., Kammer, T., Seyyedi, S., Weisbrod, M., Schwartz, A., Brix, G.: Functional magnetic resonance imaging of category-specific cor-tical activation: Evidence for semantic maps. Cognitive Brain Research 6, 309–319 (1998)
17. Tucker, G.R.: A global perspective on bilingualism and bilingual education, Technical Report EDO-FL-99-4, Center for Applied Linguistics (1999)
18. Warrington, E.K., McCarthy, R.A.: Categories of knowledge: Further fractionations and an attempted integration. Brain 110, 1273–1296 (1987)
19. Zhao, X., Li, P.: Bilingual lexical representation in a self-organizing neural network. In: Mc-Namara, D.S., Trafton, J.G. (eds.) Proceedings of the 29th Annual Meeting of the Cognitive Science Society, pp. 755–760. Cognitive Science Society, Austin (2007)

# Self-Organizing Maps with Non-cooperative Strategies (SOM-NC)

Antonio Neme[1], Sergio Hernández[1], Omar Neme[2], and Leticia Hernández[1]

[1] Non-linear Dynamics and Complex System Group, Autonomous University of
Mexico City, San Lorenzo 290, Col. Del Valle, México D.F., México
neme@nolineal.org.mx
[2] Postgraduate and Research Section. School of Economics. National Polytechnic
Institute Plan de Agua Prieta No.66, Col. Plutarco Elas Calles, México D.F., México

**Abstract.** The training scheme in self-organizing maps consists of two
phases: i) competition, in which all units intend to become the best matching unit (BMU), and ii) cooperation, in which the BMU allows its neighbor units to adapt their weight vector. In order to study the relevance of
cooperation, we present a model in which units do not necessarily cooperate with their neighbors, but follow some strategy. The strategy concept
is inherited from game theory, and it establishes whether the BMU will allow or not their neighbors to learn the input stimulus. Different strategies
are studied, including unconditional cooperation as in the original model,
unconditional defection, and several history-based schemes. Each unit is
allowed to change its strategy in accordance with some heuristics. We give
evidence of the relevance of non-permanent cooperators units in order to
achieve good maps, and we show that self-organization is possible when
cooperation is not a constraint.

## 1 Introduction

The self-organizing map (SOM) is presented as a model of self-organization of
neural connections, which is translated in the ability of the algorithm to produce
organization from disorder [1]. This property is achieved through a transformation of an incoming signal of arbitrary dimension into a low-dimensional discrete
map and by adaptively transform data in a topologically ordered fashion [2,3].

The SOM structure consists, generally, of a two-dimensional lattice of homogeneous units. Each unit maintains a dynamic weight vector which is the basic
structure for the algorithm to lead to map formation. The input space dimension is considered in the SOM by allowing each weight vector to have as many
components as dimensions in the input space. Each input vector $x$ is mapped to
the unit $i$ whose weight vector is closest to it.

The training scheme in the SOM is divided in two stages:

1. Competitive. The bets matching unit (BMU), identified as $g$ is the one whose
weight vector is the closest to the input vector:

$$BMU = argmin_g||x - w_g||$$ (1)

J.C. Príncipe and R. Miikkulainen (Eds.): WSOM 2009, LNCS 5629, pp. 200–208, 2009.

2. Cooperative The cooperation stage has been identified as fundamental for a proper map formation [4]. Within this stage, the adaptaion is diffused from the BMU $g$ to the rest of the units in the lattice through the learning equation:

$$w_i(t+1) = w_i(t) + \alpha(t)h_g^i(t)(x_j - w_i(t)) \qquad (2)$$

In which $h_g^i$ is the neighborhood function from $g$ to $i$, $\alpha$ is the learning parameter, and $x_j$ is the input vector. In this work, we study a modification to the stage 2 of training. In the original SOM, when a unit becomes the BMU it modifies the map by affecting the weight vector of those units within its neighborhood. This weight modification (cooperation) is the core of the map formation process through the SOM algorithm. Here, we describe a variation in which units may present alternatives to the cooperation scheme, refered as strategies, and still there may be a proper map formation. Strategies may be cooperative (as in the traditional SOM), non-cooperative, or based on the history of previous interactions with other BMUs.

In section 2, we introduce some relevant aspects of game theory and present the Non-Cooperative Self-Organized Map (SOM-NC), in section 3 several results are presented and analyzed, and in section 4 some conclusions are stated.

## 2   The Model

Self-organization is a property present in several structures and phenomena, ranging from biology to social situations [5]. Briefly, self-organization states that an ordered structure may be achieved from an initial, possibly disordered state, by means of local information and short-range interaction between components, with the additional feature that there is not a global unit that guides the system to global order [6].

It has been commonly stated that cooperation is the currency of self-organization. Here, we study some consequences of interactions that may not necessarily be cooperative, and give evidence that self-organization is still possible. We intend to study self-organization when unconditional cooperation between units is not a constraint. We do not intend to minimize an error measure, but to analyze the dynamics of map formation in units that do not always cooperate.

Units adapt their weight as the mechanism that leads to map formation. In the proposed model, units may also adapt the strategy they follow when they become BMU. The BMU will not necessarily affect its neighbors, but will do what its own strategy dictates. Some units will always allow their neighbors to adapt, some strategies will not allow neighborhood adaptation, and some others will decide to allow adaptation or not based on their memory, whether its neighbors allowed it to adapt or not in previous epochs.

The terms cooperation, non-cooperation and strategies are common in Game Theory (GT). GT studies possible solutions to different situations in which players may cooperate or not in order to optimize a given quantity. In cooperative games, players intend to optimize a global quantity. In a cooperative game there are two components: 1) a set of players (units), and 2) a characteristic function that specifies the goodness of a possible alliance of players [7]. This function states how good was for a given player to cooperate with other units. SOM may be seen as a cooperative game in which units cooperate in order to achieve self-organization (well-formed maps).

Although we do not intend to analize self-organizing maps in terms of GT as, for example, in [8], we borrowed some of its concepts in order to elucidate a language for presenting our model and some of its results. In the SOM algorithm, there is not an optimizing principle [9]. However, SOM may be seen as a cooperative game in which all units try to avoid a local map misformation. Units try to form a map that preserves the topology present in the data space. In that sense, each unit tries to, locally, form a map region that, when grouped to other map regions, will lead to a good topographic map. In the SOM, misformation is avoided through cooperation from BMUs to its neighbors. In other words, the BMU allows its neighbors to adapt their own weight vectors to resemble the input vector, and thus approximate the the data distribution in the feature space.

In SOM-NC each unit $i$ is assigned a strategy $\sigma_i$. Table 1 shows the strategy code, and a short description. Each strategy dictates to the BMU if it will allow its neighbors to adapt its weight vector (C or 1) or not (D or 0). If a BMU's strategy states that it will defect from its neighbors (D), then all units within its neighborhood will not adapt their weight vectors (see eq. 3). On the other hand, if a unit has a strategy that cooperates (C), then when it becomes the BMU all units within its neighborhood will be able to adapt their weight vectors.

Some of these strategies need a memory of interaction. Each unit $i$ maintains a record of what every other unit $j$ has done to it during the training period, $\Omega_i^j$. If unit $i$ has been allowed by BMU $j$ to adapt its weight vector say, two times, and then $j$ defect from $i$, then $\Omega_i^j = \{C, C, D\}$. The strategy that governs $i$ may take into account this memory to decide if it will cooperate with $j$ or not. Unit $i$ will present the same strategy for every unit $j$ in the lattice, but the final decision may be affected by $\Omega_i^j$, which is not necessarily the same for all $j$.

The strategy $\sigma_i(t)$ specifies if neuron $i$ will allow its neighbors to adapt their weight vector ($\sigma_i(t) = 1$) or not ($\sigma_i(t) = 0$) at time $t$. Derived from GT, we will call the former case the cooperation scheme whereas the latter is the defect scheme. A BMU will always cooperate with itself, in despite of its strategy. So,

**Table 1.** The studied strategies ($\sigma_i(t)$). BMU $i$ will decide to cooperate or defect from unit $i$ as a function of its own strategy and memory of previous interactions.

| Strategy id | Description |
|---|---|
| C | $i$ will always cooperate (unconditional cooperator, $\sigma_i(t) = 1$) |
| D | $i$ will always defect (unconditional defector, $\sigma_i(t) = 0$) |
| T | $i$ will do as it was done in the last interaction (tit-for-tat, $\sigma_i(t) = \Omega_i^j(t-1)$) |
| R | Random (the same probability of cooperation and defect) |
| A | Alternating C and D ($\sigma_i(t) = C, \sigma_i(t+1) = D, ...$) |
| M | $i$ will do whatever $j$ has done to $i$ most frequently ($\sigma_i(t) =$ more frequent in $\Omega_i^j$) |
| N | $i$ will do whatever $j$ has done to $i$ less frequently ($\sigma_i(t) =$ less frequent in $\Omega_i^j$) |

the adaptation equation now considers the strategy of unit $i$, and adaptation is possible only when $\sigma_i(t) = 1$:

$$w_i(t+1) = w_i(t) + \alpha(t)h_g^i(t)\sigma_i(t)(x - w_i(t)) \tag{3}$$

One of the strategies that is based on history is the so-called tit for tat (TFT). Units with the TFT strategy, when selected as BMU, will start cooperating with their neighbors, but from the second interaction on they will do as they were done: if $i$ becomes the BMU, and its strategy is TFT, then $\sigma_i(t) = \Omega_i^g(t-1)$, which means that $i$ will allow $g$ to adapt its weight vector only if the last time $g$ became BMU, it allowed $i$ to do so, otherwise, modification of $g$'s weight vector is not permitted.

Units may modify their strategy. The heuristic that leads to strategy shifting is as follows. Let $w_i$ be the weight vector for unit $i$ and let $w_h$ the average weight vector from $i$'s neighbors. If $|w_i - w_h| \leq \theta$, where $\theta$ is a threshold, then $i$ will change its strategy: $\sigma_i(t+1) \neq \sigma_i(t)$.

The basis of the proposed heuristic lies in the fact that a unit whose weight vector is very different from the weight vector of the units within its neighborhood is perturbing the proper map formation because of its wrong strategy.

In GT terms, the characteristic function that evaluates the benefit of an alliance may be identified with the difference threshold $\theta$. If unit $i$ perceives that its strategy does not benefit the local map quality as its weight vector is very different from that of their neighbors, then it will shift to another strategy.

Another control parameter for strategy shift is the periodicity $r$. It states that units may try to shift its strategy only every $r$ epochs, if $\theta$ is exceeded.

Once unit $i$ is able to change its strategy, it has to decide what strategy to adopt. Three alternatives were explored. In the first one, it will change its strategy to the more common strategy within its neighborhood. In the second one, it will adopt that strategy from the BMU whose weight vector is closest to the input vectors mapped to it. For the third strategy shift, it will chose randomly a different strategy. These strategies shifting schemes are denoted by $s = 0, s = 1$, and $s = 2$, respectively.

## 3   Experiments

In order to verify the map quality as well as the map formation process in the SOM-NC, three data sets were studied in two different sets of experiments. The ring (two-dimensional), the iris (four-dimensional), and the ionosphere (dimension 34) data sets were studied. In both experiments, network size was $20 \times 20$ and the learning parameter $\alpha$ is set to 0.1 at the beginning and exponentially decreased to 0.0001.

In the first experiment, the strategy of each unit is determined in accordance with a probability distribution. Here, p(D), p(C), p(T), p(R), p(N), p(A), and p(M) represent the probability of each unit having strategies D, C, T, R, N, A or M, respectively. These probabilities are chosen at random (see table 1). For each data set 10000 maps were formed varying the initial probability distribution, the strategy shifting scheme, the $\theta$ threshold, the number of epochs, and initial neighborhood width.

In the second experiment, $p_{ini}(C) = 1$ (all units are cooperators) and the only possible strategy shifting scheme is that of random selection.

### 3.1  Phenomenology

In SOM, the topographic error (TE) is a common measure of self-organization [10], defined as the average number of cases in which the BMU and second-best matching unit are not adjacent in the lattice for each input vector. We measured TE to compare the quality between the maps formed by SOM-NC with maps formed by SOM.

To study SOM-NC, each map was formed under some constraints: 1) the initial strategy probability distribution, refered as $p_{ini}(i)$, $i \in \{C, T, D, A, R, M, N\}$, 2) the strategy shift scheme ($s$), 3) the periodicity of strategy shifting ($r$), 4) the number of epochs, and 5) initial neighborhood width. As units may shift their strategy, there is also a final strategy probability distribution: $p_{final}(i)$, $ii \in \{C, T, D, A, R, M, N\}$ All these features define the paramater space.

In this first set of experiments, we tried to find a relation between the parameter space and maps with low TE. The method we applied to seek for this relation is that of decision trees. In this method, each input vector contains the description of an object. Every input vector is associated with a label, or class, and the method tries to identify what variables and conditions are sufficient to properly relate input vectors to classes. The algorithm seeks to summarize labeled data into a set of simple decision rules based on the mutual information function [11].

Here, we applied C4.5 to the feature vector that describes the conditions in which each map was formed, plus the final probability distribution. The class of each entry is constructed through the TE: those maps with lower TE than the TE shown by the original SOM model will be of class L, whereas those maps with higher TE will be of class H. A decision tree was obtained for each data set and for each of the strategies shifting schemes.

Partial trees for the three data sets are shown in fig. 1. For the three data sets, the final probability for C units is highly informative. Low classification errors were achieved for maps with low TE (L).

For the ring and iris data sets, the strategy shifting scheme $s$ was not informative, that is, good map formation does not depend on how a new strategy is chosen. However, for the ionosphere data set, the random strategy shifting never formed maps with lower errors than those shown by SOM. Once again, there was a tendency for the cooperative strategy $C$ to diffuse to all the units so the final strategy distribution is strongly biased towards it. If $p_{final}(C)$ is high, then the formed maps are good ones (L).

For all $i \in \{T, C, D, R, A, M, N\}$ the correlation between $p_{ini}(i)$ and TE was obtained, and for the three data sets $p_{ini}(T)$ is the most correlated (0.38, 0.42, 0.44, respectively). Also, the correlation for the final probabilities and TE was also calculated and $p_{fin}(T)$ was the highest for the studied data sets (0.51, 0.5,

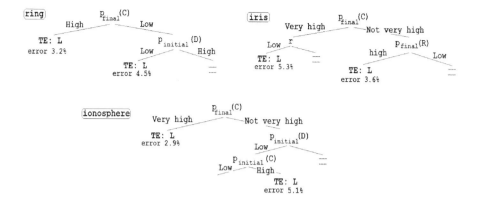

**Fig. 1.** Classification trees obtained by C4.5. $p_{final}(C)$ is highly informative for the three data sets. L class is refered to conditions that lead to maps with lower errors than that of the SOM, whereas H class refers to maps with higher error than the SOM.

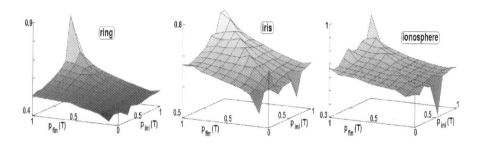

**Fig. 2.** TE as a function of final and initial probability of T units $(p_{final}(T), p_{ini}(T))$

0.6). Fig. 2 shows TE as a function of $p_{ini}(T)$ and $p_{final}(T)$ for the three data sets and for the same 10000 maps detailed in the previous analysis.

In the original SOM all units have C strategy for all the training process. In SOM-NC there are several strategies and units may change their own strategy, so it is important to study how relevant is the strategy shift in the well-formed maps. Table 2 shows the difference of the final and initial distribution for each strategy for 500 maps with the lowest TE and for all the 10000 maps. It is observed that $p_{fin}(C) > p_{ini}(C)$, but this difference is higher for the group with low TE than the difference observed for the 10000 maps, which means that a shift to a C strategy leads to properly formed maps. In the well-formed maps, the reduction of units with strategy D is clear, whereas for the 10000 maps, there is not a clear reduction at all. It may be inferred that the remotion of strategy D is also benefical for map formation.

Although C units are relevant for map unfolding, it is possible to achieve a proper map formation when other strategies are also present. As a second experiment, consider a lattice in which $p_{ini}(C) = 1$, which corresponds to the

**Table 2.** Difference of the final and initial distributions for the seven strategies for both, a) the 500 maps with the lowest TE, b) and for the 10000 maps, for each data set $(p_{fin}(i) - p_{ini}(i))$

| Data set (best 500 maps) | T | C | D | R | A | M | N |
|---|---|---|---|---|---|---|---|
| ring | -0.107 | 0.083 | -0.012 | 0.045 | 0.053 | -0.010 | -0.041 |
| iris | -0.046 | 0.056 | -0.009 | 0.017 | 0.034 | -0.016 | -0.025 |
| ion. | -0.128 | 0.0161 | -0.038 | 0.015 | 0.020 | -0.032 | 0-017 |
| Data set (All 10000 maps) | T | C | D | R | A | M | N |
| ring | -0.041 | 0.010 | -0.003 | 0.005 | 0.023 | -0.010 | 0.027 |
| iris | -0.023 | 0.004 | -0.002 | 0.005 | 0.013 | -0.005 | 0.019 |
| ion. | -0.044 | 0.009 | -0.007 | 0.001 | 0.024 | -0.008 | 0.039 |

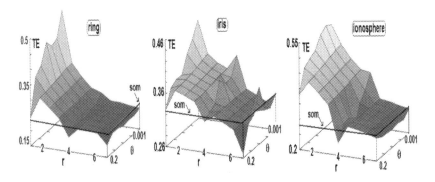

**Fig. 3.** TE as a function of $\theta$ and periodicity $(r)$, when $p_{ini}(C) = 1$ and $s = 2$. It is shown the TE coresponding to the original SOM, for 300 epochs.

original SOM scheme. The only possible strategy shifting scheme is the random one ($s = 2$). In order to study the map formation under this constraint, we varied both $\theta$ and $r$, and plotted the TE of the achieved map in fig. 3. It is shown the average TE obtained by 100 maps formed for the specifed $\theta$ and $r$.

It is observed that TE decreases when $r$ increases, and $\theta$ becomes relevant only when $r$ is low. The weight unfolding for one of the maps with low TE and $p_{ini}(C) = 1$ is shown in fig. 4. It is also shown the corresponding strategy for each unit in the lattice. This map was achieved with $r = 1$ and $\theta = 0.005$. Units change their strategies when their weight vector differs from that of their neighbors by more than $\theta$. Several units have strategies different from C, and still the formed map presents a very low TE (see fig. 4). In general good maps (low TE) are formed if $r > 1$.

## 3.2    Analysis

If $p(D) = 1$, then the model would be equivalent to the $k-$means algortihm, as the permanent defect strategy would be equivalent to a neighbordood of 0. However, units with defective strategy (D) are important for proper map

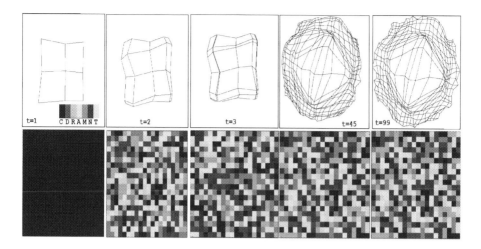

**Fig. 4.** Map unfolding and strategy for each unit in the lattice, for $p_{ini}(C) = 1$, $r = 1$, and $\theta = 0.005$

formation. Units that are located in the neighborhood of two or more units consistently selected to become BMUs may as well become BMU themselves and affect the map formation. A unit whose strategy is D, will not affect the map formation, as only its weight vector will be modified. D units may be affected by BMUs, but they are not able to affect other units when become BMUs.

The existence of units with strategy T is also relevant. If it interacts with C units, then it is equivalent to the case in which all of them cooperate. However, when a T unit interacts with a D unit, it will not affect it, as stated in the tit-for-tat strategy (see table 1).

The relevance of units with strategies R, A, M, and N is not completely clear, but if they are not considered, the formed maps do not present low TE. Units with R strategy may redirect an improper unfold by locally perturbing (or not) the map. Once a BMU with this strategy is not locally forming a good map, it should shift the strategy.

The shift of strategy locally affects the map formation and may lead to a better unfolding. However, if this perturbation is locally incorrect, as stated by the heuristic, then another strategy should be adopted. Only when the map is locally correct, as stated by the heuristic, the strategy shift will not be applied.

## 4    Conclusions and Discussion

Self-organization is possible even when cooperation is not the only strategy followed by active units (BMUs). In the traditional SOM, the BMU allows its neighbors to adapt their weight vectors in accordance with the input vectors. Here, we have shown that map formation is possible when some BMUs do not allow their neighbors to adapt, or when they decide to do so based on the memory of previous interactions.

The strategy each unit follows to cooperate or not with their neighbors is important to achieve good maps. Each unit may change its strategy. The heuristic that leads to this shift is that if the unit's weight vector is not similar to that of its neighbors, then its strategy is wrong and is perturbing the proper map formation.

For the three studied data sets, proper self-organization is possible when units try to shift to another strategy every epoch and the difference threshold is very low. That is, if some unit detects that its weight vector is not very similar to the average neighbor's weight vector, then it will shift to another strategy. The shift scheme resembles the evaluation characteristic function in game theory, which states the convenience of cooperative alliances. Here, a unit may decide not to cooperate with some other units and thus avoiding certain alliances.

# References

1. Cottrell, M., Fort, J.C., Pagés, G.: Theoretical aspects of the SOM algorithm. Neurocomputing 21, 119–138 (1998)
2. Kohonen, T.: Self-Organizing maps, 3rd edn. Springer, Heidelberg (2000)
3. Ritter, H.: Self-Organizing Maps on non-euclidean Spaces Kohonen Maps. In: Oja, E., Kaski, S. (eds.), pp. 97–108 (1999)
4. Flanagan, J.: Sufficiente conditions for self-organization in the SOM with a decreasing neighborhood function of any width. In: Conf. of Art. Neural Networks. Conf. pub. No. 470 (1999)
5. Camazine, S., Deneuburg, J., Franks, N., Sneyd, J., Theraulaz, G., Bonabeau, E.: Self-organization in biological systems. Princeton University Press, Princeton (2001)
6. Haken, H.: Information and self-organization. Springer, Heidelberg (2000)
7. Osborne, M., Rubinstein, A.: A course in Game Theory. MIT Press, Cambridge (1994)
8. Herbert, J., Yao, J.: GTSOM: Game Theoretic Self-organizing Maps. In: Trens in neural computation. Springer, Heidelberg (2007)
9. Erwin, Obermayer, K., Schulten, K.: Self-organizing maps: Ordering, convergence properties and energy functions. Biol. Cyb. 67, 47–55 (1992)
10. Bauer, H., Herrmann, M., Villmann, T.: Neural Maps and Topographic Vector Quantization. Neural Networks 12(4-5), 659–676 (1999)
11. Quinlan, R.: Improved use of continuous attributes in c4.5. Journal of Artificial Intelligence Research 4, 77–90 (1996)

# Analysis of Parliamentary Election Results and Socio-Economic Situation Using Self-Organizing Map

Pyry Niemelä and Timo Honkela

Helsinki University of Technology, Centre of Adaptive Informatics
P.O. Box 5400, 02015 TKK, Finland
timo.honkela@tkk.fi

**Abstract.** The complex phenomena of political science are typically studied using qualitative approach, potentially supported by hypothesis-driven statistical analysis of some numerical data. In this article, we present a complementary method based on data mining and specifically on the use of the self-organizing map. The idea in data mining is to explore the given data without predetermined hypotheses. As a case study, we explore the relationship between parliamentary election results and socio-economic situation in Finland between 1954 and 2003.

## 1   Introduction

In this article, we examine the possibility of exploring the relationship between the results of parliamentary elections over an extended period time and a number of political and societal variables that might influence these results. The data consists of (1) the results of the parliamentary elections between 1954 and 2003, (2) data indicating the parties in the government by the time of and before each election, and (3) a number of socio-economic variables such as unemployment rate. We are interested in finding relevant relationships between these variables (for related research within political science, see e.g., [6,16,18]). Rather than focusing on a set of specific hypotheses, we wish to explore if potentially useful relationships can be found by exploring a larger number of variables concurrently. Section 2 introduces the method, related research and the data collection. Some interesting and useful connections were found and they are reported in Section 3. We are aware that some of the conclusions are preliminary and would require additional data or more detailed qualitative analyses. On the other hand, it also seems that the SOM provides additional insight that would be difficult to gain e.g. by plain inspection of the original data, by calculating correlations between the variables, or by fitting some linear model over the data. There are some alternative methods such as multidimensional scaling that could be considered

J.C. Príncipe and R. Miikkulainen (Eds.): WSOM 2009, LNCS 5629, pp. 209–218, 2009.
© Springer-Verlag Berlin Heidelberg 2009

but the SOM is a good choice especially when the trustworthiness is considered [13,20].

## 2   Data and Method

The aim is to study the effect of socio-economic conditions on parties' approval ratings in Finnish parliamentary elections from the year 1954 to the year 2003. During chosen period there were no wars or other highly exceptional circumstances in Finland. In addition, it is easy to obtain reliable and comparable societal data from this period.

### 2.1   Data

The data consists of three parts. There are eleven variables of the elections, twelve variables of national economic conditions and ten variables of government responsibilities. The election data variables are the proportion of votes cast for the nine most important parties and the group of other parties and the turnouts in Finnish Parliamentary elections. The abbreviations, the English and Finnish names, and the former names of the parties are listed here:

- *KESK*: Centre Party of Finland, Suomen keskusta, until 1962 the Agrarian Union, in 1983 including Liberal Party.
- *SDP*: Social Democratic Party of Finland, Suomen sosiaalidemokraattinen puolue.
- *KOK*: National Coalition Party, Kansallinen kokoomus.
- *LEFT*: Left Alliance, Vasemmistoliitto, until 1987 the Democratic League of the People of Finland, in 1987 including Democratic Alternative.
- *GREENS*: Green League, Vihreä Liitto, in 1987 not as a party of its own.
- *KD*: Christian Democrats in Finland, Suomen kristillisdemokraatit, until 1999 Christian League of Finland.
- *RKP*: Swedish People's Party, Ruotsalainen kansanpuolue.
- *PS*: True Finns, Perussuomalaiset, in 1962 and 1966 the Small Holders Party of Finland and until 1995 the Finnish Rural Party.
- *LIB*: Liberals, Liberaalit, until 1966 the Finnish People's Party, until 1999 Liberal Party.

The proportion of votes cast and voting turnout are based on the elections data of Statistics Finland[1]. They are presented in Figure 1. The variables of government responsibilities contain information if the party has been in government or in opposition during the elections. The government data is based on the Finnish Minister Database MIKO published by the Finnish Government [4].

National economic conditions are analyzed using four measurements. The Change of Cost of Living Index (COLI) is used to measure inflation. The Unemployment Rate (UNEM) has significant influence on the daily life of the voters.

---

[1] http://www.stat.fi/index_en.html

| Year | KESK | SDP | KOK | LEFT | GREEN | KD | RKP | PS | LIB | OTH. | T.OUT |
|------|------|-----|-----|------|-------|-----|-----|-----|-----|------|-------|
| 1954 | 24.1 | 26.2 | 12.8 | 21.6 | | | 6.8 | | 7.9 | 0.6 | 82.9 |
| 1958 | 23.1 | 23.2 | 15.3 | 23.2 | | | 6.5 | | 5.9 | 2.8 | 78.3 |
| 1962 | 23.0 | 19.5 | 15.0 | 22.0 | | | 6.1 | 2.2 | 6.3 | 5.9 | 86.1 |
| 1966 | 21.2 | 27.2 | 13.8 | 21.1 | | 0.5 | 5.7 | 1.0 | 6.5 | 2.9 | 86.1 |
| 1970 | 17.1 | 23.4 | 18.0 | 16.6 | | 1.1 | 5.3 | 10.5 | 6.0 | 2.0 | 83.2 |
| 1972 | 16.4 | 25.8 | 17.6 | 17.0 | | 2.5 | 5.1 | 9.2 | 5.2 | 1.2 | 81.9 |
| 1975 | 17.6 | 24.9 | 18.4 | 18.9 | | 3.3 | 4.7 | 3.6 | 4.3 | 4.3 | 80.1 |
| 1979 | 17.3 | 23.9 | 21.7 | 17.9 | | 4.8 | 4.3 | 4.6 | 3.7 | 1.8 | 81.9 |
| 1983 | 17.6 | 26.7 | 22.1 | 13.5 | | 3.0 | 4.9 | 9.7 | | 2.5 | 81.2 |
| 1987 | 17.6 | 24.1 | 23.1 | 13.6 | 4.0 | 2.6 | 5.6 | 6.3 | 1.0 | 2.1 | 76.2 |
| 1991 | 24.8 | 22.1 | 19.3 | 10.1 | 6.8 | 3.1 | 5.5 | 4.8 | 0.8 | 2.7 | 71.0 |
| 1995 | 19.8 | 28.3 | 17.9 | 11.2 | 6.5 | 3.0 | 5.1 | 1.3 | 0.6 | 6.3 | 70.6 |
| 1999 | 22.4 | 22.9 | 21.0 | 10.9 | 7.3 | 4.2 | 5.1 | 1.0 | 0.2 | 5.0 | 66.8 |
| 2003 | 24.7 | 24.5 | 18.6 | 9.9 | 8.0 | 5.3 | 4.6 | 1.6 | 0.3 | 2.5 | 67.6 |

**Fig. 1.** Proportion of votes cast for different parties and voting turnout in Parliamentary elections in 1954-2003 (%) (Statistics Finland 2004, p. 11 and p. 15)

The Change of Gross Domestic Product per Capita (CGDP) and the Change of Total Consumption per Capita (CCONSUM) are good measures for economic growth. These four monetary values are transformed into constant prices of the year 2000. For each measurement, there are three variables included in the data: the first at elections year (marked with COLI(T), UNEM(T), CGDP(T), and CCONSUM(T)), the second at a year before elections (marked with COLI(T-1), etc.) and the third at two years before elections (marked with COLI(T-2), etc.). The Change of Cost of Living Index, the Change of Gross Domestic Product per Capita and the Change of Total Consumption per Capita are based on the data provided by the Statistics Finland [17]. The Unemployment Rate is based on Keinänen's unemployment and employment statistics [8].

## 2.2   Method and Earlier Work

The Self-Organizing Map (SOM) [9,10] has been used in a wide range of areas such as medicine, economics or in the analysis of industrial processes. In a paper closely related to this one, Kaski and Kohonen studied the socio-economic status of the countries in the world based on World Bank data [7]. Deboeck and Kohonen have edited a book that shows many examples of uses of the SOM in the area of finance [3]. Länsiluoto et al. conducted analysis of economic and competitive environment in the formulation of corporate strategies using the SOM [12]. Tuia et al. have used the SOM combined with Ward's classification to classify the municipalities of Western Switzerland, interpret the socio-economic landscape of the region [19]. Lendasse et al. [14] used the SOM in forecasting electricty consumption. In our work, we did not aim, for instance, to predict the results of next elections but rather provide a basis for interpreting the results and their relation to some socio-economic variables.

In the 2004 municipal and EU elections in Finland the candidates were mapped according to their answers to a number of questions. The voters could then use a web site of a major commercial broadcasting company that had an interactive version of the map. The users could answer the same questions that were presented to the candidates and then see where the like-minded candidates are positioned on the map.[1,2]

The SOM organizes data in such a way that the variables with largest variance affect the result most. In the case of relative approval ratings, no scaling is required as the figures are comparable as such. However, if the data consists of variables the status of which varies, scaling is needed. In the analysis described in Section 3, we have both relative approval ratings and some societal variables such as unemployment rate. In this study, data was normalized by dividing every variable with its variance. More detailed discussion on data normalization is provided e.g. in [10].

The study also shows how the value of different variables is distributed on the map. For interdisciplinary understandability we refer to these distributions as variable maps where the methodological community has often used the term component plane. In this study, we implemented the SOM using Matlab 7.0.1 with the SOM Toolbox 2.0.

## 3   Results

In the following, we present the results of the SOM-based analysis. We show the overall analysis of the relationship between the election years (Section 3.1) and discuss some specific findings (Section 3.2).

### 3.1   Overall Results

Figure 2 presents the parliamentary election years organized by the SOM algorithm. It is possible to see that the parliamentary election years form roughly a kind of chain. Consecutive election years are typically close to each other on the map. This feature most likely reflects the idea that societal changes happen gradually.

On the right side of Figure 2, there is a distance map. The distance map reflects distances between the locations of the map, with a lighter shade of gray denoting a relatively longer distance. On Figure 3, there are the variable maps that shows how the value of different variables is distributed on the map. The elections of the years 1954, 1962 and 1966 form their own tight cluster. This cluster corresponds to the post war years when the combined popularity of the left wing parties and the Agrarian Union was high. The elections of the years 1958 and 1970 are slightly apart from this cluster. In the 1970 election, Finnish Rural Party (PS in the variable maps) climbed up to 10.5 per cent, being only at 1.0 per cent four years earlier. The party had been founded by a former

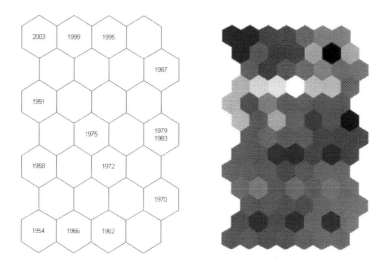

**Fig. 2.** Parliamentary election years organized by the SOM algorithm

member of Agrarian Union. The Agrarian Union, at that time the Centre Party (KESK) lost popularity: it dropped from 21.2 per cent to 17.1 per cent. This change is clearly seen on the map. In 1958 election, the inflation rate a year before the election was high (variable COLI(T-1)) as well as the popularity of the Democratic League of the People of Finland (LEFT).

The elections of the years 1972 and 1975 form a cluster of their own. The results of the 1972 elections were quite similar to those of the 1970 elections, except that the National Coalition Party (KOK) fell from the second place to the fourth in popularity. The variable maps show clearly that in that period the inflation rate was exceptionally high.

The elections of the years 1979 and 1983 form their own tight cluster. After a long period of being in the government, the Centre Party had lost much of its earlier popularity whereas the popularity of the National Coalition Party was increasing. The main issues in the 1979 parliamentary election were unemployment and taxation. In 1979, the unemployment rate was not as high as it is nowadays (see UNEM variable maps) but it had clearly grown from the earlier years.

The elections of the years 1987 and 1991 are distant from the main clusters. This period of time was particularly turbulent in Finnish 20th century history from the economical point of view. In October 1991, Finland and other EFTA member countries agreed to form a European Economic Area (EEA) with the EU from 1993 leading into Finland's EU membership in 1995. In the 1987 elections, the National Coalition Party (KOK) and conservative Prime Minister took office in 1987, heading a coalition government that included the Social Democrats. This left the Center party as the opposition for the first time since independence. The economic collapse of the USSR in 1991 caused a severe recession in Finland due to severely decreasing exports to Russia. Another factor causing the recession seems to be the liberalization of foreign loan policies. Korhonen describes this change as follows: "Perhaps the most significant change came with the granting of permission in 1986-1987 to raise long-term loans from abroad. The largest fundamental change occurred at the start of 1991, when the old comprehensive restriction was finally repealed; from then on, all foreign exchange dealings not specifically subject to approval by the Bank of Finland were unrestricted. Now foreign exchange restrictions remained only on the raising of loans abroad by private individuals and comparable corporate entities, and these were in turn lifted in October 1991 in accord with the spirit of the EEA Agreement."[11] This change lead into excessive foreign loan taking that appeared to contribute to the strong overall structural change including, among others, a strong increase of unemployment. However, when we consider the SOM analysis, these kind of conclusions cannot be drawn from the analysis results and diagrams alone. On the other hand, we have taken some time dynamics into account in the analysis by including "delayed variables": for instance, in addition to considering the unemployment rate at the year of election, we have also included the rate one and two years earlier.

The most recent elections of the years 1995, 1999 and 2003 form the cluster that is clearly apart from other elections. The most distinctive aspects include high level of unemployment, low popularity of the Left Alliance (LEFT), high popularity of the Greens (GREENS), and low level of turnout (TURNOUT). The low of turnout has raised questions about the passivity of the voters. However, at least two possible conclusions could be made. It could be, like often mentioned, that the politics has become more distant to the citizens due various reasons, one of which could be the EU membership and the relatively lower importance of national legislation. On the other hand, the economical recovery from the recession period of early 1990s may have increased the general feeling of satisfaction. This is not supported, though, by the variable maps that indicate the high degree of negative correlation between unemployment rate and turnout.

## 3.2   Specific Findings

It is commonly believed that being in the government will cause a popularity reduction in the next election. Figure 4 shows that this is true for the four largest parties: Centre Party (KESK), Social Democratic Party (SDP), National Coalition Party (KOK) and Left Alliance (LEFT). This observation is not however valid for the other parties.

There is a strong negative correlation between the Centre Party (KESK) and inflation (COLI(T), COLI(T-1) and COLI(T-2)). During high unemployment the popularity of the Centre Party has been decreasing and during low unemployment it has been increasing. The Centre Party's position as the largest party

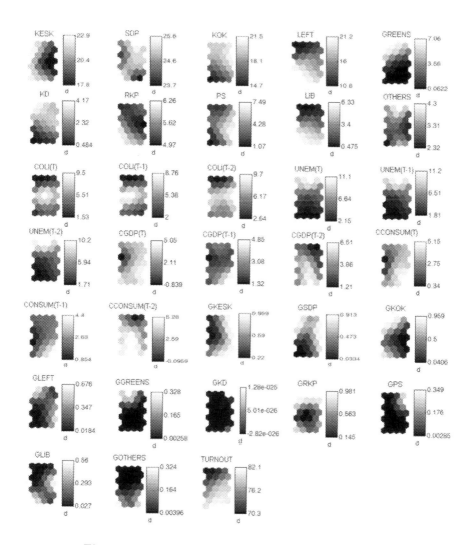

**Fig. 3.** The variable maps of all variables used in this study

that has been many times in the government could cause these findings. Voters have punished it because of unfavorable economic situations or developments. This interpretation is in harmony with a study made by Lewis-Beck with French data[15]. The study shows that increasing unemployment and inflation result in popularity reduction for the French president and prime minister.

The popularity of the National Coalition Party (KOK) has the same feature as the popularity of the Centre Party. During high unemployment it has been decreasing and during low unemployment it has been increasing. During the existence of the Green League (GREENS) the approval ratings of the Social Democratic Party and the Greens have had negative correlation. The popularity of the Left Alliance (LEFT) has been decreasing within the whole period of the study.

A change that took place in the late 1970s is clearly discernable. Many dependences between variables changed their features. Correlations turned from negative to positive and vice versa. For example, turnout has a positive correlation with the Change of Gross Domestic Product per Capita (CGDP(T), CGDP(T-1) and CGDP(T-2)) in the 1950s and 1960s. In the 1990s and 2000s, there is, on the contrary, a negative correlation. Earlier economic growth has potentially provided possibilities to be politically active and later it has made people negligent.

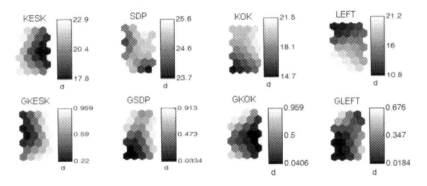

**Fig. 4.** Being in the government causes popularity reductions for the four largest parties in the next elections

## 4     Conclusions and Discussion

We have explored the relationship between parliamentary election results and political and societal situation in Finland. The data consisted of the parliamentary election results in Finland, the parties in the government by the time of and before each election, and a number of socioeconomic variables. We have used the self-organizing map algorithm as the data mining and visualization method. Using the method, we have been able to show how the parliamentary election results seem to reflect both the political and societal conditions with a large number of specific findings discussed above. We suggest that this approach can be used as a method that can serve as a bridge between qualitative and quantitative methods (see also [5]). The specific findings can serve as hypotheses that can be further studied with other statistical methods. In summary, the main ideas for the use of the data mining methodology includes the possibility of obtaining an overall picture, visual detection of correlations, and formation of hypothesis for further analysis. Future research possibilities include adding other potentially relevant variables and a more detailed analysis of time dependent phenomena.

# References

1. Berg, M., Kaipainen, M., Kojo, I.: Enhancing Usability of the Similarity Map for More Accessible Politics. In: Proceedings in 8th ERCIM Workshop, User Interfaces For All, Vienna, Austria (2004)
2. Berg, M., Marttila, T., Kaipainen, M., Kojo, I.: Exploring Political Agendas with Advanced Visualizations and Interface Tools e-Service Journal 4(2), 47–63 (2006)
3. Deboeck, G., Kohonen, T. (eds.): Visual Explorations in Finance with Self-Organizing Maps. Springer, Berlin (1998)
4. The Finnish Government n.d., MIKO-ministeritietojärjestelmä (The Finnish Minister Database),
   http://www.valtioneuvosto.fi/hakemisto/ministerikortisto/raportti.asp
   (retrieved November 13, 2005)
5. Janasik, N., Honkela, T., Bruun, H.: Text Mining in Qualitative Research: Application of an Unsupervised Learning Method. Organizational Research Methods (2008), http://orm.sagepub.com/cgi/content/abstract/1094428108317202v1
6. Johansson, K.M., Raunio, T.: Partisan responses to Europe: comparing Finnish and Swedish political parties. Eur. J. of Political Research 39(2), 25–49 (2001)
7. Kaski, S., Kohonen, T.: Exploratory data analysis by the self-organizing map: Structures of welfare and poverty in the world. In: Neural Networks in Financial Engineering, pp. 498–507. World Scientific, Singapore (1996)
8. Keinänen, P.: Työttömyys (Unemployment). In: Andreasson, K., Helin, V. (eds.) Suomen vuosisata, pp. 74–79. Statistics Finland, Helsinki (1999)
9. Kohonen, T.: Self-organized formation of topologically correct feature maps. Biological Cybernetics 43, 59–69 (1982)
10. Kohonen, T.: Self-Organizing Maps, 3rd edn. Springer, Berlin (2001)
11. Korhonen, T.: Finnish monetary and foreign exchange policy and the changeover to the euro. Bank of Finland discussion papers (2001)
12. Länsiluoto, A., Eklund, T., Back, B., Vanharanta, H., Visa, A.: Industry Specific Cycles and Companies' Financial Performance - Comparison with Self-Organizing Maps. Benchmarking. An International Journal 11(4), 267–286 (2004)
13. Lee, J.A., Lendasse, A., Verleysen, M.: Nonlinear projection with curvilinear distances: Isomap versus curvilinear distance analysis. Neurocomputing 57, 49–76 (2004)
14. Lendasse, A., Lee, J.A., Wertz, V., Verleysen, M.: Forecasting electricity consumption using nonlinear projection and self-organizing maps. Neurocomputing 48(1-4), 299–311 (2002)
15. Lewis-Beck, M.S.: Economic Conditions and Executive Popularity: The French Experience. American Journal of Political Science 24, 306–333 (1980)
16. Martikainen, P., Martikainen, T., Wass, H.: The effect of socioeconomic factors on voter turnout in Finland: A register-based study of 2.9 million voters. European Journal of Political Research 44(5), 645–669 (2005)

17. Statistics Finland 2004, Parliamentary elections 2003, elections 2004:1, Edita Prima Oy, Helsinki, p. 8, p. 11 and p. 15 (2004)
18. Raunio, T., Tiilikainen, T.: Finland in the European Union. Frank Cass, London (2003)
19. Tuia, D., Kaiser, C., Cunha, A., Kanevski, M.: Socio-economic Data Analysis with Scan Statistics and Self-organizing Maps. In: Proceedings of the international conference on Computational Science and its Applications, pp. 52–64. Springer, Berlin (2008)
20. Venna, J., Kaski, S.: Local multidimensional scaling. Neural Networks 19(6), 889–899 (2006)

# Line Image Classification by NG×SOM: Application to Handwritten Character Recognition

Makoto Otani, Kouichi Gunya, and Tetsuo Furukawa

Kyushu Institute of Technology, Kitakyushu 808-0196, Japan
otani-makoto@edu.brain.kyutech.ac.jp
http://www.brain.kyutech.ac.jp/~furukawa

**Abstract.** A method for generating a self-organizing map of line images is proposed. In the proposed method, called the NG×SOM, a set of data distributions is represented by a product space organized by a set of neural gas networks (NGs) and a self-organizing map (SOM). In this paper, it is assumed that the line images dealt with by the NG×SOM have the same, yet unknown, topology. Thus the task of the NG×SOM is to generate a map of line images with the same topology, in which the images are continuously and naturally morphed from one into another. We applied the NG×SOM to a handwritten character recognition task. The results obtained show that this method is effective, particularly when the number of training data is small.

## 1 Introduction

Shape is one of the most important basic visual clues for recognizing objects. It is, however, not easy to deal with shapes directly, owing to the difficulty in representing shapes numerically without losing information. The most usual approach for classifying a set of shapes using a conventional SOM is to transform each object shape into a numerical vector that describes the shape features. This approach is called *shape description*. A large number of features for shape description have been proposed, such as area-to-square-perimeter ratio, bending energy, moments and so on [1]. Using shape description, the SOM is expected to represent the continuous change in shape features in the map space, for example, from a round shape to a jagged shape. In other words, an intermediate point between two shapes in the map space is expected to represent the intermediate shape features. It is worth emphasizing that the phrase 'intermediate shape features' does not mean 'intermediate shape', because these shape features only describe the properties relevant to shape, and do not represent the entire shape information. Thus it can happen that different shapes with similar features are mapped to the same point in the SOM. Therefore, to obtain a map of shapes representing a continuous morphing from one shape into another, an appropriate *shape representation*, that preserves shape information, needs to be utilized.

One of the most popular approaches in the shape representation field is to represent the contour or skeleton by a manifold [1,2]. To achieve this, a SOM or another similar algorithm is employed to represent the manifold [3,4,5,6,7]. In this approach, each dot in the contour or skeleton is regarded as a data point in the $x$–$y$ space covered by the SOM. Thus the shape information is represented by a joint vector (more precisely, a

J.C. Príncipe and R. Miikkulainen (Eds.): WSOM 2009, LNCS 5629, pp. 219–227, 2009.
© Springer-Verlag Berlin Heidelberg 2009

tensor of rank 2) of the reference vectors of the SOM. If we have $n$-object shapes, then we can obtain $n$-joint reference vectors organized by $n$-SOMs. It is, therefore, expected that a map of the shapes can be obtained, if a meta-SOM is able to deal with a set of SOMs by regarding the joint vectors as a dataset.

Furukawa proposed an extension of the SOM called the $SOM^2$ or 'SOM of SOMs', that has the ability of dealing with a set of SOMs as a dataset [8,9]. In a $SOM^2$, a set of SOMs at the lower level (child SOMs or 1st-SOMs) organize maps of a set of datasets, while a SOM at the upper level (parent SOM or 2nd-SOM) generates a map of the 1st-SOMs. Since these processes proceed in parallel and have an affect on one another, the 1st-SOM maps are gradually homologized, and the 2nd-SOM represents the continuous change in these 1st-maps. It has also been shown that a $SOM^2$ can organize a map of contours using the 1st-SOMs with a circular topology.

In the case of contour representation, it is known that contours have a one-dimensional closed topology, whereas such prior knowledge is not available in the case of skeleton representation. The most typical example of this is line image classification, such as handwritten character recognition, where each character has its own inherent topology. Therefore, the 1st-SOMs of the $SOM^2$ need to be replaced by another type of vector quantization technique, that does not have the topological restriction.

The purpose of this paper is to propose an algorithm for line image classification based on the $SOM^2$. In the proposed method, a set of line images are modeled by the same number of neural gas (NG) networks instead of the 1st-SOMs. The 2nd-SOM then classifies these NGs. The resulting architecture is called the NG×SOM. Although the NG×SOM algorithm was first proposed by Furukawa [8,9], the algorithm introduced here is an improved version that considers topological preservation. The improved NG×SOM has been applied to handwritten character recognition, the results of which are presented in this paper.

## 2    Theory and Algorithm

### 2.1    Theoretical Framework

To clarify the aim of this work, let us consider some typical results produced by an NG×SOM. In Fig. 1, two maps of handwritten digits are presented. Each NG×SOM represents the same digit in various handwritten shapes, that continuously morph from one into another. A thick box indicates the best matching unit (BMU) of the training data, while the other images are all interpolated by the NG×SOM. The dots constituting the digits are reference vectors of the NGs. For the purpose of this paper, the task of an NG×SOM is defined as follows.

- Organize a self-organizing map of line images, in which the images are continuously morphed.
- The line images input into each NG×SOM are assumed to have the same (yet unknown) topology. This means that the NG×SOM is expected to represent the continuous shape change in the same character, e.g., a handwritten 'A' into another handwritten 'A'. Morphing from one character into another character (e.g., from 'A' into 'B') is not considered. This postulate is required to give a theoretical definition of the distance measure between two images.

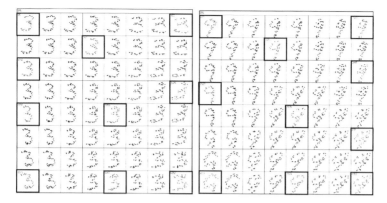

**Fig. 1.** Maps of handwritten digits '3', and '9' organized by two NG×SOMs. Thick boxes indicate the best matching units of the training data, while the other images are all interpolated.

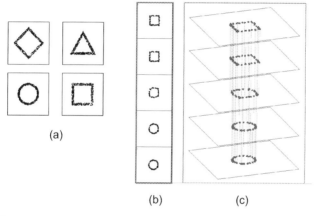

**Fig. 2.** One-dimensional NG×SOM map representing shapes morphing from a square into a circle. (a) Given line images. (b) Organized map. (c) Reference vectors of the NGs with the same index form a 'fiber'. Fibers are indicated by dotted lines.

To achieve this, the NG×SOM is required to solve the following tasks:

- Ascertaining key points in the given line images.
- Ascertaining the correspondence of key points in the given images.
- Ordering the given images so that the key points move continuously.
- Interpolating between given line images by tracing movements of the key points.

In an NG×SOM, such key points are represented by the reference vectors of the NGs. Reference vectors with the same index are connected to one another by a string, known as a *'fiber'*, which represents the continuous movement of the key point. Thus the morphing from one image into another is represented by a bundle of fibers as illustrated in Fig. 2.

## 2.2   Architecture of the NG×SOM

Fig. 3 shows the architecture of the NG×SOM, which consists of a set of NGs called '1st-NGs' and a meta-SOM called the '2nd-SOM'. The task of the 1st-NGs is to represent the shapes of the given line images, while the task of the 2nd-SOM is to organize a map of shapes represented by the 1st-NGs. The 1st-NGs and 2nd-SOM are updated reciprocally due to their ability of affecting one another.

Now suppose that we have $I$ line images, consisting of black dots on a white background. The $i$-th image is regarded as a dataset $X^i = \{\mathbf{x}^{ij}\} = \{(x^{ij}, y^{ij})\}$. Here $(x^{ij}, y^{ij})$ denotes the position of the $j$-th black dot in the $i$-th image. More generally, it is also possible to add some local feature information around the dots, such as line orientation, color, and so on. By letting $\mathbf{z}(x, y)$ be the local feature at $(x, y)$, each dot is represented by a $D$-dimensional vector $\mathbf{x}^{ij} = (x^{ij}, y^{ij}, \mathbf{z}(x^{ij}, y^{ij})) \in \mathbb{R}^D$.

The task of the 1st-NGs is to model each data distribution using $I$ 1st-NGs, all with the same structure, i.e., the same number of reference vector units. Let the 1st-NGs have $L$ reference vectors, which are denoted by $\mathbf{v}^{il} \in \mathbb{R}^D$ ($l = 1, \ldots, L$). A tensor $\mathbf{V}^i = (\mathbf{v}^{i1}, \ldots, \mathbf{v}^{iL}) \in \mathbb{R}^{L \times D}$ is referred to as the '*NG tensor*', which represents the $i$-th 1st-NG.

Suppose further that the 2nd-SOM has $K$ reference vector units $\{\mathbf{W}^1, \ldots, \mathbf{W}^K\}$, where $\mathbf{W}^k \in \mathbb{R}^{L \times D}$. Thus $\mathbf{W}^k$ is also regarded as a tensor of rank 2, consisting of $L$ reference vectors $\mathbf{W}^k = (\mathbf{w}^{k1}, \ldots, \mathbf{w}^{kL})$. Here $\mathbf{W}^k$ is referred to as the '*reference NG tensor*'. The 2nd-SOM organizes a map of the 1st-NGs by regarding $\{\mathbf{V}^i\}$ as a dataset. A *fiber* is defined as a string connecting reference vectors of the 2nd-SOM that have the same index. Thus the $l$-th fiber is represented by $\mathbf{F}^l = (\mathbf{w}^{1l}, \ldots, \mathbf{w}^{Kl}) \in \mathbb{R}^{K \times D}$. Here $\mathbf{F}^l$ is referred to as the '*fiber tensor*'. The 2nd-SOM forms a nonlinear product space represented by the reference NG tensors and the fiber tensors, in other words, a stack of sections $\{\mathbf{W}^k\}$ and a bundle of fibers $\{\mathbf{F}^l\}$. It is worth noting that every reference NG tensor acts as an ordinary reference vector unit in the 2nd-SOM, and the NG×SOM *does not* consist of NG modules.

The algorithm for the NG×SOM is summarized below [8,9]. The $i$-th 1st-NG learns the dot distribution of the $i$-th line image, and updates the NG tensor $\mathbf{V}^i$. The 2nd-SOM learns a set of NG tensors by regarding them as ordinary data vectors. After the

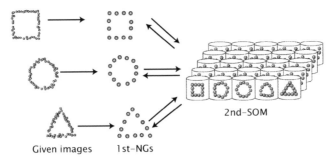

Given images     1st-NGs     2nd-SOM

**Fig. 3.** Architecture of the NG×SOM. The given images are modeled by the same number of 1st-NGs. The 2nd-SOM organizes a map of the 1st-NGs by regarding them as data vectors. The BMUs in the 2nd-SOM are fed back to the corresponding 1st-NGs as the initial state for the next iteration.

reference NG tensors $\{\mathbf{W}^k\}$ in the 2nd-SOM are updated, the best matching reference NG tensor is copied to the corresponding NG tensor, which becomes the initial state of the next iteration of the 1st-NG learning. Thus the 1st-NGs and the 2nd-SOM are updated in parallel due to their affecting one another. By iterating these processes, the key points represented by the 1st-NGs are gradually homologized, and a continuous map of the NGs is organized in the 2nd-SOM.

### 2.3   Improved NG×SOM Algorithm

In the original NG×SOM proposal [8,9], the normal NG algorithm [10,11] is adopted without any modifications. Thus each 1st-NG learns the data distribution independently, and the 2nd-SOM then orders these. In the NG×SOM proposed in this paper, an essential improvement is made to consider topology preservation.

In the improved algorithm, both the 1st-NGs and the 2nd-SOM are connected by fibers. Thus the $l$-th fiber represented by the 1st-NGs is defined as the fiber tensor $\mathbf{G}^l = (\mathbf{v}^{1l}, \ldots, \mathbf{v}^{Il}) \in \mathbb{R}^{I \times D}$. This means that the set of 1st-NGs can be regarded as a unified 1st-NG, the reference units of which are the fiber tensors $\{\mathbf{G}^l\}$. Hereafter the $\{\mathbf{G}^l\}$ are referred to as the '*reference fiber tensor*' of the unified 1st-NG.

By introducing the concept of fibers, the algorithm for the 1st-NGs can be modified at several points. The first modification involves introducing the concept of the neighborhood function to the 1st-NGs as well. In the original NG×SOM algorithm, the learning weight of each reference vector is determined independently for 1st-NGs. This means that even if $\mathbf{v}^{il_1}$ and $\mathbf{v}^{il_2}$ are neighbors in the $i$-th NG (i.e., $\mathbf{v}^{il_2}$ is given a large learning weight when $\mathbf{v}^{il_1}$ becomes the winner of the data), this is not necessarily the same in other NGs. In the improved algorithm, the learning weights are determined by the neighborhood function as in the SOM. The neighborhood relations are the same for all 1st-NGs, because the distance is defined between two fibers. Now let $d_f(l_1, l_2)$ denote the distance *from* the $l_1$-th fiber *to* the $l_2$-th fiber. In this paper, $d_f(l_1, l_2)$ is determined as follows.

$$d_f(l_1, l_2) \triangleq \frac{1}{K} \sum_{k=1}^{K} \mathrm{rank}(\mathbf{w}^{kl_1}, \mathbf{w}^{kl_2}) \tag{1}$$

Here $\mathrm{rank}(\mathbf{w}^{kl_1}, \mathbf{w}^{kl_2})$ gives the rank of $\mathbf{w}^{kl_2}$ from $\mathbf{w}^{kl_1}$, i.e., $\mathrm{rank}(\mathbf{w}^{kl_1}, \mathbf{w}^{kl_2}) = n$ if $\mathbf{w}^{kl_2}$ is the $n$-th nearest neighbor of $\mathbf{w}^{kl_1}$. Note that $d_f(l_1, l_2) = 0$ if, and only if, $l_1 = l_2$. (Strictly speaking, $d_f(\cdot, \cdot)$ is a quasi-distance measure, since $d_f(l_1, l_2) \neq d_f(l_2, l_1)$.) The learning weight is then determined by using the neighborhood function and this distance table. Unlike in the SOM, the distance table $d_f(l_1, l_2)$ is also updated at every iteration of the NG×SOM algorithm; if the distance table is fixed through learning, the algorithm resembles that for the SOM[2] instead of for the NG×SOM.

The second modification controls the winning rate. By using the NG algorithm, every reference vector wins more or less an equal number of data points after learning. (Strictly speaking, the magnification factor should also be considered.) This means that the winning rates of the 1st-NGs are controlled individually in the original algorithm, so that each reference vector unit becomes a BMU more or less equally. In the improved

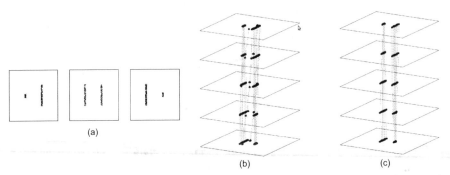

**Fig. 4.** Winning rate control. (a) Three line images given to the NG×SOM. (b) Nonexistent strokes appearing in the NG×SOM using the original algorithm. (c) Continuous change in strokes is well represented using the improved algorithm.

algorithm, the winning rates of the fibers ($\mathbf{G}^l$), not the reference vectors ($\mathbf{v}^{il}$), are considered. Thus the winning rates of the reference vectors will not necessarily be equal, as long as the winning rates are almost equal for all fibers. The reason for this improvement is that the key points (i.e., the reference vectors) are expected to be assigned to the same stroke (line) in the images. For example, the reference vectors assigned to the horizontal stroke of the letter 'A' are expected to represent the same stroke for all images of 'A', regardless of whether the stroke is long or short.

Fig. 4 shows the difference between the original and the improved NG×SOM algorithms. Using the original algorithm, the assigned stroke of each reference vector is changed depending on the stroke length. As a result, nonexistent strokes appeared between two existing strokes in the 2nd-SOM (Fig. 4 (b)). This phenomenon did not occur when using the improved algorithm, and the assigned strokes were consistent for all 1st-NGs.

## 2.4   Algorithm for the Improved NG×SOM

Taking the above points into consideration, the improved algorithm is described below.

*Step 1: Learning process of the 1st-NGs*
In step 1, each 1st-NG is updated by the following equations.

$$l_{ij}^*(t) = \arg\min_{l} \|\mathbf{x}^{ij} - \tilde{\mathbf{v}}^{il}(t)\| \tag{2}$$

$$\beta_{ij}^l(t) = \frac{\bar{p}^{l_{ij}^*}}{p_i^{l_{ij}^*}} \exp\left[-\frac{d_f^2(l, l_{ij}^*(t))}{2\sigma_1^2(t)}\right] \tag{3}$$

$$\mathbf{v}^{il}(t) = (1 - \varepsilon)\tilde{\mathbf{v}}^{il}(t) + \varepsilon \frac{\sum_{j=1}^J \beta_{ij}^l(t)\mathbf{x}^{ij}}{\sum_{j'=1}^J \beta_{ij'}^l(t)} \tag{4}$$

Here $\tilde{\mathbf{v}}^{il}(t)$ is the initial state of $\mathbf{v}^{il}$ at calculation time $t$, which is obtained from step 4 in the preceding iteration. $\sigma_1(t)$ is the neighborhood size of the 1st-NGs, and $\bar{p}^l$ and $p_i^l$

are the winning rates of the $l$-th reference fiber and the $l$-th reference vector in the $i$-th 1st-NG, respectively. Thus,

$$p_i^l \triangleq \frac{\sum_{j=1}^{J} \delta(l_{ij}^*, l)}{J} \tag{5}$$

$$\bar{p}^l \triangleq \frac{\sum_{i=1}^{I} \sum_{j=1}^{J} \delta(l_{ij}^*, l)}{IJ}. \tag{6}$$

Here $\delta(\cdot, \cdot)$ is Kronecker's delta function.

*Step 2: Learning process of the 2nd-SOM*

In step 2, the 2nd-SOM is updated by the conventional batch SOM algorithm, whilst regarding $\{\mathbf{V}^i(t)\}$ as a dataset.

$$k_i^*(t) = \arg\min_k \|\mathbf{V}^i(t) - \mathbf{W}^k(t)\| \tag{7}$$

$$\alpha_i^k = \exp\left[-\frac{d_s^2(k, k_i^*(t))}{2\sigma_2^2(t)}\right] \tag{8}$$

$$\mathbf{W}^k(t) = \frac{\sum_{i=1}^{I} \alpha_i^k(t)\mathbf{V}^i(t)}{\sum_{i'=1}^{I} \alpha_{i'}^k(t)} \tag{9}$$

Here $d_s(k_1, k_2)$ gives the distance between two reference NG tensors in the map space.

*Step 3: Updating the distance table*

In step 3, the distance table $d_f(l_1, l_2)$ is updated according to (1).

*Step 4: Feedback from the 2nd-SOM to the 1st-NGs*

Finally, the best matching reference NG tensors are copied to the corresponding 1st-NGs. Thus,

$$\tilde{\mathbf{V}}(t + 1) = \mathbf{W}^{k_i^*}(t). \tag{10}$$

The above four processes are iterated whilst reducing the neighborhood size.

## 3   Application to Handwritten Character Recognition

### 3.1   Method

To investigate the ability of the NG×SOM, we performed a recognition experiment on handwritten digits using the NIST handwritten character database (NIST special database 19) [12]. Before applying the NG×SOM, all images were processed by a median filter to remove noise. A set of angle filters was then used to describe the local features (Fig. 5(a)). The output from these filters is the ratio of the "black pixels in the fan shape area" to "all the pixels in the fan-shape area". Twelve filters, tuned every 30°, were used to transform every dot into a 14-dimensional data vector.

To recognize digits, we used 10 NG×SOMs corresponding to '0',…,'9'. After generating 10 maps of handwritten digits from the training dataset, the recognition rate was measured for 10,000 test data which were not used during the training phase. The recognition rates were measured whilst changing the number of training data. To compare the effectiveness, two different methods were employed. One is the simple nearest neighbor method using all training data as reference patterns. The other method uses a set of conventional SOMs, each of which represents pixel images of the corresponding digit. In both methods, each pixel image was transformed into a data vector with (64 pixel)×(64 pixel)×(12 angle filters)= 49, 152 dimensions. The number of reference patterns was 64 patterns/digit for the NG×SOMs and SOMs, which is equal to the number of training patterns in the case of the simple nearest neighbor method.

Fig. 5 (b) shows the recognition rate of the NG×SOM, the conventional SOM and the simple nearest neighbor method. Some of the representative maps generated by the NG×SOMs are shown in Fig. 1 and Fig. 6 (b), while a map organized by the conventional SOM is shown in Fig. 6 (a). In these figures, the number of training data was 100, i.e., 10 patterns/digit. In spite of the small number of training data, the average recognition rate of the NG×SOM was sufficiently high (94.69%). When the number of

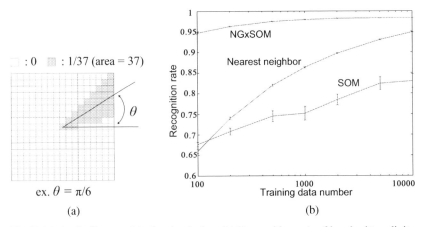

**Fig. 5.** (a) Angle filter used in the simulation. (b) Recognition rate of handwritten digits.

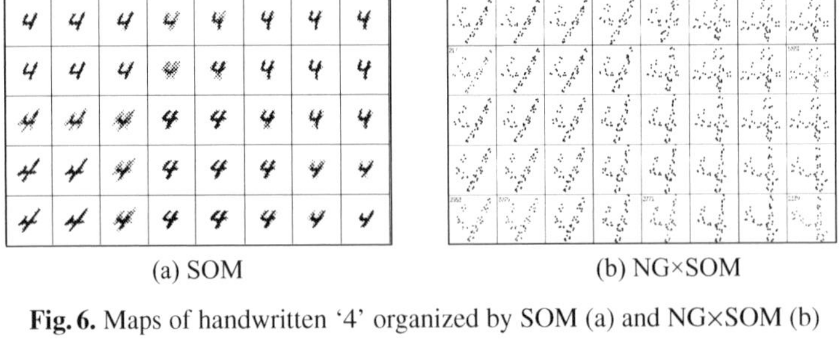

(a) SOM                    (b) NG×SOM

**Fig. 6.** Maps of handwritten '4' organized by SOM (a) and NG×SOM (b)

training data was increased, the recognition rate of NG×SOM also improved to 98.32%. The reason for the high recognition rate is that the NG×SOM interpolates 'intermediate digit shapes' from the training images.

## 4    Conclusion

In this paper, we proposed a novel shape classification method using an NG×SOM. The simulation results showed this method's high ability for shape representation and classification. This ability was also confirmed by the high recognition rate of handwritten characters.

In this paper, we assumed that the topology of shapes was homogeneous. It would be interesting to investigate how to deal with shapes with heterogeneous topologies and also natural images with similar topological structures, such as face images. These issues will be explored in future studies.

## References

1. Loncaric, S.: A survey of shape analysis techniques. Pattern Recognition 31, 983–1001 (1998)
2. Zhang, D., Lu, G.: Review of shape representation and description techniques. Pattern Recognition 37, 1–19 (2004)
3. Parui, S.K., Datta, A., Pal, T.: Shape approximation of arc patterns using dynamic neural networks. Signal Processing 42, 221–225 (1995)
4. Datta, A., Pal, T., Parui, S.K.: A modified self-organizing neural net for shape extraction. Neurocomputing 14, 3–14 (1997)
5. Datta, A., Parui, S.K.: Skeletons from dot patterns: A neural network approach. Pattern Recognition Letters 18, 335–342 (1997)
6. Kumar, G.S., Kalra, P.K., Dhande, S.G.: Curve and surface reconstruction from points: an approach based on self-organizing maps. Applied Soft Computing 5, 55–66 (2004)
7. Huang, D., Yi, Z.: Shape recovery by a generalized topology preserving SOM. Neurocomputing 72, 573–580 (2008)
8. Furukawa, T.: SOM of SOMs: Self-organizing map which maps a group of self-organizing maps. In: Duch, W., Kacprzyk, J., Oja, E., Zadrożny, S. (eds.) ICANN 2005. LNCS, vol. 3696, pp. 391–396. Springer, Heidelberg (2005)
9. Furukawa, T.: SOM of SOMs. Neural Networks (in press)
10. Martinetz, T.M., Berkovich, S.G., Schulten, K.J.: "Neural-gas" network for vector quantization and its application to time-series prediction. IEEE Transaction on Neural Networks 4, 558–567 (1993)
11. Cotrell, M., Hammer, B., Hasenfuß, A., Villmann, T.: Batch and median neural gas. Neural Networks 19, 762–771 (2006)
12. Grother, P.J.: NIST special database 19, Handprinted forms and characters database, National Institute of Standards and Technology (1995)

# Self-Organization of Tactile Receptive Fields: Exploring Their Textural Origin and Their Representational Properties

Choonseog Park, Heeyoul Choi, and Yoonsuck Choe

Department of Computer Science and Engineering
Texas A&M University
College Station, TX 77843-3112
{cspark13,hchoi}@cs.tamu.edu, choe@tamu.edu

**Abstract.** In our earlier work, we found that feature space induced by tactile receptive fields (TRFs) are better than that by visual receptive fields (VRFs) in texture boundary detection tasks. This suggests that TRFs could be intimately associated with texture-like input. In this paper, we investigate how TRFs can develop in a cortical learning context. Our main hypothesis is that TRFs can be self-organized using the same cortical development mechanism found in the visual cortex, simply by exposing it to texture-like inputs (as opposed to natural-scene-like inputs). To test our hypothesis, we used the LISSOM model of visual cortical development. Our main results show that texture-like inputs lead to the self-organization of TRFs while natural-scene-like inputs lead to VRFs. These results suggest that TRFs can better represent texture than VRFs. We further analyzed the effectiveness of TRFs in representing texture, using kernel Fisher discriminant (KFD) and the results, along with texture classification performance, confirm that this is indeed the case. We expect these results to help us better understand the nature of texture, as a fundamentally tactile property.

## 1 Introduction

Humans process sensory information from different specialized modalities (e.g., vision, touch, and hearing), yet relatively little is known about how specific input stimuli affect the cortical organization. Textural patterns have been studied as important cues that help form the sensory cortex [1]. In our earlier work, tactile representation was found to be better than vision-based ones in texture tasks [2]. Given computational models based on visual receptive fields (VRFs) [3] and tactile receptive fields (TRFs) (Fig. 1) [4], those based on TRFs showed a significantly superior texture boundary detection performance compared to those based on VRFs (t-test: $n = 100, p < 0.03$) [2]. This suggests that TRFs are intimately related with texture-like input, and that texture is fundamentally tactile.

In this paper, we investigate how TRFs can self-organize and if texture-like input play a key role. Our main hypothesis is that TRFs can be self-organized

J.C. Príncipe and R. Miikkulainen (Eds.): WSOM 2009, LNCS 5629, pp. 228–236, 2009.

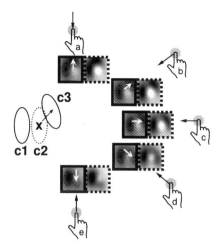

**Fig. 1.** Tactile Receptive Fields (TRFs). TRFs found in the somatosensory area 3b are similar to visual receptive fields (VRFs) (marked C1 and C2, representing inhibitory and excitatory blobs) but there is an added dynamic inhibitory component (marked C3). C3's position relative to the fixed components change, centered at "X", depending on the direction of scan of the tactile surface, e.g., the finger tip (right). The arrow on the finger tip shows the scan direction; the solid outline box shows how the dynamic inhibitory component is shifted (white arrow) in the opposite direction of the scan; and the dotted outline box shows the resulting TRF shape. Adapted from [5] (also see [4]).

using the a visual cortical development model by simply exposing it to texture-like inputs. In order to test our hypothesis, we used the LISSOM (Laterally Interconnected Synergetically Self-Organizing Map) model which was originally developed to model the self-organization of the visual cortex [6]. However, the LISSOM model is actually a more general model of how the cortex (in general) organizes to represent correlations in the sensory input, regardless of the input modality. Thus, LISSOM should work equally well in modeling the development of non-visual sensory modalities (e.g., see [7]).

Our main results show that texture-like inputs lead to the self-organization of TRFs while natural-scene-like inputs lead to VRFs. This result proposes that TRFs could have become accommodated to (surface) textures with a regular repetition of pattern, while VRFs adjusted to handle natural scenes containing various objects and backgrounds that do not repeat over space. We further analyzed the effectiveness of the TRFs and VRFs in representing texture, using kernel Fisher discriminant analysis (KFD) [8]. The results confirmed that TRFs are better suited for textures than VRFs.

The rest of this paper is organized as follows. Section 2 describes the process and results of self-organization using LISSOM. In section 3, a manifold analysis (KFD) for the TRF and VRF feature space is given. Section 4 discusses issues arising from our work, followed by the conclusion in section 5.

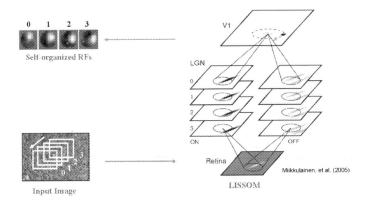

**Fig. 2.** Self-organization process with LISSOM. Given a large image, motion of the gaze window results in a sequence of inputs being generated on the LISSOM retina, which in turn activates the lateral geniculate nucleus (LGN) ON/OFF sheets, one by one, depending on the sheet's built-in delay. After projecting the activities from the LGN ON/OFF sheets, V1 (the primary visual cortex) self-organizes its RFs and lateral connections (excitatory and inhibitory). LISSOM figure adapted from [6].

**Fig. 3.** Sample Input Patterns. The top row shows natural scenes and the bottom row textures used in our experiments. Note that the texture set has texture elements at varying scales. Adapted from [5].

## 2    Self-Organization of the Tactile Receptive Fields

In order to investigate the developmental origin of TRFs, we used the Topographica neural map simulator package (http://topographica.org) [9,6]. Topographica implements a superset of the LISSOM model.

Fig. 2 shows the experimental process we followed to develop self-organized RFs. We generated input stimulus that are natural-scene-like or texture-like, while sampling across the input image with the retina. Fig. 3 shows the inputs we used: natural-scenes and textures.

Given an image, we randomly picked an initial location and moved the gaze window in a random direction along a straight line at a fixed interval. Moving input on an image following a scanning direction are presented on the retina in discrete time steps, like frames of a movie. At each time step t, all LGN cells calculate their activities with lag $t$ one after another as a scalar product of a fixed weight vector (standard on-center/off-surround and vice versa) and input response on the retinal sheet. Each V1 neuron computes its initial response like that of an LGN cell. After the initial response, the V1 activity settles through short-range excitatory and long-range inhibitory lateral interaction. Note that for the texture input, the above process simulates the somatosensory pathway, starting with the texture image standing in for raw mechanoreceptor array activations. After the activity has settled, the connection weights of each V1 neuron are modified according to the normalized Hebbian learning rule. The weakest connections are eliminated periodically, resulting in self-organized patterns similar to those observed in the cerebral cortex. See [6] for details.

For the simulation reported in this paper, four 24×24 LGN-ON cell sheets and four 24×24 LGN-OFF cell sheets received input from a 48×48 retinal sheet, and a 48×48 V1 sheet was used to self-organize the RFs. The learning parameters were the same as in the basic LISSOM model in Topographica [6] with small modification of several scaling factors for low-contrast inputs of images as described in the appendices of [6].

Fig. 4 shows the self-organized RFs of six representative V1 neurons trained with texture-like input and natural-scene-like input after 20,000 training iterations. The self-organized RFs produced from LISSOM with the texture input set are visualized in Fig. 4a. The neurons developed spatiotemporal RFs strongly resembling tactile RFs found in the somatosensory cortex (Fig. 1) [4]: excitatory (bright) and inhibitory (dark) components of each neuron consists of ring and blob-like features. Note that these RF shapes arise not because circular texture elements dominate the texture input set we used. There are interesting variations as well, such as the last three columns in Fig. 4a. In those RFs, the polarity is reversed, i.e., instead of an excitatory region in the middle and inhibitory region in the surround, these RFs have an inhibitory region in the middle and the excitatory region in the surround.

On the other hand, RFs self-organized based on natural-scene-like inputs show a significantly different pattern. Nearly all neurons in Fig. 4b developed spatiotemporal RFs strongly selective for both direction and orientation. The receptive fields consist of excitatory (bright) and inhibitory (dark) lobes according to the preferred orientation and direction of the neuron, showing spatiotemporal preference. That is, each neuron is highly responsive to a line with a particular orientation moving in a direction perpendicular to that orientation. Such properties of the receptive fields are similar to those of the receptive fields of neurons found experimentally in the visual cortex [10].

The overall layout (i.e., map organization) of the RFs developed in these simulations is shown in Fig. 5 (roughly every 3rd neuron is plotted, horizontally and vertically). The texture input set we used show texture elements at varying scales, however, on closer observation, the size of the receptive field (15 × 15) is usually smaller than the round or oval features in the texture input set. So, these receptive fields are not direct memorization of the dominant features in the texture input

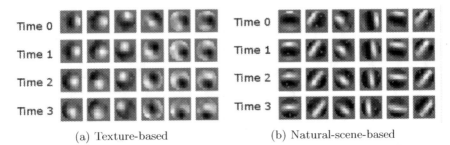

(a) Texture-based                    (b) Natural-scene-based

**Fig. 4.** RFs Resulting from Self-Organization on the Natural Scene Input Set. Six spatiotemporal RFs from (a) the texture (b) the natural-scene based experiments are shown. Each column corresponds to an individual neuron's RF, and each column represents the different time-lag. In (a), the RF shapes resemble the ring-like shape of tactile RFs found in [4] (also see Fig. 1). In (b) we can see that the pattern moves in a direction perpendicular to the orientation preference as in [10]. Adapted from [5].

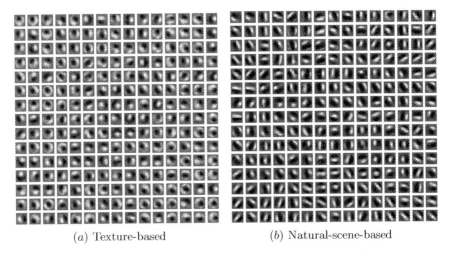

(a) Texture-based                    (b) Natural-scene-based

**Fig. 5.** RFs Resulting from Self-Organization on (a) the Texture Input Set and (b) the Natural-Scene Input Set. From the 48 × 48 cortex, only 15 × 15 are plotted (roughly every 3rd RF) for a detailed view of the RFs. The RFs in (a) mostly resemble tactile RFs while the RFs in (b) mostly resemble visual RFs. Adapted from [5].

set. Note that Fig. 5 only shows the first frame among the total of four (note that these are spatiotemporal RFs). Fig. 5b shows the map trained with natural inputs, and here we can see most RFs have a oriented Gabor-like property, just like in the visual cortex [6]. Fig. 6 shows the orientation selectivity histograms for the two maps: texture-based and natural-scene-based. The natural-scene based map (i.e., the "visual" map) shows a much higher orientation selectivity.

The results show that exposure to texture-like input can drive a general cortical learning model to develop RFs that resemble tactile RFs, while exposure

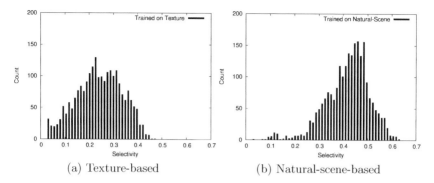

(a) Texture-based                    (b) Natural-scene-based

**Fig. 6.** Selectivity in Orientation Maps. The orientation selectivity histogram are shown for the two $48 \times 48$ V1 sheets (maps): (a) texture-based map, and (b) natural-scene-based map. As we can already see from Fig. 5, the map trained with natural scenes show much higher selectivity (peak near 0.45), compared to the case with textures (peak near 0.25). Note that higher selectivity means that RFs are more sharply tuned to one specific orientation (i.e., RFs are more slender).

to natural-scene-like input leads to visual RFs. The significance of this results is that it shows an intimate connection between texture and the tactile modality.

## 3   Manifold Analysis of RF Response

The responses of the RFs are represented in high-dimensional feature spaces, and it is hard to interpret. An effective approach for analyzing the characteristics of the responses is to assume that the responses of each RF lie on a non-linear low-dimensional manifold embedded in the high dimensional feature space. Each embedded manifold is spanned by a few dominant factors. In order to find the dominant factors of the features, we applied kernel Fisher discriminant (KFD) [11] to the feature spaces of the RF response. Here, we briefly review KFD.

KFD is a generalized version of Fisher discriminant analysis (or linear discriminant analysis, LDA) using kernel trick as in support vector machines or kernel principal component analysis [12]. The basis function in the feature space can be obtained by maximizing the ratio of the within-class scatter matrix in the feature space to the between-class scatter matrix in the feature space, as in LDA. Let $\mathcal{X}_i = \{x_1^i, x_2^i, ..., x_{l_i}^i\}, (i = 1, ..., C)$, be samples from $C$ classes and $\mathcal{X} = \bigcup_i^C \mathcal{X}_i$. Suppose $\Phi(\cdot)$ is a nonlinear mapping function to the feature space, then the within-class scatter matrix in feature space, $S_W^\Phi$, is given by

$$S_W^\Phi = \sum_{i=1}^{C} \sum_{x \in \mathcal{X}_i} (\Phi(x) - m_i^\Phi)(\Phi(x) - m_i^\Phi)^T, \tag{1}$$

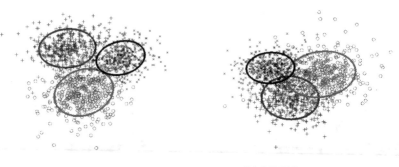

(a) TRF feature space          (b) VRF feature space

**Fig. 7.** Kernel Fisher discriminant (KFD) feature spaces for TRF and VRF responses. KFD analysis of (a) TRF and (b) VRF responses to texture-like input are shown. In each plot, response samples from three different textures, projected on the 1st and 2nd KFD axes, are shown. The ellipses show the $1.5 \times \sigma$ equidistance trace from the class centers. We can see that the classes in (a) are more separable than those in (b).

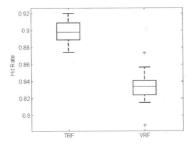

**Fig. 8.** Comparison of texture classification rate based on TRF response (left) and VRF response (right) to texture-like input is shown. The box plot shows the quartile, median, and the upper quartile, while the whiskers show 1.5 times the interquartile range ("+" marks outliers, $n = 30$). TRF-based response shows higher texture classification performance.

where $m_i^\Phi = \frac{1}{l_i} \sum_{j=1}^{l_i} \Phi(x_j^i)$. The between-class scatter matrix in feature space is given by $S_B^\Phi = S_T^\Phi - S_W^\Phi$, where the total scatter matrix in feature space, $S_T^\Phi$, is given by

$$S_T^\Phi = \sum_{x \in \mathcal{X}} (\Phi(x) - m^\Phi)(\Phi(x) - m^\Phi)^T, \tag{2}$$

where $m^\Phi = \frac{1}{|\mathcal{X}|} \sum_{i=1}^{C} l_i m_i^\Phi$ and $|\mathcal{X}|$ is the sample size.

We applied KFD to the responses of TRF and VRF on texture-like inputs (three textures were from Fig. 3). Fig. 7 shows the two different embedded manifold (TRF-based and VRF-based) in two-dimensional space. We used the square root function as the kernel function for both cases. The figure shows that the TRF responses give clusters that are more separable across texture classes than those based on VRF responses.

In order to further quantify the merit of the different RF types in dealing with texture, we measured the classification performance on the KFD results. We ran the experiment for 30 times and for each experiment 50% of data set was randomly chosen as training data and the rest as testing data. As a classifier, k-nearest neighbor (kNN) was used. Fig. 8 shows the boxplot of the classification rate for both RFs on texture-like inputs. The averages were 89.8% (for TRF-based) and 83.4% (for VRF-based), respectively. We can see that TRF is better than VRF in texture classification task. Another interesting thing is that the standard deviation in the TRF case ($= 0.0121$) is less than that of the VRF case ($= 0.0156$), which means that the performance of the TRF-based representation is more stable than that of the VRF. We also conducted a similar experiment, this time on natural-scene inputs, but the results were inconclusive, i.e., both TRFs and VRFs showed an equal level of (high) performance in the scene classification task. We are currently investigating the cause, since we expected VRFs to be better than TRFs for this task.

## 4   Discussion and Conclusion

The main contribution of this work is to have shown a developmental and a functional relationship between tactile RFs and texture. We have shown that texture-like input can drive the self-organization of tactile RFs, and tactile RFs are more effective in dealing with texture than visual RFs. The novelty of our result is not that it showed changed RF organization due to altered stimulus statistics, since that is already well-established (see [6] for a review). The novelty of our work is more specific, by explicitly linking texture to tactile RF development. The results in this paper further confirm our initial insight on the nature of texture: texture as a surface property in 3D [13]. From a computational perspective, it is also interesting to note that the TRF response distribution shows a power-law property, which is known to indicate sparse representations (see [5] for the data). Sparse coding is known to provide an efficient representation for natural scenes and receptive field characteristics, similar to those found in the primary visual cortex [14]. Finally, it would be interesting to apply our finding in the investigation of visuo-tactile integration in the blind. The use of texture as the stimulus can help tease out the common functional processes in the two different modalities (cf. [15]).

To conclude, the main objective of this work was to confirm the relationship between tactile RFs and texture. The results suggest that tactile RFs can be self-organized by texture-like input using a general cortical development model (LISSOM) initially inspired by the visual cortex, and that the representations from tactile RFs are better than vision-based ones for texture tasks. We expect our results to help us better understand the nature of texture, as a fundamentally tactile property.

**Acknowledgments.** Part of the results in section 2 (Figs. 4&5) has been accepted for presentation in [5]. This research was funded in part by NIH/NIMH (#1R01-MH66991) and NIH/NINDS (#1R01-NS54252).

# References

1. Knierim, J.J., van Essen, D.C.: Neuronal responses to static texture pattern in area v1 of the alert macaque monkey. Neurophysiol 67, 961–980 (1992)
2. Bai, Y.H., Park, C., Choe, Y.: Relative advantage of touch over vision in the exploration of texture. In: Proceedings of the 19th International Conference on Pattern Recognition, Tampa, FL (in press) (2008)
3. Jones, J.P., Palmer, L.A.: An evaluation of the two-dimensional Gabor filter model of simple receptive fields in cat striate cortex. Neurophysiology 58(6), 1233–1258 (1987)
4. DiCarlo, J.J., Johnson, K.O.: Spatial and temporal structure of receptive fields in primate somatosensory area 3b: Effects of stimulus scanning direction and orientation. Neuroscience 20, 495–510 (2000)
5. Park, C., Bai, Y.H., Choe, Y.: Tactile or visual?: Stimulus characteristics determine RF type in a self-organizing map model of cortical development. In: Proceedings of 2009 IEEE Symposium on Computational Intelligence for Multimedia Signal and Vision Processing, Nashville, TN, pp. 6–13 (2009)
6. Miikkulainen, R., Bednar, J.A., Choe, Y., Sirosh, J.: Computational Maps in the Visual Cortex. Springer, New York (2005)
7. Wilson, S.: Self-organisation can explain the mapping of angular whisker deflections in the barrel cortex. Master's thesis, The Universiy of Edinburgh, Scotland, United Kindom (2007)
8. Khurd, P., Baloch, S., Gur, R., Davatzikos, C., Verma, R.: Manifold learning techniques in image analysis of high-dimensional diffusion tensor magnetic resonance images. In: The IEEE Conference on CVPR (2007)
9. Bednar, J.A., Choe, Y., De Paula, J., Miikkulainen, R., Provost, J., Tversky, T.: Modeling cortical maps with Topographica. Neurocomputing, 1129–1135 (2004)
10. DeAngelis, G.C., Ghose, G.M., Ohzawa, I., Freeman, R.D.: Functional microorganization of primary visual cortex: Receptive-field analysis of nearby neurons. Neuroscience 19, 4046–4064 (1999)
11. Mika, S., Ratsch, G., Weston, J., Schölkopf, B., Müller, K.: Fisher discriminant analysis with kernels. In: Proceedings of IEEE Neural Networks for Signal Processing Workshop, pp. 41–48 (1999)
12. Schölkopf, B., Smola, A.J.: Learning with Kernels. MIT Press, Cambridge (2002)
13. Oh, S., Choe, Y.: Segmentation of textures defined on flat vs. layered surfaces using neural networks: Comparison of 2D vs. 3D representations. Neurocomputing 70, 2245–2255 (2007)
14. Olshausen, B.A., Field, D.J.: Emergence of simple-cell receptive field properties by learning a sparse code for natural images. Nature 381, 607–609 (1996)
15. Grant, A.C., Thiagarajah, M.C., Sathian, K.: Tactile perception in blind braille readers: A psychophysical study of acuity and hyperacuity using gratings and dot patterns. Perception and Psychophysics 62, 301–312 (2000)

# Visualization by Linear Projections as Information Retrieval

Jaakko Peltonen

Helsinki University of Technology,
Department of Information and Computer Science,
P.O. Box 5400, FI-02015 TKK, Finland
jaakko.peltonen@tkk.fi

**Abstract.** We apply a recent formalization of *visualization as informa-tion retrieval* to linear projections. We introduce a method that optimizes a linear projection for an information retrieval task: retrieving neighbors of input samples based on their low-dimensional visualization coordinates only. The simple linear projection makes the method easy to interpret, while the visualization task is made well-defined by the novel information retrieval criterion. The method has a further advantage: it projects input features, but the input neighborhoods it preserves can be given separately from the input features, e.g. by external data of sample similarities. Thus the visualization can reveal the relationship between data features and complicated data similarities. We further extend the method to kernel-based projections.

**Keywords:** visualization, information retrieval, linear projection.

## 1 Introduction

Linear projections are widely used to visualize high-dimensional data. They have the advantage of easy interpretation: each axis in the visualization is a simple combination of original data features, which in turn often have clear meanings. Linear projections are also fast to apply to new data. In contrast, nonlinear projections can be hard to interpret, if a functional form of the mapping is available at all. Some nonlinear methods also need much computation or approximation of the mapping to embed new points. Kernel-based projections are a middle ground between linear and nonlinear projections; their computation is linear in the kernel space, and their interpretability depends on the chosen kernel.

The crucial question in linear visualization is what criterion to use for find-ing the projection. Traditional answers include preservation of maximum variance as in principal component analysis (PCA); preservation of an indepen-dence structure as in independent component analysis; preservation of distances and pairwise constraints as in [1]; or maximization of class predictive power as in linear discriminant analysis, informative discriminant analysis [2], neighborhood components analysis [3], metric learning by collapsing classes [4], and others.

When the linear projection is intended for visualization, the previous criteria are insufficient, as they are only indirectly related to visualization. One must

J.C. Príncipe and R. Miikkulainen (Eds.): WSOM 2009, LNCS 5629, pp. 237–245, 2009.

first formalize what is the task of visualization, and what are good performance measures for the task. This question has recently been answered in [5], where the task of visualization is formalized as an *information retrieval task*, and goodness measures are derived which are generalizations of *precision* and *recall*. Based on the goodness measures one can form an optimization criterion, and directly *optimize the goodness of a visualization in the information retrieval task*; however, so far this approach has only been used for nonlinear embedding where output coordinates are directly optimized without any parametric mapping [5,6].

We introduce a novel method for linear and kernel-based visualization called Linear Neighbor Retrieval Visualizer (LINNEA): we apply the formalization of visualization as an information retrieval task, and optimize precision and recall of such retrieval. A useful property is that the input features being projected and the distances used to compute the input neighborhoods *can be given separately*: for example, features can be word occurrence vectors of text documents and distances can be distances of the documents in a citation graph. In special cases, LINNEA is related to the methods *stochastic neighbor embedding* [7] and *metric learning by collapsing classes* [4], but it is more general; it can be used for unsupervised and supervised visualization, and allows the user to set the tradeoff between precision and recall of information retrieval. We show by preliminary experiments that LINNEA yields good visualizations of several data sets.

## 2    Visualization as Information Retrieval

We briefly summarize the novel formalization of visualization introduced in [5].

The task is *visualization of neighborhood or proximity relationships* within a high-dimensional data set. For a set of input points $\mathbf{x}_i \in \mathbb{R}^{d_o}$, $i = 1, \ldots, N$, a visualization method yields output coordinates $\mathbf{y}_i \in \mathbb{R}^d$, which should reveal the neighborhood relationships. This is formalized as an *information retrieval task*: for any data point, the visualization should allow the user to retrieve its neighboring data points in the original high-dimensional data. Perfect retrieval from a low-dimensional visualization is usually not possible, and the retrieval will make two kinds of errors: not retrieving a neighbor decreases *recall* of the retrieval, and erroneously retrieving a non-neighbor decreases *precision*.

To apply the information retrieval concepts of *precision* and *recall* to visualization, in [5] they are generalized to continuous and probabilistic measures as follows. For each point $i$, a *neighborhood probability distribution* $p_{i,j}$ over all other points $j$ is defined; in [5] an exponentially decaying probability based on input distances $d(\mathbf{x}_i, \mathbf{x}_j)$ is used. In this paper we allow the $d(\mathbf{x}_i, \mathbf{x}_j)$ to arise from any definition of distance between points $i$ and $j$. The *retrieval of points from the visualization* is also probabilistic: for each point $i$ a distribution $q_{i,j}$ is defined which tells the probability that a particular nearby point $j$ is retrieved from the visualization. The $q_{i,j}$ are defined similarly to the $p_{i,j}$, but using Euclidean distances $||\mathbf{y}_i - \mathbf{y}_j||$ between visualization coordinates $\mathbf{y}_i$. This yields

$$p_{i,j} = \frac{e^{-d^2(\mathbf{x}_i,\mathbf{x}_j)/2\sigma_i^2}}{\sum_{k \neq i} e^{-d^2(\mathbf{x}_i,\mathbf{x}_k)/2\sigma_i^2}} , \qquad q_{i,j} = \frac{e^{-||\mathbf{y}_i-\mathbf{y}_j||^2/2\sigma_i^2}}{\sum_{k \neq i} e^{-||\mathbf{y}_i-\mathbf{y}_k||^2/2\sigma_i^2}} \qquad (1)$$

where $\sigma_i$ are scale parameters which can be set by fixing the entropy of the $p_{i,j}$ as suggested in [5]. Since $p_{i,j}$ and $q_{i,j}$ are probability distributions, it is natural to use Kullback-Leibler divergences to measure how well the retrieved distributions correspond to the input neighborhoods. The divergence $D_{KL}(p_i, q_i) = \sum_{j \neq i} p_{i,j} \log(p_{i,j}/q_{i,j})$ turns out to be a *generalization of recall* and $D_{KL}(q_i, p_i)$ turns out to be a *generalization of precision*. The values of the divergences are averaged over points $i$ which yields the final goodness measures.

## 3 The Method: Linear Neighborhood Retrieval Visualizer

The generalizations of precision and recall above can be directly used as optimization goals, but as both precision and recall cannot usually be maximized together, the user must set a tradeoff between them. Given the tradeoff a single cost function can be defined and visualizations can be directly optimized in terms of the cost function. In the earlier works [5,6] this approach was used to compute a nonlinear embedding, that is, the output coordinates $\mathbf{y}_i$ of data points were optimized directly. In this paper we instead consider a parametric, linear projection $\mathbf{y}_i = \mathbf{W}\mathbf{x}_i$ where $\mathbf{W} \in \mathbb{R}^{d \times d_0}$ is the projection matrix. We wish to optimize $\mathbf{W}$ so that the projection is good for the information retrieval task of visualization. We call the method Linear Neighborhood Retrieval Visualizer (LINNEA). We use the same cost function as in [5], that is,

$$E = \lambda \sum_i D_{KL}(p_i, q_i) + (1 - \lambda) \sum_i D_{KL}(q_i, p_i)$$

$$= \sum_i \sum_{j \neq i} \left[ -\lambda p_{i,j} \log q_{i,j} + (1 - \lambda) q_{i,j} \log \frac{q_{i,j}}{p_{i,j}} \right] + const. \quad (2)$$

where the tradeoff parameter $\lambda$ is to be set by the user to reflect whether precision or recall is more important. We simply use a conjugate gradient algorithm to minimize $E$ with respect to the matrix $\mathbf{W}$. The gradient $\frac{\partial E}{\partial \mathbf{W}}$ is

$$\sum_{i,j \neq i} \left[ \lambda(p_{i,j} - q_{i,j}) + (1 - \lambda) q_{i,j} \left( D_{KL}(q_i, p_i) - \log \frac{q_{i,j}}{p_{i,j}} \right) \right] \frac{(\mathbf{y}_i - \mathbf{y}_j)(\mathbf{x}_i - \mathbf{x}_j)^T}{\sigma_i^2}$$

$$(3)$$

which yields $O(N^2)$ computational complexity per gradient step.

*Optimization details.* In this paper we simply initialize the elements of $\mathbf{W}$ to uniform random numbers between 0 and 1; more complicated initialization, say by initializing $\mathbf{W}$ as a principal component analysis projection, is naturally possible. To avoid local optima, we use two simple methods. Firstly, in each run we first set the neighborhood scales to large values, and decrease them after each optimization step until the final scales are reached, after which we run 40 conjugate gradient steps with the final scales. Secondly, we run the algorithm from 10 random initializations and take the result with the best cost function value.

## 3.1    Kernel Version

We now present a kernel version of LINNEA. Instead of simple linear projections we optimize projections from a kernel space: we set $\mathbf{y}_i = \mathbf{W}\Phi(\mathbf{x}_i)$ where $\Phi(\cdot)$ is some nonlinear transformation to a potentially infinite-dimensional space with inner products given by a kernel function $k(\mathbf{x}_i, \mathbf{x}_j) = \Phi(\mathbf{x}_i)^T\Phi(\mathbf{x}_j)$. As usual, the kernel turns out to be all we need and knowing $\Phi$ is not required.

The task is the same as before: to optimize the projection (visualization) so that it is good for information retrieval according to the cost function (2).

It is reasonable to assume that the rows $\mathbf{w}_l^T$ of $\mathbf{W}$ can be expressed as linear combinations of the $\Phi(\mathbf{x}_i)$, so that $\mathbf{w}_l = \sum_m a_m^l \Phi(\mathbf{x}_m)$ where $a_m^l$ are the coefficients. Then the projection has the simple form

$$\mathbf{y}_i = \left[ \sum_m a_m^1 \Phi(\mathbf{x}_m), \ldots, \sum_m a_m^d \Phi(\mathbf{x}_m) \right]^T \Phi(\mathbf{x}_i) = \mathbf{A}\mathbf{K}(\mathbf{x}_i) \qquad (4)$$

where the matrix $\mathbf{A} \in \mathbb{R}^{d \times N}$ contains the coefficients $A(l, m) = a_m^l$ and $\mathbf{K}(\mathbf{x}_i) = [k(\mathbf{x}_1, \mathbf{x}_i), \ldots, k(\mathbf{x}_N, \mathbf{x}_i)]^T$. As before, the coordinates $\mathbf{y}_i$ can be used to compute the neighborhoods $q_{i,j}$, the cost function, and so on.

To optimize this kernel-based projection, it is sufficient to optimize the cost function with respect to the coefficient matrix $\mathbf{A}$. We can again use a standard conjugate gradient method: the gradient with respect to $\mathbf{A}$ is the same as equation (3), except that $\mathbf{x}_i$ and $\mathbf{x}_j$ are replaced by $\mathbf{K}(\mathbf{x}_i)$ and $\mathbf{K}(\mathbf{x}_j)$. Since $\mathbf{A}$ has $N$ columns, the computational complexity becomes $O(N^3)$ per gradient step.

## 3.2    Properties of LINNEA

A crucial property of LINNEA is that the input features $\mathbf{x}_i$ being projected and the distances $d(\mathbf{x}_i, \mathbf{x}_j)$ used to compute the input neighborhoods can be given separately. At simplest $d(\mathbf{x}_i, \mathbf{x}_j)$ can be the Euclidean distance $\|\mathbf{x}_i - \mathbf{x}_j\|$, but it can also be based on other data: for example, $\mathbf{x}_i$ can be word occurrence vectors of text documents and $d(\mathbf{x}_i, \mathbf{x}_j)$ can be distances of the documents in a citation graph (we test this example in Section 4). When distances are directly computed from input features, the projection is unsupervised. When distances are given separately the projection is supervised by the distances; then the projection is optimized to allow retrieval of neighbors which are based on the separately given distances, so it reveals the relationship between the features and the distances.

Note that a visualization based on the distances only, say multidimensional scaling computed from citation graph distances between documents, would not provide any relationship between the visualization and the features (document content); in contrast, the LINNEA visualization is directly a projection of the features, which is optimized for retrieval of neighbors based on the distances.

If we set $\lambda = 1$ in (2), that is, we maximize recall, this yields the cost function of *stochastic neighbor embedding* (SNE; [7]); thus LINNEA includes a linear version of SNE as a special case. More generally, the cost function of LINNEA implements a flexible user-defined tradeoff between precision and recall.

Another interesting special case follows if the input neighborhoods are derived from class labels of data points. Consider a straightforward neighborhood $p_{i,j}$: for any point $i$ in class $c_i$, the neighbors are the other points from the same class, with equal probabilities. It is easy to show that if we set $\lambda = 1$ in the cost function (that is, we maximize recall), this is equivalent to maximizing

$$\sum_i \sum_{j \neq i} \delta_{c_i, c_j} \log \frac{e^{-||\mathbf{y}_i - \mathbf{y}_j||^2 / 2\sigma_i^2}}{\sum_{j' \neq i} e^{-||\mathbf{y}_i - \mathbf{y}_{j'}||^2 / 2\sigma_i^2}} \tag{5}$$

where $\delta_{c_i, c_j} = 1$ if the classes $(c_i, c_j)$ are the same and zero otherwise, and for simplicity classes are assumed equi-probable. This is the cost function of *metric learning by collapsing classes* (MCML; [4]) which was introduced as a supervised, linear version of SNE. LINNEA includes MCML as a special case. We thus give a new interpretation of MCML: it maximizes *recall of same-class points*.

Note that LINNEA yields meaningful solutions for the above kind of straightforward input neighborhoods because the mapping is a linear projection of input features. In contrast, methods that freely optimize output coordinates could yield trivial solutions mapping all input points of each class to a single output point. To avoid trivial solutions, such methods can e.g. apply topology-preserving supervised metrics as in [6]; such complicated metrics are not needed in LINNEA.

In summary, LINNEA can be used for both supervised and unsupervised visualization; it is related to well-known methods but is more general, allowing the user to set the tradeoff between the different costs of information retrieval.

## 4   Experiments

In this first paper we do not yet make thorough comparisons between LINNEA and earlier methods. We show the potential of LINNEA in four experiments; we use principal component analysis (PCA) as a baseline. We use the non-kernel version of LINNEA (Section 3), and set the user-defined tradeoff between precision and recall to $\lambda = 0$ (favoring precision only) in experiments 1-3 and $\lambda = 0.1$ in experiment 4. Other parameters were defaults from the code of [5].

*Experiment 1: Extracting the relevant dimensions.* We first test LINNEA on toy data where the visualization can perfectly recover the given input neighborhoods. Consider a spherical Gaussian cloud of 500 points in the Hue-Saturation-Value (HSV) color space, shown in Fig. 1 (left). One cannot represent all three dimensions in one two-dimensional visualization, and without additional knowledge one cannot tell which features are important to preserve, as the shape of the data is the same in all directions. However, if we are also given pairwise distances between points, they determine which features to preserve. Suppose those distances have secretly been computed based only on Hue and Value; then the correct visualization is to take those two dimensions, ignoring Saturation.

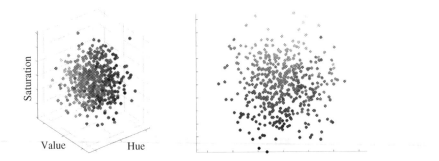

**Fig. 1.** Projection of points in the Hue-Saturation-Value color space. **Left:** the original three-dimensional data set is a Gaussian point cloud; coordinates correspond to Hue, Saturation (grayishness-colorfulness) and Value (lightness-darkness) of each dot. Pairwise input distances were computed from Hue and Value only. **Right:** LINNEA has correctly found the Hue and Value dimensions in the projection and ignored Saturation.

We optimized a two-dimensional projection with LINNEA; we gave the HSV components of each data point as input features, and computed input neighborhoods from the known pairwise distances. As shown in Fig. 1 (right), LINNEA found the Hue-Value dimensions and ignored Saturation, as desired; the weight of Saturation in the projection directions is close to zero.

*Experiment 2: S-curve.* We visualize data set having a simple underlying manifold: 1000 points sampled along a two-dimensional manifold, embedded in the three-dimensional space as an S-shaped curve as shown in Fig. 2 (left). No external pairwise distances are given and input neighborhoods are computed from the three-dimensional input features. The task is to find a visualization where original neighbors on the manifold can be retrieved well. Unfolding the manifold would suffice; however, a linear projection cannot unfold the nonlinear S-curve perfectly. The PCA solution in Fig. 2 (middle) leaves out the original Z-axis, which is suboptimal for retrieving original neighbors as it leaves visible only one of the coordinates of the underlying two-dimensional manifold. The LINNEA result in Fig. 2 (right) emphasizes directions Z and Y; this shows the coordinates of the underlying manifold well and allows retrieval of neighbors of input points.

*Experiment 3: Projection of face images.* We visualize a data set of human faces ([8]; available at http://web.mit.edu/cocosci/isomap/datasets.html). The data set has 698 synthetic face images in different orientations and lighting directions; each image has 64 × 64 pixels. We first find linear projections of the face images using the pixel images as input features, without giving any additional knowledge. As shown in Fig. 3 (top left), PCA reveals part of the data structure, but the result is unsatisfactory for retrieving neighboring faces, since PCA has clumped together the back-lit faces. In contrast, as shown in Fig. 3 (top right), LINNEA spreads out both front-lit and back-lit faces. The projection directions can be interpreted as linear filters of the images. For PCA the filter on the

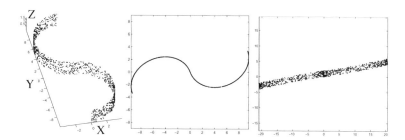

**Fig. 2.** Projections of an S-curve. **Left:** the original three-dimensional data. **Middle:** PCA neglects the original Z-direction. **Right:** LINNEA finds a projection where neighbors on the underlying manifold can be retrieved well from the visualization.

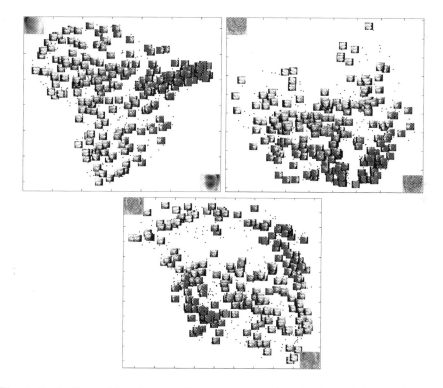

**Fig. 3.** Projections of face images. **Top:** unsupervised projections of the pixel images by PCA (left) and LINNEA (right). The linear projection directions can be interpreted as linear filters of the images, which are shown for each axis. **Bottom:** a supervised projection by LINNEA. Pairwise distances were derived from known pose/lighting parameters of the faces. LINNEA has optimized projections of the pixel images, for retrieving neighbors having similar pose/lighting parameters. See the text for more analysis.

horizontal axis roughly responds to a left-facing head; the filter on the vertical axis roughly detects left-right lighting direction. The LINNEA filters are complicated; more analysis of the filters is needed in future work.

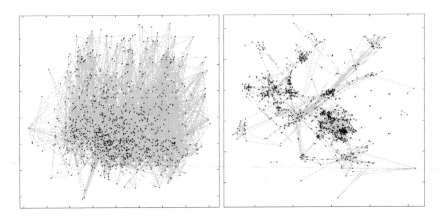

**Fig. 4.** Projections of scientific documents. Documents are shown as dots and citations between two documents are shown as lines. **Left:** PCA projection of document content vectors does not reveal citation neighborhoods well. **Right:** projection by LINNEA shows clusters of citing documents and connections between clusters.

*Projection to retrieve known pose/lighting neighbors.* For the face data the pose and lighting parameters of the faces are available. We can then compute pairwise input distances based on these parameters, and use LINNEA to find a supervised visualization of the pixel images that best allows retrieval of the pose/lighting neighbors of each face. The LINNEA projection is shown in Fig. 3 (bottom). The face images are arranged quite well in terms of the pose and lighting; the top left–bottom right axis roughly separates left and right-facing front-lit faces, and the top right–bottom left axis roughly separates left and right-facing back-lit faces. The corresponding filters are somewhat complicated; the filters on the vertical axis and horizontal axis seem to roughly detect edges and lighting direction respectively. The underlying pose/lighting space is three-dimensional and cannot be represented exactly by a two-dimensional mapping, thus the filters are compromises between representing several aspects of pose/lighting. Note that running e.g. PCA on the known pose/lighting parameters would not yield filters of the pixel images, thus it would not tell how pixel data is related to pose/lighting; in contrast, LINNEA optimizes filters for retrieval of pose/lighting neighbors.

*Experiment 4: Visualization of scientific documents.* We visualize the CiteSeer data set which contains scientific articles and their citations. The data set is available at http://www.cs.umd.edu/projects/linqs/projects/lbc/index.html. Each article is described by a binary 3703-dimensional vector telling which words appeared in the article; we used these vectors as the input features. To reduce computational load we took the subset of 1000 articles having the highest number of inbound plus outbound citations. We provide separate pairwise input distances, simply taking the graph distance in the citation graph: that is, two documents where one cites the other have distance 1, documents that cite the same other document have distance 2, and so on. As a simplification we assumed citation to be a symmetric relation, and as a regularization we upper bounded the graph distance to 10. We use LINNEA (with $\lambda = 0.1$) to optimize a two-dimensional visualization where neighbors according to this graph distance can be best retrieved. The result is shown in Fig. 4. In the baseline PCA projection (left subfigure) citations are spread all over the data with little visible structure, whereas the LINNEA projection (right subfigure) shows clear structure: clusters where documents cite each other, and citation connections between clusters. For this data each feature is a word; unfortunately the identities of the words are unavailable. In general one can interpret the projection directions given by LINNEA by listing for each direction the words having the largest weights.

## 5   Conclusions

We introduced a novel method for visualization by linear or kernel based projection. The projection is optimized for information retrieval of original neighbor points from the visualization, with a user-defined tradeoff between precision and recall. The method can either find projections for input features as such, or find projections that reveal the relationships between input features and separately given input distances. The method yields good visualization of several data sets.

**Acknowledgments.** The author belongs to the Adaptive Informatics Research Centre and to Helsinki Institute for Information Technology HIIT. He was supported by the Academy of Finland, decision 123983, and in part by the PASCAL2 Network of Excellence. He thanks Samuel Kaski for very useful discussion.

## References

1. Cevikalp, H., Verbeek, J., Jurie, F., Kläser, A.: Semi-supervised dimensionality reduction using pairwise equivalence constraints. In: Proc. VISAPP 2008, pp. 489–496 (2008)
2. Peltonen, J., Kaski, S.: Discriminative components of data. IEEE Trans. Neural Networks 16(1), 68–83 (2005)
3. Goldberger, J., Roweis, S., Hinton, G., Salakhutdinov, R.: Neighbourhood components analysis. In: Proc. NIPS 2004, pp. 513–520. MIT Press, Cambridge (2005)
4. Globerson, A., Roweis, S.: Metric learning by collapsing classes. In: Proc. NIPS 2005, pp. 451–458. MIT Press, Cambridge (2006)
5. Venna, J., Kaski, S.: Nonlinear dimensionality reduction as information retrieval. In: Proc. AISTATS 2007 (2007)
6. Peltonen, J., Aidos, H., Kaski, S.: Supervised nonlinear dimensionality reduction by neighbor retrieval. In: Proc. ICASSP 2009 (in press, 2009)
7. Hinton, G., Roweis, S.T.: Stochastic neighbor embedding. In: Proc. NIPS 2002, pp. 833–840. MIT Press, Cambridge (2002)
8. Tenenbaum, J.B., de Silva, V., Langford, J.C.: A global geometric framework for nonlinear dimensionality reduction. Science 290 (December 2000)

# Analyzing Domestic Violence with Topographic Maps: A Comparative Study

Jonas Poelmans[1], Paul Elzinga[2], Stijn Viaene[1,3], Guido Dedene[1,4], and Marc M. Van Hulle[5]

[1] K.U.Leuven, Faculty of Business and Economics, Naamsestraat 69,
3000 Leuven, Belgium
[2] Police Amsterdam-Amstelland, James Wattstraat 84,
1000 CG Amsterdam, The Netherlands
[3] Vlerick Leuven Gent Management School, Vlamingenstraat 83,
3000 Leuven, Belgium
[4] Universiteit van Amsterdam Business School, Roetersstraat 11
1018 WB Amsterdam, The Netherlands
[5] K.U.Leuven, Laboratorium voor Neuro- en Psychofysiologie
Campus Gasthuisberg 3000 Leuven, Belgium
{Jonas.Poelmans,Stijn.Viaene,Guido.Dedene}@econ.kuleuven.be
Paul.Elzinga@amsterdam.politie.nl
marc@neuro.kuleuven.be

**Abstract.** Topographic maps are an appealing exploratory instrument for discovering new knowledge from databases. During the recent years, several variations on the Self Organizing Maps (SOM) were introduced in the literature. In this paper, the toroidal Emergent SOM tool and the spherical SOM are used to analyze a text corpus consisting of police reports of all violent incidents that occurred during the first quarter of 2006 in the police region Amsterdam-Amstelland (The Netherlands). It is demonstrated that spherical topographic maps provide a powerful instrument for analyzing this dataset. In addition, the performance of the toroidal Emergent SOM is compared to that of the spherical SOM, and it turned out to be superior to that of an ordinary classifier, applied directly to the data.

**Keywords:** Topographic maps, domestic violence, knowledge discovery in databases, Emergent SOM, BLOSSOM.

## 1 Introduction

According to the department of Justice of the Netherlands, domestic violence can be characterized as serious acts of violence committed by someone of the domestic sphere of the victim. Violence includes all forms of physical assault. The domestic sphere includes all partners, ex-partners, family members, relatives and family friends of the victim. Family friends are those persons who have a friendly relationship with the victim and who regularly meet the victim in his/her home [1].

J.C. Príncipe and R. Miikkulainen (Eds.): WSOM 2009, LNCS 5629, pp. 246–254, 2009.
© Springer-Verlag Berlin Heidelberg 2009

Research has proven that domestic violence is a largely underestimated problem in our modern society [2,3,4,5]. Pursuing an effective policy against offenders is one of the top priorities of the police organization of the region Amsterdam-Amstelland in the Netherlands. Of course, in order to pursue an effective policy against offenders, being able to swiftly recognize cases of domestic violence and label reports accordingly is of the utmost importance. Still this has proven to be problematic. In the past, intensive audits of the police databases related to filed reports have established that many reports tended to be wrongly classified as domestic or as non-domestic violence cases. One of the conclusions was that there was a need for an in-depth investigation of this problem area.

In the current paper, we develop an application in the problem area of topographic maps [7], which are particularly suited for high-dimensional data visualization. Two recent tools will be considered, the Emergent SOM and the spherical SOM, and their performances compared. The remainder of this paper is composed as follows. In section 2, we discuss the essentials of topographic map theory and in particular the Emergent SOM and Spherical SOM. In section 3, we elaborate on the dataset. In section 4, the results of the comparative analysis of the toroidal ESOM (using the Databionics tool) and the Spherical SOM (using the BLOSSOM tool) are presented. Section 5 concludes the paper.

## 2    Topographic Map Essentials

From a practitioner's point of view, topographic maps are an especially appealing technique for knowledge discovery in databases [15]. It performs a non-linear mapping of a high-dimensional space to a low-dimensional one, usually a two-dimensional one. It offers the user a useful tool for exploring the dataset [12]. It can be used to detect clusters and it maintains the neighborhood relationships that are present in the input space. It also provides the user with an idea of the complexity of the dataset, the distribution of the dataset (e.g. spherical) and the amount of overlap between the different classes. The lower-dimensional data representation is also an advantage when constructing classifiers.

### 2.1    Emergent SOM

An Emergent Self Organizing Map (ESOM) is a very recent type of topographic map [8]. It is argued to be especially useful for visualizing sparse, high-dimensional datasets, yielding an intuitive overview of its structure [10]. An Emergent SOM differs from a traditional SOM in that a very large number of neurons (at least a few thousand) are used [9]. Alfred Ultsch argues that the topology preservation of the traditional SOM projection is of little use when using small maps: the performance of a small SOM is almost identical to that of k-means clustering, with k equal to the number of nodes in the map [8]. An additional advantage of an ESOM is that it can be trained directly on the available dataset without first having to go through a feature selection procedure [11]. ESOM maps can be created and used for data analysis by means of the publicly available *Databionics ESOM Tool*. This tool allows the user to construct both flat and unbounded (i.e., toroidal) ESOM maps.

## 2.2  Spherical SOM

In a spherical SOM, the neurons are arranged on a sphere. Recently, several spherical self-organizing topographic maps have been introduced in the literature [6]. These maps are spherical or toroidal and, thus, not bounded as in the case of e.g. the traditional SOM and its many versions, and thus should not suffer from the border effect. The border effect is a phenomenon which occurs in flat maps because the number of neighborhood neurons of a neuron at the border of the map is smaller than the number of neighborhood neurons of a neuron at the center of the map [14]. This might cause distortions of the map, e.g. leading to a too small area for cluster detection near the edges of the map. The spherical SOM tool used here is BLOSSOM [13].

## 3  Dataset

The dataset consists of 4146 police reports describing all violent incidents from the first quarter of 2006. All domestic violence cases from that period are a subset of this dataset. Unfortunately, many of these 4146 police reports did not contain the reporting of a crime by a victim, which is necessary for establishing domestic violence. This happens for example when a police officer was sent to an incident and later on wrote a report in which he/she mentioned his/her findings, while the victim did not make an official statement to the police. Therefore, we only retained the 2288 documents in which the victim reported a crime to a police officer. From these 2288 documents, we removed the follow-up reports referring to previous cases. This filtering process resulted in a set of 1794 reports. From these reports, the person who reported the crime, the suspect, the persons involved in the crime, the witnesses, the project code and the statement made by the victim to the police were extracted. Of these 1794 reports, 462 were cases of domestic violence; the others not. These data were used to generate the 1794 html-documents that were used during the research.

We also have at our disposal a thesaurus – a collection of terms – that was obtained by performing word frequency analyses on these police reports. The relevant terms that occurred most often were retrieved and added to the initially empty thesaurus. This resulted in a set of 123 terms. In the categorical dataset, it is indicated for each police report which ones of these terms appear in the report. In the continuous dataset, the relevance of each term is indicated for each police report by means of a continuous value between 0 and 1. This value was calculated on the basis of the number of times the term appeared in the report.

For each police report, some additional information is available. This information includes whether or not the suspect of the criminal offence is known, the gender of the victim, the age of the victim, whether the perpetrator and victim lived at the same address, etc.

## 4  Experiment

In a first step, a toroidal ESOM map was trained on the basis of these 2 datasets, in order to discover the distribution of the dataset. In the map displayed in Fig. 1, the best matching (nearest-neighbor) nodes are labeled in the two classes for the given

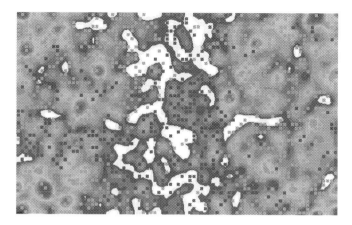

**Fig. 1.** Toroid ESOM map trained on the categorical dataset with all features

test data set (red for domestic violence, green for non-domestic violence). By analyzing the ESOM map based on the categorical dataset, we found that there is one large domestic violence cluster running vertically through the center of the map, and one less clearly demarcated domestic violence cluster running to the left. The latter continues over the edge of the map and has an outlier on the right of the map. Therefore, it seems natural to use a spherical or toroidal SOM for visualizing this dataset.

For both tools, it was possible to train a map directly on the entire dataset with more than 123 features. However, in order to prevent distortions on the map caused by irrelevant and redundant features, it was chosen to apply feature selection. A heuristic feature selection procedure called minimal-redundancy-maximal-relevance (mRMR), as described in [16], was considered. The aim was to select the 50 most relevant features. To obtain the optimal feature set, an SVM, a Neural Network, a kNN (with k=3) and a Naïve Bayes classifier were used to measure the classification performance for an increasing number of features. The classification performance was plotted as a function of the number of features and it was decided to retain the best 18 features.

For the ESOM, a SOM with a lattice containing 50 rows and 82 columns of neurons was used (50x82=4100 neurons in total). The weights were initialized randomly by sampling a Gaussian with the same mean and standard deviation as the corresponding features. A Gaussian bell-shaped kernel with initial radius of 24 was used as a neighborhood function. Further, an initial learning rate of 0.5 and a linear cooling strategy for the learning rate were used. The number of training epochs was set to 20. Both a map with a toroidal topology of the neurons as well as a flat topology were used. For BLOSSOM, a network consisting of 642 neurons was used. The weights were initialized randomly. A Gaussian kernel with initial radius $\pi$ was used as a neighborhood function. Further, an initial learning rate of 0.9 and a linear cooling strategy for the learning rate were used. The number of training epochs was set to 50.

The BLOSSOM map trained on the categorical dataset is displayed in Fig. 2. The BLOSSOM map trained on the continuous dataset is displayed in Fig. 3. The toroidal ESOM map trained on the categorical dataset is displayed in Fig. 4. The flat ESOM map trained on the categorical dataset is displayed in Fig. 5.

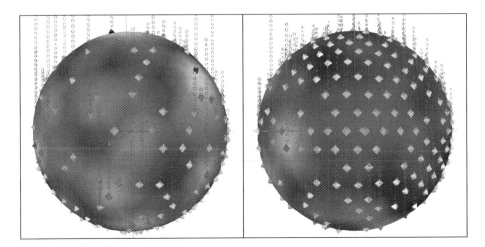

**Fig. 2.** Two views of the BLOSSOM map trained on the categorical dataset with 18 features. The grayscales on the surface indicate local densities (white= high density). The small tetrahedrons indicate the nearest-neighbor neurons for the two types of labels; "x" indicates a domestic violence case, "o" a non-domestic violence case.

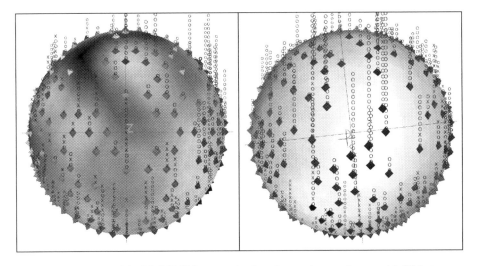

**Fig. 3.** Two views of the BLOSSOM map trained on the continuous dataset with 18 features

Finally, a kNN classifier was built for the ESOM and BLOSSOM maps. For BLOSSOM, k was set to 1. In order to obtain the misclassification error of the BLOSSOM map, the Euclidean distance of each input vector to each weight vector was measured. For each weight vector (corresponding to a node of the map) it was calculated how many of the domestic and non-domestic violence cases had this weight vector as a best match. If the node dominantly contained domestic violence cases, it was labeled as a domestic violence node and the non-domestic violence cases

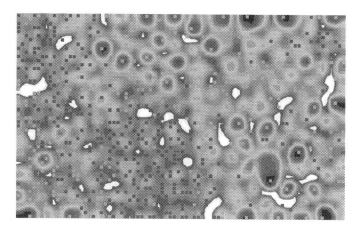

**Fig. 4.** Flat ESOM map trained on the categorical dataset with 18 features

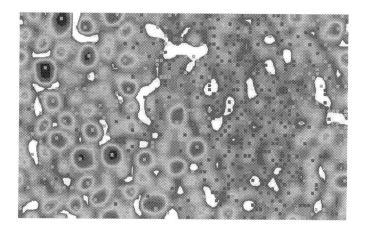

**Fig. 5.** Toroid ESOM map trained on the categorical dataset with 18 features

that best matched to this node were considered to be wrong classifications. Because the ESOM map contained about 2 times as many best-matched neurons as the BLOSSOM map (680 vs. 316), k was set to 2 for the ESOM map. A best-matched neuron is a neuron for which there exists at least one input vector for which the Euclidean distance to the weight vector of this node is minimal.

## 5   Analysis and Results

The ESOM tool was first applied to the dataset containing all features for quickly obtaining an overview of the structure of the dataset. We observed that some of the clusters continued over the edges of the map, thereby making BLOSSOM an interesting candidate tool. A problem with the ESOM map is that the density profile of the map does not match the uniform distribution of the labeled data vectors.

Moreover, there is no ridge in the map that separates the domestic- from the non-domestic violence cases. Therefore, the 'watershed' technique will not lead to a correct identification of the classes. This problem was not solved by lowering the number of features. Nevertheless, much more density variations can be observed in Fig. 5. BLOSSOM was problematic in that not all labels are visible on the map. Many of the labels at the upper side of the map were impossible to see because of the small window size used by the tool.

By examining the BLOSSOM map shown in Fig. 2, one can conclude that there are no clearly demarcated clusters of domestic violence cases available in the categorical dataset. This was also the case for the spherical map trained on the dataset containing all 123 features. However, several clearly demarcated clusters of non-domestic violence cases can be observed in the map of Fig. 2. These clusters correspond to the white (= high density) areas of the map.

We found that these clusters correspond to types of incidents that can be clearly distinguished from other types of incidents. In burglary cases e.g., the suspect is typically not known, neither a description of the suspect is provided, and one or more locations inside the house are mentioned. These typical characteristics result in a grouping of such cases by the BLOSSOM tool. From the map displayed in Fig. 3, one can conclude that this is also the case for the continuous dataset. However, it is conspicuous that the latter contains much less density variations.

Another interesting result is that the map provides a good division between domestic and non-domestic violence cases. Many of the best matched nodes dominantly contain either domestic or non-domestic violence cases. This indicates that there is only a small amount of overlap between them. The observed overlap probably indicates that the feature set is not sufficiently refined to discriminate between the two classes. However, it should be considered that some cases might have been wrongly classified by police officers. The latter might be due to the vagueness of the domestic violence definition.

When the flat ESOM map is compared to the toroidal ESOM map, one concludes that the toroidal map provides a better visualization of the dataset. The border effect is clearly present in the flat map resulting in undesired distortions of the map. Most of the observed clusters are located at the border of the map, which makes them smaller in area, and the large group of domestic violence cases is less clearly demarcated from the non-domestic violence cases.

Finally, the results of the nearest neighbor classifiers based on the ESOM and BLOSSOM maps are displayed in table 1 and 2. These values are averages of the accuracy obtained during 40 runs of each method.

**Table 1.** Classification performance on the categorical dataset

|  | Overall accuracy | False Positive Rate | False Negative Rate |
|---|---|---|---|
| BLOSSOM 1NN | 90.4% | 2.9% | 29% |
| Toroid ESOM 2NN | 90.8% | 8% | 12.6% |
| Flat ESOM 2NN | 91.4% | 6.8% | 13.9% |
| Toroid ESOM 1NN | 95% | 4% | 7.6% |
| Flat ESOM 1NN | 95.1% | 4% | 7.4% |

An interesting result is the striking difference in performance of the traditional kNN classifer (65%) and the kNN classifier based on the spherical BLOSSOM map (90,4%). This is due to the topographic map being a model of the data distribution: it forms an approximation of the data manifold, offering interpolating facilities, and it spends more neural hardware at clusters in the data, leading to a modeling of the local density.

**Table 2.** Classification performance on the continuous dataset

|  | Overall accuracy | False Positive Rate | False Negative Rate |
|---|---|---|---|
| BLOSSOM 1NN | 87% | 3.7% | 40% |
| Toroid ESOM 2NN | 88.7% | 8.6% | 20.6% |
| Flat ESOM 2NN | 88.9% | 8.8% | 18% |
| Toroid ESOM 1NN | 94.4% | 0.3% | 21% |
| Flat ESOM 1NN | 94.7% | 0.3% | 20% |

From table 1 and table 2, one may conclude that the overall accuracy of the 1NN classifier based on the BLOSSOM map and the overall accuracy of the 2NN classifier based on the ESOM map are almost equal. However, a clear difference can be observed in the false positive rates and the false negative rates. The false positive rate (i.e. the number of non-domestic violence cases that were incorrectly classified as domestic violence, divided by the number of non-domestic violence cases contained in the dataset) for the BLOSSOM map more than twice as good as the false positive rate for the ESOM map. The opposite is true for the false negative rate (i.e. the number of domestic violence cases that were not classified as such by the NN classifier, divided by the number of domestic violence cases contained in the dataset). Surprisingly, there is almost no difference in classification performance for the flat and the toroidal ESOM map. Although the former map contains many undesired distortions, this does not result in a lower classification accuracy. Another interesting result is that, although the overall classification accuracy on the continuous dataset is only slightly worse, there is a very large difference between the false negative rates for both datasets. Since false negatives are critical, the ESOM map is better suited for our case than BLOSSOM. Finally, it should be noted that more complex classifiers such as the SVM did not perform better than the ESOM or BLOSSOM, and that the currently used system for our case is a multi-layer perceptron, which does not provide any insight into the problem (since it is a black-box), and its performance is around 80% only.

## 6   Conclusions and Future Work

In this paper, the usefulness of two recent SOM tools for studying an interesting police dataset was showcased. By applying the ESOM tool, it was possible to discover that the distribution of the dataset is spherical. By consequence, the spherical SOM tool BLOSSOM seemed natural to apply. By using this spherical SOM technique, interesting results for exploratory purposes of the data were discovered. Finally, a comparison between the ESOM and the BLOSSOM maps was performed by means of a nearest-neighbor classifier. However, it should be noted that a full-fledged benchmarking of the ESOM and BLOSSOM tools, using the full dimensionality of 123 features, is beyond the scope of this paper and is a topic for future research.

The authors are grateful to the Amsterdam-Amstelland Police, for providing us with the data. Jonas Poelmans is aspirant of the Research Foundation – Flanders.

# References

[1] Keus, R., Kruijff, M.S.: Huiselijk geweld, draaiboek voor de aanpak. Directie Preventie, Jeugd en Sanctiebeleid van de Nederlandse justitie (2000)

[2] Watts, C., Timmerman, C.: Violence against women: global scope and magnitude. The Lancet 359(9313), 1232–1237 (RMID 1155557)

[3] Waits, K.: The criminal Justice System's response to Battering: Understanding the problem, forging the solutions. Washington Law Review 60, 267–330 (1984-1985)

[4] Minleer-Black, C.: Domestic violence: Findings from a new British Crime Survey self-completion questionnaire. Home Office Research Study, London (1999)

[5] Vincent, J.P., Jouriles, E.N.: Domestic violence. Guidelines for research-informed practice. Jessica Kingsley Publishers, London (2000)

[6] Ritter, H.: Non-Euclidean Self-Organizing Maps, pp. 97–109. Elsevier, Amsterdam (1999)

[7] Kohonen, T.: Self-Organized formation of topologically correct feature maps. Biological Cybernetics 43, 59–69 (1982)

[8] Ultsch, A., Moerchen, F.: ESOM-Maps: Tools for clustering, visualization, and classification with Emergent SOM. Technical Report Dept. of Mathematics and Computer Science, University of Marburg, Germany, No. 46 (2005)

[9] Ultsch, A., Hermann, L.: Architecture of emergent self-organizing maps to reduce projection errors. In: Proc. ESANN 2005, pp. 1–6 (2005)

[10] Ultsch, A.: Density Estimation and Visualization for Data containing Clusters of unknown Structure. In: Proc. GfKI 2004 Dortmund, pp. 232–239 (2004)

[11] Ultsch, A.: Maps for visualization of high-dimensional Data Spaces. In: Proc. WSOM 2003, Kyushu, Japan, pp. 225–230 (2003)

[12] Ultsch, A., Siemon, H.P.: Kohonen's Self Organizing Feature Maps for Exploratory Data Analysis. In: Proc. Intl. Neural Networks Conf., pp. 305–308 (1990)

[13] Tokutaka, H., BLOSSOM Software Tool, http://www.somj.com

[14] Nakatsuka, D., Oyabu, M.: Application of Spherical SOM in Clustering. In: Proc. Workshop on Self-Organizing Maps (WSOM 2003), pp. 203–207 (2003)

[15] Van Hulle, M.: Faithful Representations and Topographic Maps from distortion based to information based Self-Organization. Wiley, New York (2000)

[16] Peng, H., Long, F., Ding, C.: Feature Selection Based on Mutual Information: Criteria of Max-Dependency, Max-Relevance, and Min-Redundancy. IEEE Transactions on pattern analysis and machine intelligence 27(8) (2005)

# On the Finding Process of Volcano-Domain Ontology Components Using Self-Organizing Maps

J.R.G. Pulido[1], M.A. Aréchiga[1],
E.M.R. Michel[1], G. Reyes[2], and V. Zobin[2]

[1] Faculty of Telematics, University of Colima, México
{jrgp,mandrad,ramem}@ucol.mx
[2] Volcanic observatory (RESCO), University of Colima, México
{gard,zobin}@ucol.mx

**Abstract.** Monitoring volcanic activity is a task that requires people from a number of disciplines. Infrastructure, on the other hand , has been built all over the world to keep track of these living earth entities, ie volcanoes. In this paper we present an approach that merges a number of computational tools and that may be incorporated to existing ones to predict important volcanic events. It mainly consists of applying artificial learning, ontology, and software agents for the analysis, organization, and use of volcanic-domain data for the communities of people, living nearby volcanoes, benefit. This proposal allows domain experts to have a view of the knowledge contained in and that can be extracted from the Volcanic-Domain Digital Archives (VDDA). Specific-domain knowledge components with further processing, and by embedding them into the digital archive itself, can be shared with and manipulated by software agents. In this first study, we deal with the issue of applying Self-Organizing Maps (SOM), to volcano-domain signals originated by the activity of the Volcano of Colima, Mexico. By applying this algorithm we have generated clusters of volcanic activity and can readily identify families of important events.

## 1   Introduction

Every day the activity of quiet a few volcanoes in the world attract the attention of the goverment and scientists. This activity varies in intensity. Volcanic eruptions are, in most cases, one of the most deadly natural disasters in the world. In the worst case, whole areas are devastated by erupting volcanoes, including communities living near by. A number of computational architectures and resources have been set up all around the world to monitor, forecast, and alert people regarding volcano activity. This paper is a first approach to the problem of analysing volcanic seismology signals from the computational perspective, in particular applying Self-Organizing Maps.

The next generation of volcano domain computational tools require that the huge amount of information generated by volcanoes and contained into VDDA is structured [8, 4, 19]. In the last few years a number of proposals on how to represent knowledge via ontology languages have paraded [5, 13]. Now that OWL has become a standard [18], the real challenge, in the context of semantics, has started. Eventually, the knowledge contained into VDDA will become semantic knowledge, ie software agents will be able

J.C. Príncipe and R. Miikkulainen (Eds.): WSOM 2009, LNCS 5629, pp. 255–263, 2009.

**Table 1.** Seismic signal sample file

| TipEvent | EZV4 | EZV5 | Lat. | Long. | Mag. | Prof. | VelAp. | #E | Archivo |
|---|---|---|---|---|---|---|---|---|---|
| ve | 408 | 416 | 19.519 | -103.629 | 3.8 | 0.8 | 17.97 | 6 | 02030131.rss |
| lp | 38 | 46 | 19.528 | -103.612 | 1.0 | 2.8 | 14.68 | 6 | 02030202.rss |
| ve | 380 | 385 | 19.525 | -103.607 | 3.7 | 2.9 | 11.57 | 6 | 02030240.rss |
| lp | --- | 25 | 19.831 | -103.526 | 0.6 | 15.0 | 10.43 | 3 | 02030255.rss |
| lp | 26 | 26 | 19.815 | -103.489 | 0.7 | 15.0 | 12.09 | 3 | 02030257.rss |
| lp | 34 | 24 | 19.826 | -103.512 | 0.8 | 15.0 | 10.31 | 3 | 02030258.rss |
| rf | 75 | --- | ------ | ------- | --- | --- | ----- | 1 | 02030640.rss |
| lp | 12 | 12 | 19.827 | -103.516 | -0.1 | 15.0 | 11.00 | 3 | 02030813.rss |
| ve | 401 | 410 | 19.525 | -103.628 | 3.7 | 1.7 | 17.00 | 6 | 02031045.rss |

to understand, manipulate, and even carry out inferencing and reasoning tasks for us. Converting such as digital archives into semantic ones is to take much longer if no semi-automatic approaches are taken into account to carry out this enterprise. This is what our paper is all about, a step forward towards the realization of semantic volcano-domain digital archives.

The remainder of this paper is organized as follows. Some related work is presented in section 2, including volcanology signal processing, ontology, and artificial learning concepts. Our approach is described in section 3. Some results are reported in section 4. The paper concludes in section 5 with thoughts on the approach we have applied to analyse volcanic-domain data and some future work.

## 2   Related Work

One of the most important aspects of monitoring volcano activity is forecasting. An important number of research papers on this area are found in the literature [21, 15, 23, 7]. On the other hand, in the context of *semantics*, perhaps the most important aspect is related to mapping unstructured data into software agent enable knowledge [1, 4].

### 2.1   Volcanology

A volcano is a vent in the crust of earth from which melted rock, gas, steam and ash from inside earth sometimes burst. Explosion sequences are common hazards at many volcanoes. Statistical analyses of such sequences form the basis of forecasting models and reveal underlying processes. Some mathematical models have been proposed to describe th behavior of processes occurring within volcanoes [3]. However, no single statistical model describes what is going on exactly in there, ie interexplosion repose intervals in vulcanian systems [16]. Artificial learning techniques, such as the one described in this paper may help here.

*The Volcano of Colima.* This is an andesitic 3860m high stratovolcano. It is one of the most active in Mexico and in the world. Located in the western part of the country (103.62°W, 19.514°N) has had significant eruptive activity over the last five centuries [2]. The seismic stations of the Telemetric Seismic Network (RESCO) systematically record the seismic signals produced by the Volcano of Colima [23].

**Table 2.** A excerpt of a volcano ontology

```
<owl:Class rdf:about="#Volcano">
    <rdfs:subClassOf rdf:resource="#TopographicalRegion"/>
    <rdfs:subClassOf rdf:resource="#VolcanicSystem"/>
.
.
.
    <rdfs:subClassOf>
      <owl:Restriction>
        <owl:onProperty rdf:resource="#primarySubstance"/>
        <owl:someValuesFrom rdf:resource="&substance.owl;#Magma"/>
      </owl:Restriction>
    </rdfs:subClassOf>
  </owl:Class>
```

## 2.2  Signal Processing

The analysis and processing of signals of volcano-domain signals requires the avail-ability of a formal description. This mathematical description may be referred to as the signal model. In Table 1, an excerpt of volcano signal sampling file is presented, some data are omitted.

*Deterministic vs random.*  Any observed data representing a physical phenomenon can be broadly classified as being either deterministic or nondeterministic. *Deterministic* refers to the fact that past, present and future values of a signal are known by means of a mathematical expression. *Random* signals, on the other hand, evolve in time in an unpredictable manner, ie a physical phenomenon, such a volcanic process, cannot be described by an explicit mathematical relationship, because each observation is unique, ie random.

*Stationary vs nonstationary.*  Nondeterministic (random) data may be further classified. *Stationary* random data refers to the fact that a set of *moments* and joint-moments of the datasets are time-invariant. For instance, if we take the mean-value, ie first moment, of the random process at time $t_1$ and $t_2$ we find that these values do not vary signifi-cantly in time. *Nonstationary*, on the other hand, refers to random processes where the moments, and usually also the joint-moments, vary in time.

*Volcanic processes.*  Volcanoes are then complex, dynamic, nonlinear systems [14]. These systems are the results of a number of subsystems interacting. A sample record of random data is also known as a sample function $x(t, s)$. The set of sample functions of the phenomenon is a random process denoted as $X(t, S)$ where $t$ represents time, $S$ the sample space of all possible sample functions, and $s$ is a value of the sample space, $s \in S$. For the sake of simplicity, the variables $s, S$ are usually dropped, so the random process is denoted as $X(t)$ and a single realization as $x(t)$.

A vast source of information for research can be found in [22]. In this book, the properties of volcano-tectonic earthquakes are described. A methodology and some applications for predicting eruptions are discussed. A classification of volcanic earth-quakes is also presented. A study of volcanic explosions carried out onto four volcanoes is described in [20]. This study focuses on applying several basic statistical techniques to small-scale events in trying to find clustering properties. An important software tool for volcanic-domain data is visualization. In [9] a study that explores these techniques is presented. Researchers in the geoscience areas consider increasingly important using visualization and clustering software tools as an useful device to analyse data.

### 2.3    The Purpose of Ontology

Scientists among disciplines require a framework in order to be able to interact with each other. Ontology is a framework that makes it possible for people and software agents to communicate in a consistent, complete, and distributed way. Even more, we are able to encode, for a particular domain, say a volcano-domain, the following:

- entities, objects, processes, and concepts.
- relationships of entities, objects, processes, and concepts.
- relationships across discipline areas.
- domain-dependant axioms.
- multilingual knowledge of the domain.
- assumptions, parameter settings, experimental conditions as well.

In Table 2, an excerpt of a *volcano* ontology written in OWL [18] and defined by the Semantic Web for Earth and Environmental[1] Terminology is presented. Some superclasses are shown. From this, a taxonomy can then be derived and viceversa in a semi-automatic way by means of appropriate ontology software tools.

### 2.4    Artificial Learning

Creating a volcano domain taxonomy scheme may help improve existing forecasting software systems. Perhaps the most well-known SOM project is [11], where the results of applying it as a document organization, searching and browsing system, to a set of about 7 million electronic patent abstracts was described. In this case, a document map is presented as a series of HTML pages facilitating exploration. A specified number of best-matching points are marked with a symbol that can be used as starting points for browsing. Documents are grouped using self-organizing maps, and then a graphical real-world metaphor is used to present the documents to users.

## 3    Our Approach

As we have mentioned, by merging a number of computational tools, we are able to deliver enhanced forecasting software systems. In the first stage of our approach, we focus on applying artificial learning to VDDA in order to create a volcano-domain taxonomy that may be of help for the creation of a basic volcano-domain ontology. Our system can be regarded as a set of software tools that helps in the semi-automatic construction of

---

[1] http://sweet.jpl.nasa.gov

domain-specific ontologies, in particular by clustering together a number of elements of the following sets 1) set of objects (entities, concepts), 2) set of functions (for example *is-a*), 3) set of relations (*has* for instance). Of course, finding functions and relations is a domain-dependant task. In our case, we are just looking for objects, ie families of seismic signals to create a basic volcano-domain ontology. Domain experts are always needed in order to validate the ontology components that have been identified.

## 3.1   Constructing Ontologies

The obvious source of information for constructing a volcano-domain ontology is the data contained in the digital archives themselves. Datasets can be regarded as high dimensional vector spaces and can be represented either in a tabular form as shown in the following table:

| $X$ | $v_1$ | $\cdots$ | $v_m$ |
|---|---|---|---|
| $x_1$ | $a_{11}$ | $\cdots$ | $a_{1m}$ |
| $\vdots$ | $\vdots$ | $\ddots$ | $\vdots$ |
| $x_n$ | $a_{1n}$ | $\cdots$ | $a_{nm}$ |

or in a mathematical way as follows:

$$x_j = \sum_k a_{jk} e_k \tag{1}$$

where $\{v_1, \cdots, v_n\}$ are n-dimensional *variables*, and $\{x_1, \cdots, x_n\}$ are m-dimensional *samples*, $e_k$ is the unit vector and $a_{jk}$ is the value of $v_j$ in $x_k$. For the volcano-domain ontology construction process, it is important to identify knowledge components and not to start from scratch. A good ontology assures scientists that software agents can reason properly about the domain knowledge and, for instance, forecast important events on our behalf. It is very important to bear in mind that a domain expert, ie a volcanologist, must be always part of the taxonomy creation team for validating the ontology.

## 3.2   Visualizing Ontology Components

By using SOM we are able to visualize, clustered together, volcano-domain ontology components. Self-Organizing Maps can be viewed as a model of unsupervised learning and an adaptive knowledge representation scheme. Adaptive means that at each iteration a unique sample is taken into account to update the weight vector of a neighbourhood of neurons [10]. Adaptation of the model vectors take place according to the following equation:

$$m_i(t+1) = m_i(t) + h_{ci}(t)[x(t) - m_i(t)] \tag{2}$$

where $t \in \mathcal{N}$ is the discrete time coordinate, $m_i \in \Re^n$ is a node, and $h_{ci}(t)$ is a neighbourhood function. The latter has a central role as it acts as a smoothing kernel defined over the lattice points and defines the stiffness of the surface to be fitted to the data points. This function may be constant for all the cells in the neighbourhood and zero elsewhere. A common neighbourhood kernel that describes a natural mapping and that is used for this purpose can be written in terms of the Gaussian function:

$$h_{ci}(t) = \alpha(t) \exp(-\frac{||r_c - r_i||^2}{2\sigma^2(t)}) \tag{3}$$

where $r_c, r_i \in \Re^2$ are the locations of the winner and a neighbouring node on the grid, $\alpha(t)$ is the learning rate ($0 \leq \alpha(t) \leq 1$), and $\sigma(t)$ is the width of the kernel. Both $\alpha(t)$ and $\sigma(t)$ decrease monotonically.

Our system consists of two applications: Spade and Grubber [17]. The former pre-processes data and creates a dataspace suitable for training purposes. The latter is fed with the dataspace and produces knowledge maps that allow us visualize ontology components contained in the digital archive. As we have mentioned, in a semantic context, they may later be organized as a set of *Entities*, *Relations*, and *Functions*. Problem solvers use this triad for inferring new data from the domain [1, 6] and carrying out reasoning. The major steps of our approach are as follows:

1. Produce a *dataspace*. A dataset is created with the individual vector spaces from the domain by *spade*. In some cases, when the dataset already exists, spade carries out a pre-processing validation task, ie merging sample files, removing headers from files.
2. Construct the SOM. Once the dataset is a valid one, a second software tool, *grubber*, is fed and trained with the dataset and ontology maps are then created.

## 4   Results

In this section the results of the experiments we have carried out are presented. Three stages have been clearly identified regarding volcanic activity: 1) preliminary seismic activity stage 2) period from the first to the largest explosion stage 3) post-explosion activity stage. For our study we have used data from a second stage. From the bulk of data, including long-period (LP), rockfall, explosion, only three kind of events have been select, namely hybrid, volcano-tectonic, and tremor. In particular, apparent velocity (kms/s), type of event, and coda (duration of the event in secs) and magnitude (Mb) of prior eruption events were used (cf.table 1). By simple observation, volcano-domain experts have been able to classify volcanic events [22]. In figure 1 a visual representation of the events studied is presented. These events usually occur before erupting activity. However, they may also occur during erupting stages.

It took over 56 minutes on a Pentium 4 CPU 2Gb RAM computer training 22x22 SOM ($1.4 \times 10^6$ iterations) fed with a 4k sample dataset selected from a 40k bulk dataset. Background knowledge is very important. Domain experts, ie volcanologists, have to validate our findings in order to create a good ontology. Browsing the SOMs gives us a clear idea and helps us understand what the domain is all about.

Volcano-domain ontology components can be visualized clustered together from the knowledge maps created. We start with a randomly initialized map and after a training process, clusters of volcano-domain ontology components can be readily identified from the map. After the maps are trained through repeated presentations of all the samples in the collection, a labelling phase is carried out. Neighbouring nodes that contain

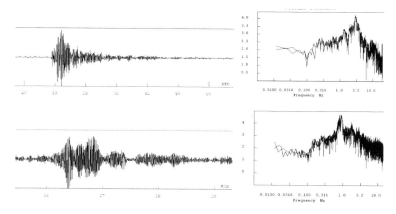

**Fig. 1.** Volcano-tectonic event (above). Tremor event (below). On the left, the time-series. On the right, the frequency-series.

the same winning elements merge to form concept regions. The resulting maps represent areas where neighbouring elements are similar to each other. The software interface we have created allows us to relate information from the samples in such a way that each node has a feature that relates to its corresponding subfeatures. This can be seen as we browse the maps and help us understand the clusters that have been formed. A brief definition of the clusters found is given as follows.

*Volcano-tectonic.* VT events are usually the result of pressure from magma cracking solid rock. Also known as high-frequency and type-A events. VT seismograms look just like typical earthquake seismograms. VT activity is an early sign of volcanoes becoming active. These are generated 1-20km down the volcano. They occur in *swarms* and generally have a 1-5Hz frequency, a 0-2.2Mb magnitude, an apparent velocity 0-50kms, and a duration of days to years.

*Tremor.* Sudden changes in pressure within magma filled cracks and channels cause this kind of events. These are long duration low frequency surface wave occurrences. These are a good indicator of forthcoming eruptions. Volcanic tremors have a irregular sinusoid form compared with earthquakes of the same magnitude. They have a 0.5-3Hz frequency, and a duration of minutes to months.

*Hybrid.* Volcanoes produce a variety of pre-eruptive seismic signals. VT activity sometimes generate tremors and viceversa. This kind of seismic signal contains a mixture of VT and tremor data. Hybrid activity comprises a class of signals usually having high-frequency onsets followed by low-frequency. This kind of event has a 1.5-4.5Hz frequency, a magnitude 0-2.5Mb, an apparent velocity 0-100kms, and a duration of hours to weeks.

These preliminary results were surprisingly close to our intuitive expectations. The created clusters correspond to hybrid, volcano-tectonic, and tremor, consistent to other studies [21, 23]. However, as can be seen from figure 2 some other new clusters have emerged. After this, some other ontology tools such as editors can be used to organize

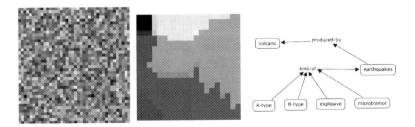

**Fig. 2.** After a training process, a randomly initialized SOM (left) becomes a categorized one (middle). Then an ontology can be derived with the help of a domain-expert (right).

this knowledge. Then, it can be embedded into the digital archive where it was extracted from by means of any of the ontology languages that exist.

## 5   Conclusions

The vast amount of data generated by volcanoes has eventually to be transformed into semantic data such that software agents are able to carry out, on our behalf, inference and reasoning tasks, including forecasting. The acquisition and representation of knowledge needs to take into account the complexity that is often present in domains. Volcano-domain experts are always needed in order to assure the quality of the ontology created. In this paper we have presented a novel approach that generates clusters of volcanic activity and readily help us create basic volcano-domain ontologies, and later identify situations of risk for predicting important events. However, some more research is required in order to fine tune the semi-automatic specific-domain ontology creation process.

In further stages of this research we are to analize some more specific families of volcanic seismic signals, in particular extrusions and temor families, as well as the ones emerging from our study.

## References

[1] Berners-Lee, T., et al.: The Semantic Web. Scientific American 284(5), 34–43 (2001)
[2] Breton, M., et al.: Summary of the historical eruptive activity of volcán de Colima, México 1519-2000. Journal of Volcanology and Geothermal Research 117(1-2), 21–46 (2002)
[3] Connor, C., et al.: Exploring links between physical and probabilistic models of volcanic eruptions: The soufriere hills volcano, montserrat. Geophysical research letters 30(13) (2003)
[4] Crow, L., Shadbolt, N.: Extracting focused knowledge from the Semantic Web. Int. J. Human-Computer Studies 54, 155–184 (2001)
[5] Gómez, A., Corcho, O.: Ontology languages for the Semantic Web. IEEE Intelligent Systems (2002)
[6] Gómez, A., et al.: Knowledge maps: An essential technique for conceptualisation. Data & Knowledge Engineering 33, 169–190 (2000)

[7] Jenkins, S., et al.: Multi-stage volcanic events: A statistical investigation. Journal of Volcanology and Geothermal Research 161(4), 275–288 (2007)

[8] Pulido, J.R.G., et al.: A novel approach to the analysis of volcanic-domain data using self-organizing maps: A preliminary study on the volcano of colima. Research in computer science 40, 49–59 (2008)

[9] Kadlec, B., et al.: Visualization and analysis of multi-terabyte geophysical datasets in an interactive setting with remote webcam capabilities. In: Yin, X., et al. (eds.) Computational earthquake physics: simulations, analysis and infrastructure PART II, pp. 2455–2465. Birkhauser-Verlag, Basel (2006)

[10] Kohonen, T.: Self-Organizing Maps, 3rd edn. Information Sciences Series. Springer, Berlin (2001)

[11] Kohonen, T., et al.: Self organization of a massive text document collection. In: Oja, E., Kaski, S. (eds.) Kohonen Maps, pp. 171–182. Elsevier Sci., Amsterdam (1999)

[12] Legrand, S., Pulido, J.R.G.: NLP agents and document clustering in knowledge management: the semantic web case. In: Raisinghani, M. (ed.) Handbook of Research on Global Information Technology, pp. 476–498. IGI Global Publishing (2008)

[13] Martin, P., Eklund, P.: Embedding knowledge in web documents. Computer Networks 31, 1403–1419 (1999)

[14] Melnik, O., Sparks, R.: Nonlinear dynamics of lava dome extrusion. Nature 402, 37–41 (1999)

[15] Reyes, G., Cruz, S.: Experience in the short-term eruption forecasting at volcán de Colima, México, and public response to forecasts. Journal of Volcanology and Geothermal Research 117(1–2), 121–127 (2002)

[16] Watt, S., et al.: Vulcanian explosion cycles: Patterns and predictability. Geology 35(9), 839–842 (2007)

[17] Pulido, J.R.G., et al.: Identifying ontology components from digital archives for the semantic web. In: IASTED Advances in Computer Science and Technology (ACST), pp. 1–6 (2006) (CD edition)

[18] Pulido, J.R.G., et al.: Ontology languages for the semantic web: A never completely updated review. Knowledge-Based Systems 19(7), 489–497 (2006)

[19] Pulido, J.R.G., et al.: Artificial learning approaches for the next generation web: part I. Ingeniería Investigación y Tecnología, UNAM (CONACyT) 9(1), 67–76 (2008)

[20] Varley, N., et al.: Applying statistical analysis to understand the dynamics of volcano explosions. In: Mader, H., et al. (eds.) Statistics in volcanology, pp. 57–76. Geological society for IAVCEI, London (2006)

[21] Zamora, A., et al.: The 1997-1998 activity of Volcán de Colima, western México: Some aspects of the associated seismic activity. Pure and applied geophysics 164(1), 39–52 (2007)

[22] Zobin, V.: Introduction to volcanic seismology. Elsevier, Amsterdam (2003)

[23] Zobin, V., et al.: Comparative characteristics of the 1997-1998 seismic swarms preceding the november 1998 eruption of volcán de Colima, México. Journal of Volcanology and Geothermal Research 117(1–2), 47–60 (2002)

# Elimination of Useless Neurons in Incremental Learnable Self-Organizing Map

Atsushi Shimada and Rin-ichiro Taniguchi

Department of Advanced Information Technology, Kyushu University
744, Motooka, Nishi-ku, Fukuoka 819–0395, Japan
{atsushi,rin}@limu.ait.kyushu-u.ac.jp
http://limu.ait.kyushu-u.ac.jp

**Abstract.** We propose a method to eliminate unnecessary neurons in Variable-Density Self-Organizing Map. We have defined an energy function which denotes the error of the map, and optimize the energy function by using graph cut algorithm. We conducted experiments to investigate the effectiveness of our approach.

**Keywords:** Self-Organizing Map, Elimination of Neurons, Graph Cut.

## 1 Introduction

Self-Organizing Map (SOM) proposed by Kohonen[1,2] is one of the most widely used artificial neural network algorithms which uses unsupervised learning. There are two types of learning algorithms. One is batch learning and the other is incremental learning. Generally, the batch learning algorithm is effective when all of the training samples are given preliminarily. Meanwhile, the incremental learning algorithm should be used when learning data are input into SOM sequentially. In the incremental leaning of SOM, it is very difficult to determine the number of neurons in the map since the number of training samples is usually unknown. In addition, a new training sample affects the weights of neurons. Therefore, the SOM tends to forget training data previously given. Shimada *et al.* proposed an incremental learning algorithm which resolves these problems[3]. In this algorithm, new neurons are inserted into the map when the number of neurons is not enough to learn a new training sample (see detailed algorithm in section 2). However, the algorithm has a problem that topological relation is destroyed among the inserted neurons and the existing neurons. This is because some neurons including inserted neurons and the existing neurons become useless neurons which destroy the topological relation in the process of incremental learning. Therefore, we propose an approach to eliminate such useless neurons from the map. The elimination method is summarized as follows: 1) find candidate neurons which destroy the topology of the map, 2) calculate an error energy of the map after eliminating candidate neurons, 3) minimize the error energy according to the Graph Cuts algorithm[4].

J.C. Príncipe and R. Miikkulainen (Eds.): WSOM 2009, LNCS 5629, pp. 264–271, 2009.

## 2   Incremental Learning of Self-Organizing Map

In this section, we explain the incremental learning algorithm of Self-Organizing Map. We call the incremental learnable Self-Organizing Map "VDSOM (Variable Density Self-Organizing Map)" [3] derived from its appearance. VDSOM has three types of neurons in its map.

- Weight-fixed neuron
- Weight-quasi-fixed neuron
- Normal neuron

In the following section, we will give an explanation of these neurons. And then, we will show the algorithm of incremental learning.

### 2.1   Weight-Fixed Neuron

A weight vector of weight-fixed neuron is no longer updated in the training process. This helps SOM remember a training sample previously given. Therefore, one of the neurons is inevitably selected the weight-fixed neuron to remember the training sample. On the other hand, weight vectors of neurons surrounding a weight-fixed neuron should not be updated easily because they need to have similar weight vector with the weight-fixed neuron. We call such a neuron "weight-quasi-fixed neuron". Finally, normal neurons are far from a weight-fixed neuron. Their weight vectors are updated in the same way as a standard SOM.

### 2.2   Learning Algorithm

**Step 1.** Initializing the weights of neurons randomly.
**Step 2.** A neuron which satisfies equation (1) is selected as the winner neuron $u_c$.

$$c = \operatorname*{argmin}_{i} \left( ||\boldsymbol{x} - \boldsymbol{w}_i|| \right) \tag{1}$$

where $\boldsymbol{x}$ is an input vector of training sample, and $\boldsymbol{w}_i$ is the weight vector of neuron $u_i$.
**Step 3.** If the $u_c$ is weight-fixed neuron, eight neurons are inserted around the neuron $u_c$. Otherwise, jump to Step 5.
**Step 4.** The initial weights of the newly inserted neurons $\boldsymbol{w}_{new}$ are calculated as follow. If $u_c$ is the nearest weight-fixed neuron from the neuron $u_{new}$,

$$\boldsymbol{w}_{new} = h_{c,new}\boldsymbol{w}_c \tag{2}$$

otherwise,

$$\boldsymbol{w}_{new} = \frac{h_{c,new}\boldsymbol{w}_c + h_{f,new}\boldsymbol{w}_f}{h_{c,new} + h_{f,new}} \tag{3}$$

$f$ denotes the nearest weight-fixed neuron. $h_{a,b}$ is a neighborhood kernel defined by equation (4):

$$h_{a,b} = \exp\left( -\frac{||r_a - r_b||^2}{2\sigma^2(t)} \right) \tag{4}$$

where $r_a$ is position vector of neuron $u_a$. $\sigma(t)$ is monotonically decreasing function calculated as follow.

$$\sigma(t) = \sigma_i \left(\frac{\sigma_f}{\sigma_i}\right)^{t/T} \tag{5}$$

$T$ is the limit of iterations, $\sigma_i$ is the initial value of $\sigma(t)$ and $\sigma_f$ is the last value of $\sigma(t)$.

**Step 5.** Weight vectors are updated by equation (6). However, the weight-fixed neurons are not updated.

$$\boldsymbol{w}_i(t+1) = \boldsymbol{w}_i(t) + H(d)\alpha(t)h_{c,i}(\boldsymbol{x}(t) - \boldsymbol{w}_i(t)) \tag{6}$$

$\alpha(t)$ is also monotonically decreasing function.

$$\alpha(t) = \frac{-\alpha_0(t-T)}{T} \tag{7}$$

$\alpha_0$ is the initial value of training rate. $H(d)$ is the function which defines the distribution weight-quasi-fixed neurons.

$$H(d) = \frac{1 - \exp(-d \cdot k)}{1 + \exp(-d \cdot k)} \tag{8}$$

$d$ is the position distance between a neuron $u$ and its nearest weight-fixed neuron. $k$ is a coefficient to determine the slope of the function $H(d)$. This function is about 0 with decreasing the value of $d$. Therefore, the weights of neurons around a weight-fixed neuron are not updated easily. On the other hand, with increasing the value of $d$, the function $H(d)$ comes close to 1. Therefore, the weights of neurons are updated in a similar way to standard SOM. This function enables SOM to spread weight-fixed neurons all over the map.

**Step 6.** The same training sample is learned by Step 2 to Step 5 iteratively until the weight-fixed condition; $U < d_f$ is satisfied, where $U$ is the distance between the input vector $x$ and the weight of the winner neuron $\boldsymbol{w}_c$ and $d_f$ is a given threshold.

**Step 7.** If one of the neurons satisfies the weight-fixed condition, the next training sample is input to VDSOM.

### 2.3  Negative Effects of Incremental Learning

The VDSOM grows the map by inserting new neurons adaptively. This is very effective to learn a new training sample incrementally. However, it has some problems summarized as follows.

- The newly inserted neurons are not always useful for the map. In other words, some neurons are useless to memorize training samples previously given.
- Some neurons come to destroy the proximity of the map since the newly inserted neurons have initial weights which are not rigorously but simply calculated according to the surrounding neurons.

To solve above problems, we have to find "useless neurons" in the map and eliminate them.

# 3 Elimination of Useless Neurons

## 3.1 Labeling for Neuron

We define a topographic error of SOM[5] by following equation (9):

$$E_t = \frac{1}{K} \sum_{k=1}^{K} f(\boldsymbol{x}_k)$$

$$f(x_k) = \begin{cases} 1, & u_f^k \text{ and } u_s^k \text{ non-adjacent} \\ 0, & \text{otherwise} \end{cases} \tag{9}$$

where $K$ is the number of training samples and $x_k$ is an input vector of a training sample. We denote a neuron which has nearest weight vector for the input vector $x_k$ with $u_f^k$ and second-nearest weight vector with $u_s^k$. The $E_t$ comes close to 0 with decreasing the topographic error.

At the next step, we define two kinds of labels for each neuron $u_i$. One is a label about topographic error (equation (10)) and the other is a label about winner neuron (equation (11)).

$$l_i^E = \begin{cases} 1, & \text{all } u_i \text{ between } u_f^k \text{ and } u_s^k \text{ when } f(\boldsymbol{x}_k) \text{ is 1} \\ 0, & \text{otherwise} \end{cases} \tag{10}$$

$$l_i^W = \begin{cases} 1, & \text{if } u_i \text{ is best-match unit} \\ 2, & \text{if } u_i \text{ is second-best-match unit} \\ 0, & \text{otherwise} \end{cases} \tag{11}$$

For these labeling, neurons are divided into 6 types according to the combination of two kinds of labels ($l_i^E$ and $l_i^W$). These labels are used when an error energy of the map is calculated (see detailed explanation in the next section). Finally, we define another label about elimination.

$$l_i^D = \begin{cases} 1, & \text{if } u_i \text{ is useless} \\ 0, & \text{otherwise} \end{cases} \tag{12}$$

The labeling for $l_i^D$ is achieved after minimizing the error energy of the map. The neuron whose label is $l_i^D = 1$ will be eliminated from the map.

## 3.2 Error Energy of Self-Organizing Map

The error energy of the map is defined by equation (13).

$$P(\boldsymbol{l}^D; \boldsymbol{u}) = \sum_{i \in N} P_L(l_i^D; u_i) + \sum_{(i,j) \in M} P_S(l_i^D, l_j^D; u_i, u_j) \tag{13}$$

We denote $P_L$ the data term for neuron's label and $P_S$ the smoothing term. The $N$ and $M$ are a set of all neurons and a set of neighboring neurons respectively.

The data term is calculated by equation (14).

$$P_L(l_i^D = 0; u_i) = \begin{cases} 0, & \text{if } l_i^E = 1, l_i^W = 0 \\ 1, & \text{if } l_i^E = 0, l_i^W = 1 \\ 1/d_{f,s}, & \text{if } l_i^E = 0, l_i^W = 2 \\ \alpha, & \text{if } l_i^E = 1, l_i^W = 1 \\ \alpha/d_{f,s}, & \text{if } l_i^E = 1, l_i^W = 2 \\ 1/2, & \text{otherwise} \end{cases} \tag{14}$$

$$P_L(l_i^D = 1; u_i) = 1 - P_L(l_i^D = 0; u_i)$$

where $\alpha (0 \leq \alpha \leq 1)$ is a parameter to handle the neuron which has labels of $l_i^E = 1$ and $l_i^W \geq 1$. Such a neuron has bilateral characters; one is that the neuron is useless because of $l_i^E = 1$, the other is that the neuron is required by a training sample because of $l_i^W \geq 1$. If the value of $alpha$ is small, the neuron is easily regarded as useless. In addition, if $l_i^W = 2$, we calculate $d_{f,s}$ (position distance between best-match unit neuron and second-best-match neuron), and let the data term small by dividing itself by $d_{f,s}$.

The smoothing term is defined by equation (15).

$$P_S(l_i^D, l_j^D; u_i, u_j) = \frac{\beta}{d_{i,j}} \frac{|l_i^E - l_j^E| + |l_i^W - l_j^W|}{\ln \left( ||\boldsymbol{w}_i - \boldsymbol{w}_j||^2 + 1 + \varepsilon \right)} \tag{15}$$

where $\boldsymbol{w}_i$ is the weight vector of neuron $u_i$. The $\beta$ is a parameter to adjust the degree of incidence of smoothing term, and $\varepsilon \ll 1$.

### 3.3    Minimization of Error Energy

It is a combinational optimization problem to minimize the error energy of the map $P(\boldsymbol{l}^D; \boldsymbol{u})$ for labels of all neurons simultaneously. We use the Graph Cuts algorithm[4] to acquire the global optimum solution. We make a graph which has nodes and edges shown in Fig. 1. First, we prepare nodes corresponding to neurons in VDSOM, a terminal node (source $T$) which denotes the useful neuron and a terminal node (sink $R$) which denotes the useless neuron. Next, we make edges between nodes which have costs $q(i, j)$ calculated by equation (15).

$$q(i, j) = q(j, i) = P_S(l_i^D, l_j^D; u_i, u_j) \tag{16}$$

Edges are also made between a node and a terminal, which have costs $q(t, i)$ or $q(i, r)$.

$$q(t, i) = \lambda \cdot P_L(l_i^D = 0; u_i) \tag{17}$$

$$q(i, r) = \lambda \cdot P_L(l_i^D = 1; u_i) \tag{18}$$

Therefore, the error energy of the map (equation (13)) is acquired by calculating sum of costs.

$$P(\boldsymbol{l}^D; \boldsymbol{u}) \approx Q(T, R) = \sum_{x \in T, y \in R} q(x, y) \tag{19}$$

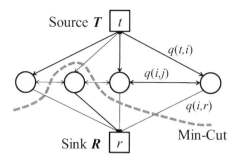

**Fig. 1.** Graph for minimizing the error energy

The Graph Cuts algorithm will give us the global optimum solution which minimize the error energy. Finally, a neuron which has the label $l_i^D = 1$ will be eliminated from the map.

## 4    Experimental Results

We generated 125 kinds of colors arbitrarily. The VDSOM learned the colors one by one incrementally. The initial map of the VDSOM consisted of $16(4 \times 4)$ neurons and we set each parameter: $\alpha_0 = 0.9, \sigma_i = 1.0, \sigma_f = 0.3, k = 0.3$. After the training process, there were 307 neurons in the map (see Fig. 2). In the Fig. 2, we painted each neuron in a color which the neuron had memorized after the training. We have also marked an area which enlarged the topographic error of the map. We can see that the color gradient is not smooth in the area.

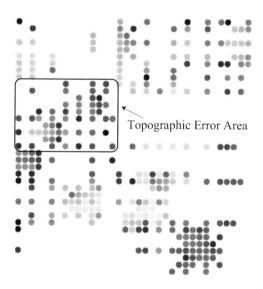

**Fig. 2.** The VDSOM after finishing the training process

Next, we eliminated some neurons which had enlarged the topographic error according to minimization of error energy. We have evaluated the topographic error $E_t$ (equation (9)) and the quantization error defined by following equation (20).

$$E_q = \frac{1}{K} \sum_{k=1}^{K} ||\boldsymbol{x}_k - \boldsymbol{w}_f^k|| \tag{20}$$

If we eliminate too much neurons, $E_t$ will become smaller. However, the $E_q$ will become larger since the number of neurons is too short to memorize all training samples. Therefore, we have to reduce $E_t$ without increasing the $E_q$. We show the changes of the $E_t$ and $E_q$ between before and after elimination of useless neurons. We can see that the parameter *beta* affected the $E_t$. The $E_t$s after elimination were smaller than the one before elimination. The error was smallest in 0.2 to 1.0 range of *beta*. On the other hand, the $E_q$s were not changed between before and after elimination. This showed that only useless neurons had been eliminated from the map by minimizing the error energy of the map. We

**Fig. 3.** The VDSOM after elimination of useless neurons

**Table 1.** Topographic error, quantization error and the number of neurons before and after elimination of useless neurons

| $\beta$ | after elimination | | | | before elimination |
|---|---|---|---|---|---|
| | 0 | 0.15 | [0.2 − 1] | 2 | − |
| $E_{t[\times 10^{-2}]}$ | 8.8 | 7.2 | 6.4 | 7.2 | 16.8 |
| $E_{q[\times 10^{-3}]}$ | 2.6 | 2.6 | 2.6 | 2.6 | 2.6 |
| $N$ | 256 | 245 | [241 − 230] | 226 | 307 |

show the map after elimination of useless neurons in Fig. 3. The color gradient becomes smooth compared with the area marked in Fig. 2.

## 5   Conclusion

We have proposed the method to eliminate useless neurons in Variable-Density Self-Organizing Map. We have defined the error energy of the map and have minimized the energy by using Graph Cuts algorithm. The method gave good results in our experiment. We are going to investigate a timing to eliminate useless neurons in our future work.

## References

1. Kohonen, T.: Self-Organization and Associative Memory. Springer, Heidelberg (1989)
2. Kohonen, T.: Self-Organizing Maps. Springer Series in Information Science (1995)
3. Shimada, A., Taniguchi, R.-I.: Variable-Density Self-Organizing Map for Incremental Learning. In: The 5th Workshop On Self-Organizing Maps (WSOM 2007) (2007)
4. Boykov, Y., Kolmogorov, V.: An Experimental Comparison of Min-Cut/Max-Flow Algorithms for Energy Minimization in Computer Vision. IEEE Transactions on Pattern Analysis and Machine Intelligence 26, 1124–1137 (2004)
5. Kiviluoto, K.: Topology Preservation in Self-Organizing Maps. In: International Conference on Neural Networks (ICNN), pp. 294–299 (1996)

# Hierarchical PCA Using Tree-SOM for the Identification of Bacteria

Stephan Simmuteit[1], Frank-Michael Schleif[1], Thomas Villmann[2]
and Markus Kostrzewa[3]

[1] Univ. Leipzig, Dept. of Medicine, 04107 Leipzig, Germany
Tel.: +49(0)3419718954; +49(0)3419718955
`stephan.simmuteit@medizin.uni-leipzig.de`,
`schleif@informatik.uni-leipzig.de`
[2] Univ. of Appl. Sc. Mittweida, Dept. of MPI, 09648 Mittweida, Germany
Tel.: +49(0)372758-1328
`thomas.villmann@hs-mittweida.de`
[3] Department of Bioanalytics, Bruker Daltonik GmbH, 28359 Bremen, Germany
`km@bdal.de`

**Abstract.** In this paper we present an extended version of *Evolving Trees* using Oja's rule. Evolving Trees are extensions of *Self-Organizing Maps* developed for hierarchical classification systems. Therefore they are well suited for taxonomic problems like the identification of bacteria. The paper focus on clustering and visualization of bacteria measurements. A modified variant of the Evolving Tree is developed and applied to obtain a hierarchical clustering. The method provides an inherent PCA analysis which is analyzed in combination with the tree based visualization. The obtained loadings support insights in the classification decision and can be used to identify features which are relevant for the cluster separation.

**Keywords:** tree som, bacteria identification, mass spectrometry, hierarchical PCA, unsupervised feature selection.

## 1 Introduction

The identification of bacteria in medical and biological environments by means of classical methods like gram stain is time consuming and frequently leads to mistakes in separation of species or even genus. These data are categorized in a taxonomical tree-structure. It can be expected that the supporting measurements reflect such a structure. Further its known that for some bacteria molecular finger prints exist [9]. Taking these two aspects into account we derive the *Hierarchical PCA-based Evolving Tree* to obtain an optimal compact encoding and tree-structured representation of such data based on *Evolving Trees* [13] and Oja-PCA learning [12].

The utilization of mass spectrometry (MS) provides a fast and reproducible way to receive bio-chemical information to identify bacteria cultured on nutrient solution. One task in this line is an appropriate classification of the high-dimensional mass spectra. This requires a reasonable classification structure to

J.C. Príncipe and R. Miikkulainen (Eds.): WSOM 2009, LNCS 5629, pp. 272–280, 2009.

achieve adequate storage and retrieval performance. It is further valuable to obtain interpretable visualizations of the data for a later expert analysis. Existing approaches are based on the direct comparison of spectra with manually selected reference spectra by means of a (pre-filtered) peak matching including their intensity as well as their mass position [9,11].

The application of MS for bacteria identification is quite new and a representation of the taxonomic (tree-) nature of bacteria is difficult. The problem of discriminating bacteria species with MS is described in [1]. Forero et al. use extracted features from images of bacteria to identify them [5]. Discrimination of bacteria can be done also by bio-markers based on MS spectra [10]. Most of those approaches are also based on the evaluation of the peak intensities. In case of bacteria even the peak intensities alone are an unsafe criterion. Further, the encoded peaks (line spectra) to be compared are huge-dimensional vectors representing a functional relation. Fast and reliable investigation of line spectra requires, on the one hand side, an adequate processing, which preserves the relevant information as good as possible. On the other hand, optimum interpretable data structures are required.

This contribution provides new aspects for efficient information-preserving representation of line spectra by a data-driven tree generation using the *Hierarchical PCA-based Evolving Trees (ET)*.

## 2 Evolving Trees and Hierarchical PCA

As mentioned above the 'natural' identification methodology in taxonomy/analysis of bacteria is tree structured. Therefore, in context of machine learning, decision trees (DT) may come into mind. However, DTs don't integrate structural data information like data shape and density in an adequate manner during tree generation. An alternative is presented by PAKKANEN ET AL. – the *evolving trees (ET)* for which we provide a formal definition later on. The ET-approach is an extension of the concept of *self-organizing maps* (SOMs) introduced by KOHONEN [6].

SOMs project high-dimensional vectorial data onto a predefined low-dimensional regular grid usually chosen as a hypercube. This mapping is topology preserving under certain conditions, i.e. in case of no violations similar data points in the data space are mapped onto the same or neighbored grid nodes. For this purpose, to each node a weight vector, also called prototype, is assigned. A data point is mapped onto this node, the prototype of which is closest according to a similarity measure in the data space, usually the Euclidean distance. This rule is called winner-take all. In this sense, all data points mapped onto the same node are called *receptive field* of this node and the respective prototype is a representative of this field.

### 2.1 Evolving Trees

Yet, the usual rectangular lattice as output structure is only mandatory. Other choices are possible depending on the task. ETs use trees as output structures and, hence, are potentially suited for mapping of vectorial data with an inherent hierarchical structure.

Suppose we consider an ET $\mathcal{T}$ with nodes $r \in R_{\mathcal{T}}$ (set of nodes) and root $r_0$ which has the depth level $l_{r_0} = 0$. A node $r$ with depth level $l_r = k$ is connected to its successors $r'$ with level $l_{r'} = k + 1$ by directed edges $\varepsilon_{r \to r'}$ with length is unit. The set of all direct successors of the node $r$ is denoted by $S_r$. If $S_r = \varnothing$ is valid, the node $r$ is called a leaf. The degree of a node $r$ is $\delta_r = \#S_r$, here assumed to be constant $\delta$ for all nodes except the leafs. A sub-tree $\mathcal{T}_r$ with node $r$ as root is the set of all nodes $r' \in R_{\mathcal{T}_r}$ such that there exists a directed cycle-free path $p_{r \to r'} = \varepsilon_{r \to m} \circ \ldots \circ \varepsilon_{m' \to r'}$ with $m, \ldots, m' \in R_{\mathcal{T}_r}$ and $\circ$ as the concatenation operation. $L_{p_{r \to r'}}$ is the length of path $p_{r \to r'}$, i.e. the number of concatenations plus 1. The distance $d_{\mathcal{T}}(r, r')$ between nodes $r, r'$ is defined as

$$d_{\mathcal{T}}(r, r') = L_{p_{\hat{r} \to r}} + L_{p_{\hat{r} \to r'}} \tag{1}$$

with paths $p_{\hat{r} \to r}$ and $p_{\hat{r} \to r'}$ in the sub-tree $\mathcal{T}_{\hat{r}}$ and $R_{\mathcal{T}_{\hat{r}}}$ contains both $r$ and $r'$ and the depth level $l_{\hat{r}}$ is maximum for all sub-trees $\mathcal{T}_{\hat{r}'}$ which contain $r$ and $r'$. A connecting path between a node $r$ and a node $r'$ is defined as follows: let $p_{\hat{r} \to r'}$ and $p_{\hat{r} \to r}$ be direct paths such that $L_{p_{\hat{r} \to r'}} \cdot L_{p_{\hat{r} \to r}}$ is $d_{\mathcal{T}}(r, r')$. Then $p_{r \to r'}$ is the reverse path $p_{r' \to \hat{r}} \cdot p_{\hat{r} \to r}$ and the node set of $P$ is denoted by $\mathcal{N}_{p_{r \to r'}}$. As for usual SOMs, each node $r$ is equipped with a prototype $\mathbf{w}_r \in \mathbb{R}^D$, provided that the data to be processed are given by $\mathbf{v} \in V \subseteq \mathbb{R}^D$. Further, we assume a differentiable similarity measure $d_V : \mathbb{R}^D \times \mathbb{R}^D \to \mathbb{R}$. The winner detection is different from usual SOM but remains the concept of winner-take-all. For a given subtree $\mathcal{T}_r$ with root $r$ the *local winner* is

$$s_{\mathcal{T}_r}(\mathbf{v}) = \arg\min_{r \in S_r} (d_V(\mathbf{v}, \mathbf{w}_r)) \tag{2}$$

If $s_{\mathcal{T}_r}(\mathbf{v})$ is a leaf then it is also the overall winner node $s(\mathbf{v})$. Otherwise, the procedure is repeated recursively for the sub-tree $\mathcal{T}_{s_{\mathcal{T}_r}}$. The *receptive field* $\Omega_r$ of a leaf $r$ (or its prototype) is defined as

$$\Omega_r = \{\mathbf{v} \in V | s(\mathbf{v}) = r\} \tag{3}$$

and the receptive field of root $r'$ of a sub-tree $\mathcal{T}_{r'}$ is defined as

$$\Omega_{r'} = \cup_{r'' \in R_{\mathcal{T}_{r'}}} \Omega_{r''} \tag{4}$$

The adaptation of the prototypes $\mathbf{w}_r$ takes place only for those prototypes, where the nodes $r$ of are leafs. The other nodes remain fixed. This learning for a randomly selected data point $\mathbf{v} \in V$ is neighborhood-cooperatively as in usual SOM:

$$\triangle \mathbf{w}_r = \epsilon h_{SOM}(r, s(\mathbf{v}))(\mathbf{v} - \mathbf{w}_r) \tag{5}$$

with $s(\mathbf{v})$ being the overall winner and $\epsilon > 0$ a small learning rate. The neighborhood function $h_{SOM}(r, r')$ is defined as a function depending on the tree distance $d_{\mathcal{T}}$ usually of Gaussian shape

$$h_{SOM}(r, r') = \exp\left(\frac{-(d_{\mathcal{T}}(r, r'))^2}{2\sigma^2}\right). \tag{6}$$

with neighborhood range $\sigma$.

Unlike for the SOM we cannot guarantee that $s(\mathbf{v})$ is the true best matching unit ($bmu$), because the tree model is subject of a stochastic optimization process.

The whole ET learning is a repeated sequence of adaptation phases according to the above mentioned prototype adaptation and tree growing beginning with a minimum tree of root $r_0$ and its $\delta$ successors as leafs. The decision, which leafs become roots of sub-trees at a certain time can be specified by the user. Subsequently for each node $r$ a counter $b_r$ is defined. This counter is increased if the corresponding node becomes a winner and the node is branched if a given threshold $\theta \in \mathbb{N}, \theta > 0$ is reached.

Possible criteria might be the variance of the receptive fields of the prototypes or the number of winner hits during the competition. The prototypes of the new leafs should be initialized in a local neighborhood of the root prototype according to $d_V$. Hence, the ET also can be taken as a special growing variant of SOM as it is known for example from [2].

Since ETs are extended variants of usual SOM one can try to transfer evaluation methods known from SOMs to ETs. Unknown samples can be identified using the ET in the following way. The ET is fully labeled by assignment of a label to each node by an analysis of the receptive fields of the corresponding sub-trees. The root node remains unlabeled. For each receptive field a common label is determined by a majority voting of the contained samples and their labels. An unknown, new item is preprocessed as described later on. For this item the $bmu$ in the tree is determined in accordance to Equation (2) and $s(\mathbf{v})$ is calculated. The label of the receptive field of $s(\mathbf{v})$ defines the label of the item.

## 2.2   Hierarchical PCA by Evolving Tree Learning Using Oja's Rule

In [12] a learning rule for neuron models has been proposed which inherently provides a principal component analyses of the represented data. This rule was recently used in [7] to get an optimal data encoding and proven to be effective in learning using neighborhood cooperativeness. We combine this approach with the learning of Evolving Trees such that the prototype representing a data cluster become the first eigenvector of this cluster. In this way a hierarchical PCA can be calculated. We replace the learning rule of Equation (5) by the following Oja based learning dynamic but keeping the neighborhood cooperativeness of ET:

$$\triangle \mathbf{w}_r = \epsilon h_{ET}\left(r, s\left(\mathbf{v}\right)\right) O \left(\mathbf{v} - O\mathbf{w}_r\right) \tag{7}$$
$$O \qquad = <\mathbf{v}, \mathbf{w_r}> \tag{8}$$

Further the winner determination of Equation (2) is changed accordingly

$$s_{T_r}\left(\mathbf{v}\right) = \arg\max_{r \in S_r} \left(<\mathbf{v}, \mathbf{w}_r>\right) \tag{9}$$

As pointed out in [12] the update for the weight vector $\mathbf{w}_r$ as defined by Equation (7) will, neglecting statistical fluctuations, tend to the dominant eigenvector $c$ of the input correlation matrix $C$ of the input data $v$ limited to the receptive field of $\mathbf{w_r}$. Using this approach we obtain eigenvectors for each cluster, at each depth level $l_r$ for each node of the tree. The first eigenvector as obtained from an analysis of the prototype $\mathbf{w_r}$ at $l_{r_o}$ is the regular first principal component of the whole data set. With increasing depth of the tree the data are clustered by the Tree-SOM approach and a hierarchical PCA analysis of the sub-clusters become available. The principal components can be used to analyse and visualize the cluster separability. Further the obtained loadings provide insights in a variance based analysis of the individual input dimensions of the clusterings such that separating features become apparent.

## 3    Evolving Tree Applied on Mass Spectra of Bacteria

The introduced *Hierarchical PCA-based Evolving Tree* is now applied to investigate MS-spectra for identification of bacteria. These data are spectra of different species of *Vibrio-* and *Listeria-*bacteria. Thereby we use the spectra in a pre-processed form of line spectra. The resulting identification is visualized and it is shown how the obtained hierarchical PCA model can be interpreted.

### 3.1    Data, Measurement and Pre-processing

The data used in the experiments are MS spectra of 56 different vibrio species and 7 different *Listeria* species. Every data-set contains about $20 - 40$ single spectra, being measurements of the same bacterium. Together there are 1452 spectra of vibrio and 231 spectra of *Listeria*. Each MS measurement is processed as described later on. Biological details on the bacteria samples can be obtained from [4].

Details on the mass spectrometry technique can be found in [8]. At the end of the measurement process one obtains for each measurement a spectrum with a mass axis in m/z respectively Dalton and an unit-less intensity for every mass. The spectrum is encoded as a high-dimensional vector (profile spectrum) of intensities, often visualized as a function of mass.

The standard pre-processing to generate a line spectrum (consisting only of peaks) is provided by the measurement system as detailed in [3]. A line spectrum typically consists of around $100 - 500$ peaks depending on the sample complexity and system mode while the profile spectra are original given as measurements with around 40 000 sample points. In order to the line spectra for our approach the input vectors of peak lists are mapped onto a global mass vector covering every appearing peak within a predefined tolerance (here 500 ppm) depending on the expected measurement accuracy.

The resulting aligned peak-lists are now located in the same data space, still very high-dimensional. For the *Listeria* data the line spectra have a dimensionality of $D = 1181$ (peak positions) whereas for the vibrio data the dimensionality is given as $D = 2382$.

### 3.2    Experiments

Euclidean distance is used to find the *bmu.* $\delta_r = 3$ for all nodes without leafs. The learning is done in accordance to the standard SOM approach, thereby the initial learning rate $\alpha_0$ is defined as $\alpha_0 = 0.2$ which is logarithmically decreased

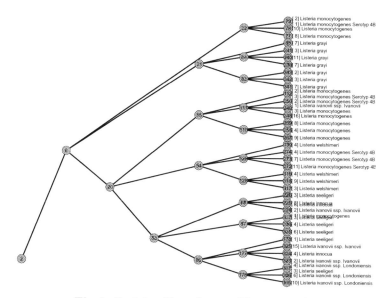

**Fig. 1.** Evolving Tree of seven *Listeria* species

during learning to a final value of $\alpha_{\mathrm{end}} = 0.01$. The neighborhood cooperation value $\sigma$ is initialized with $\sigma = 1$ and logarithmically decreased to $\sigma = 0.35$ in accordance to suggestions given in [15]. The total number of learning iterations $I$ is determined depending on the number of training samples, the desired number of clusters #C, $\delta_r$ and $\theta$ as shown in [14].

We apply the proposed methodology on the two data sets. In the first experiment an hierarchical PCA using the ET is generated for the *Listeria* and the *Vibrio* data. This is a simplified example of a bacteria identification on the genus level. In a second experiment we consider the *Listeria* bacteria only. Thereby we assume that the genus of the considered bacteria is already identified as *Listeria* using the first tree and the remaining task consists in an identification and visualization of the species and subspecies level. For both settings we generate the tree, analyze the hierarchical, local PCA visualizations and identify relevant mass positions (features) by means of PCA loadings. For simplicity we provide the plots for the *Listeria* data, only. In Figure 1 the Evolving-Tree for the *Listeria* data is shown[1]. We observe a quite clear separation of the different *Listeria* species in the tree, but also some mixed clusters occur. Especially for the *monocytogenes* data subclusters can be identified, this however is a intended effect because the *monocytogenes* group is known to be diverse. Here a single taxonomical label does not perfectly reflect the biochemical picture[2]. In Figure 5 we analyze the loadings of the local PCA of node 6, as obtained in the hierarchical PCA using the ET. We observe a cut in the loadings histogram such that $2 - 7$ dimensions can be considered to be relevant. Taking

---

[1] Here we show the subtree from the *Listeria/Vibrio*-Tree, but an individual generated tree is actually very similar, ignoring permutations.

[2] This effect becomes even more explicit for e.g. bacillus data - which are in fact multiple subgroups (genera) (not distinctly labeled in the taxonomy of bacteria).

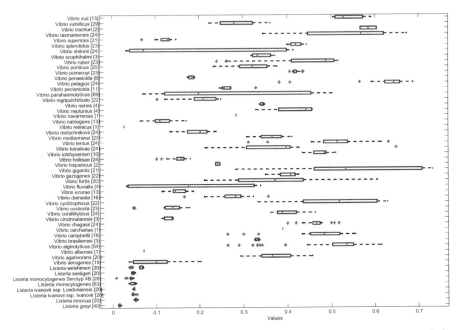

**Fig. 2.** Box plot of the pc's at the 2 node, branching the *Listeria* and most of the *Vibrio* data in an ET on the bacteria data

these input dimensions into account a Pseudo-Gelview can be generated as depicted in Figure 3 showing a top/gelview of the spectra restricted to the peaks intensities at the masses indicated by the PCA. Some peaks differentiate between *Listeria* groups by means of intensity variations, as e.g. in the first peak with moderate intensity values for the *ivanonvii*, a missing peak situation for the *grayi* and high intensities else. We noticed that in general a 0/1 encoding of the peak intensities (peak absent/present) is sufficient but for some species and subspecies the incorporation of intensity information is valuable. In addition a box plot of the projected data on the principal component as depicted in Figure 2 may provide further information on the separation potential of the hierarchical PCA based clustering. Doing a traversal through the feature loadings over the different tree levels the approach identified the following masses most relevant $4276.4Da, 4278.0Da, 5181.0Da, 9751.0Da$. The first three dimensions are relevant to separate the vibrio data from the *Listeria* and to get separations with the vibrio genus, separating different (but not all) vibrio species. The last dimension is a clear indicator for the presents of *Listeria*.

In the Table 4 the most relevant dimensions for the *Listeria* experiment identified by the hierarchical PCA at node 6 are depicted. Similar analyses can be done for the other nodes as well. It should be noted that the identified masses at a specific node are interpreted as those dimensions explaining the largest variance of the data presented in the underlying clustering. This is an unsupervised interpretation, hence the relevant dimensions may not be relevant with respect to a provided labeling. For bacteria data however we observed, that the highlighted dimensions are in general meaningful for the taxonomy as well.

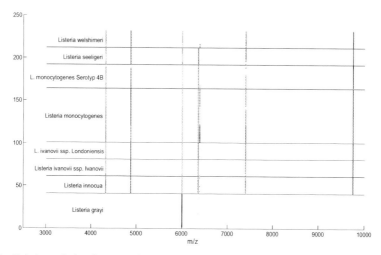

**Fig. 3.** Gelview of the *Listeria* data restricted to the identified most relevant masses

| Rank | Contribution | Dim. | Relevant Mass |
|------|-------------|------|---------------|
| 1 | 0.6859 | 2243 | 9751.11 |
| 2 | 0.5860 | 897 | 4876.13 |
| 3 | 0.2190 | 1764 | 7402.22 |
| 4 | 0.1987 | 1441 | 6362.80 |
| 5 | 0.1879 | 664 | 4323.25 |
| 6 | 0.1323 | 1449 | 6388.37 |
| 7 | 0.1074 | 1307 | 6006.82 |

**Fig. 4.** Relevant masses contributing to the first principal component in the tree node (6) pooling all *Listeria* subspecies

**Fig. 5.** Analysis of the loadings (truncated to 100) of the local PCA for node 6

## 4    Conclusions

A method for an unsupervised hierarchical PCA based analysis of bacteria spectra from mass spectrometry has been presented. One obtains a hierarchical representation of the bacteria by means of a Evolving Tree with local principal components in a hierarchical manner. This can be used to get a better interpretation of the underlying clustering. The approach is unsupervised but nicely reflects the expected taxonomical ordering of the data. The approach can be used to identify masses which are relevant for the clustering in a hierarchical way, e.g. by traversing through the different levels of the tree. If the clustering fits to an added set of meta information, as in our case, the taxonomy of bacteria the identified dimensions could be interpreted in a supervised scheme as well. The approach can be used to get highly interpretable representations of bacteria spectra and to get quick identifications with a logarithmic number of comparisons.

**Acknowledgment.** The authors are grateful to S. Klepel and T. Maier for providing the bacteria data and helpful discussions (both Bruker Daltonik Leipzig, Germany).

# References

1. Barbuddhe, S.B., Maier, T., Schwarz, G., Kostrzewa, M., Hof, H., Domann, E., Chakraborty, T., Hain, T.: Rapid identification and typing of listeria species by matrix-assisted laser desorption ionization-time of flight mass spectrometry. Applied and Environmental Microbiology 74(17), 5402–5407 (2008)
2. Bauer, H.-U., Villmann, T.: Growing a Hypercubical Output Space in a Self–Organizing Feature Map. IEEE TNN 8(2), 218–226 (1997)
3. Bruker Daltonik GmbH. Bruker BioTyper 2.0, user manual (2008), http://www.bdal.de
4. Bruker Daltonik GmbH. Bruker listeria and vibrio spectra (Dr. Markus Kostrzewa), personal communicated (2008), http://www.bdal.de
5. Forero, M.G., Sroubek, F., Cristobal, G.: Identification of tuberculosis bacteria based on shape and color. Real-time Imaging 10(4), 251–262 (2004)
6. Kohonen, T.: Self-Organizing Maps, 2nd edn. Springer Series in Information Sciences, vol. 30. Springer, Heidelberg (1995) (1997)
7. Labusch, K., Barth, E., Martinetz, T.: Learning data representations with sparse coding neural gas. In: Proc. of ESANN 2008, pp. 233–238 (2008)
8. Liebler, D.C.: Introduction to Proteomics. Humana Press (2002)
9. Maier, T., Kostrzewa, M.: Fast and reliable maldi-tof ms-based microorganism identification. Chemistry Today 25, 68–71 (2007)
10. Mazzeo, M.F., Sorrentino, A., Gaita, M., Cacace, G., Di Stasio, M., Facchiano, A., Comi, G., Malorni, A., Siciliano, R.A.: Matrix-assisted laser desorption ionization-time of flight mass spectrometry for the discrimination of food-borne microorganisms. Applied and Environmental Microbiology 72(2), 1180–1189 (2006)
11. Mellmann, A., Cloud, J., Maier, T., Keckevoet, U., Ramminger, I., Iwen, P., Dunn, J., Hall, G., Wilson, D., LaSala, P., Kostrzewa, M., Harmsen, D.: Evaluation of matrix-assisted laser desorption/ionization time-of-flight-mass spectrometry MALDI-TOF MS in comparison to 16s rrna gene sequencing for species identification of nonfermenting bacteria. J. Clinical Microbiology 46, 1946–1954 (2008)
12. Oja, E.: A simplified neuron model as a principal component analyzer. Journal of Mathematical Biology 15, 267–273 (1982)
13. Pakkanen, J., Iivarinen, J., Oja, E.: The evolving tree—a novel self-organizing network for data analysis. Neural Process. Lett. 20(3), 199–211 (2004)
14. Simmuteit, S.: Effizientes Retrieval aus Massenspektrometriedatenbanken, Diplomarbeit, Technische Universität Clausthal (February 2008)
15. Villmann, T., Der, R., Herrmann, M., Martinetz, T.: Topology Preservation in Self–Organizing Feature Maps: Exact Definition and Measurement. IEEE Transactions on Neural Networks 8(2), 256–266 (1997)

# Optimal Combination of SOM Search in Best-Matching Units and Map Neighborhood*

Mats Sjöberg and Jorma Laaksonen

Helsinki University of Technology TKK,
Department of Information and Computer Science,
P.O. Box 5400, FI-02015 TKK, Finland

**Abstract.** The distribution of a class of objects, such as images depicting a specific topic, can be studied by observing the best-matching units (BMUs) of the objects' feature vectors on a Self-Organizing Map (SOM). When the BMU "hits" on the map are summed up, the class distribution may be seen as a two-dimensional histogram or discrete probability density. Due to the SOM's topology preserving property, one is motivated to smooth the value field and spread out the values spatially to neighboring units, from where one may expect to find further similar objects. In this paper we study the impact of using more map units than just the single BMU of each feature vector in modeling the class distribution. We demonstrate that by varying the number of units selected in this way and varying the width of the spatial convolution one can find an optimal combination which maximizes the class detection performance.

## 1 Introduction

In many crucial information processing applications, such as high-level indexing and querying on multimedia data, it has proven to be very useful to have models of semantically related classes, i.e. meaningful subsets of the full dataset under study [1]. When a Self-Organizing Map (SOM) [2] is trained on a large dataset, mapping the data vectors of some semantic class to their best-matching units (BMUs) produces a distribution characterizing that particular class in the context of the full dataset. For example, when studying a database of animal images, one could map the class of objects depicting lions on a SOM trained from color features extracted from all the images. The SOM may then be used for example in an image retrieval task for detecting images of lions in a new batch of unannotated images.

The rest of this paper is organized as follows: Section 2 describes modeling of class distributions with BMUs, Section 3 smoothing in the spatial and feature domains. In Section 4 an image retrieval experiment is shown, and finally conclusions are drawn in Section 5.

---

* The research leading to these results has received funding from the European Community's Seventh Framework Programme (FP7/2007–2013) under *grant agreement* n° 216529, Personal Information Navigator Adapting Through Viewing, PinView. Mats Sjöberg has been supported by a grant from the Nokia Foundation.

J.C. Príncipe and R. Miikkulainen (Eds.): WSOM 2009, LNCS 5629, pp. 281–289, 2009.

## 2    Modeling Class Distributions with BMUs

For any database of objects, feature vectors can be extracted for analyzing the properties of the objects. If the features are selected properly they should be of moderate dimensionality, while still preserving semantically important information of the objects and their distribution. Figure 1 (left), visualizes how the original very-high-dimensional pattern space is first projected to a lower-dimensional feature space, the vectors of which are then used in training a SOM. The dark areas in the figure illustrate how a class of objects might be projected, ideally to a compact distribution in the feature space if the discriminative properties of the class are well represented in the feature extraction process.

If the best-matching units of the objects of a specific semantic class are marked with a positive impulse, the "hits" on a SOM surface form a sparse value field. When these values are summed up and properly normalized, the formed distribution can be seen as a two-dimensional discrete probability density that characterizes the object class. Such distributions were studied in an earlier article [3] in the context of our content-based retrieval system PicSOM [4], and information-theoretic measures were proposed for evaluating their properties.

Due to the topography-preserving property of the SOM, we can now expect to find more similar objects in the map areas with many nearby hits. In order to spread the values to such neighboring units the value field is, in the Pic-SOM system, low-pass filtered with a tapered kernel. This facilitates finding new unannotated objects of the same class, and also aids in visual inspection of the map distribution. It also serves to emphasize areas with many hits close-by and deemphasize areas with only a few sporadic hits. A visual example is shown in Figure 1 (right) where a class of video frames depicting scenes with "explosion or fire" have been mapped to a SOM trained from Color Layout feature vectors. Areas occupied by objects of the concept in question are shown with gray shades. Clearly the hits from this class seem to be concentrated into the bottom right corner of the map.

These class-conditional distributions or class models can be considered as estimates of the true distributions of the semantic concepts in question, not on the original feature spaces, but on the discrete two-dimensional grids defined by the used SOMs. Thereby, instead of modeling probability densities in the

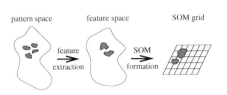
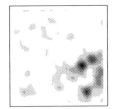

**Fig. 1.** Left: Stages in creating a class model from the very-high-dimensional pattern space through the high-dimensional feature space to the two-dimensional SOM grid. Right: An example of image class model "explosion or fire" on a Color Layout SOM.

high-dimensional feature spaces, the PicSOM system is essentially performing kernel-based estimation of discrete class densities over the SOM grid. Depending on the variance of the kernel function, these kernels will overlap and weight vectors close to each other will partially share each other's probability mass. As an example, the most representative objects of a given semantic class can be obtained by locating those SOM units, and the objects mapped to these units, that have the highest responses on the estimated class distribution.

In this paper, we study the use of more than just one BMU when mapping the members of a semantic class to a SOM. We sort all the model vectors of the map in ascending order of the distance to the input vector and apply a weighting kernel to this set, giving the highest weight to the best-matching unit, and decreasing weights according to the list rank. By varying the width of this kernel we can choose the number of nearest units selected for each input vector. We call this number the "BMU depth". For example, for BMU depth equal to three, we select the second and third best-matching units (generally with decreasing weights) in addition to the normal BMU. Thus, we use both spatial SOM surface smoothing and smoothing in the BMU depth, i.e. we spread the "hit" values both in the SOM grid and feature space domains.

To compare, the WEBSOM system [5] for interactive browsing of large text document databases, used only the BMU depth approach, not spatial smoothing. An idea similar to ours was explored in [6], where the cluster structure of the data could be visualized on different levels of detail by varying the smoothing parameter (equivalent to our BMU depth). Another related concept is to force the map convolution to follow the form of the U-matrix, i.e. the convolution span is inversely proportional to the distance between the SOM units [7]. The advantage of the proposed approach over U-matrix based weighting is computational simplicity; instead of tuning the convolution separately for each unit we need only select a small set of best-matching units. Finding BMUs is very fast, especially in the PicSOM system that implements the tree-structured SOM variant [8] which does BMU search in logarithmic time.

## 3   Smoothing in the SOM and Feature Spaces

In this paper we introduce smoothing in the feature space domain in combination with the traditional spatial SOM surface smoothing. Instead of only using the single best-matching unit, we order the list of SOM model vectors by increasing distance from the input vector. Such ordered lists can be generated off-line for each database object storing only a restricted set of the best matches.

Let us assume that we have a set R of training set objects $j$ whose membership value $r_j$ in the studied object class is known. Then

$$r_j = \begin{cases} +\rho_+ & \text{, if } j \text{ is a member of the class} \\ 0 & \text{, if } j\text{'s membership in the class is unknown ,} \\ -\rho_- & \text{, if } j \text{ is not a member of the class} \end{cases} \qquad (1)$$

where $\rho_+$ and $\rho_-$ are properly selected non-negative weights for the member and non-member samples, respectively. In PicSOM, the values of $\rho_+$ and $\rho_-$ have been inverses of the number of positive and negative samples and consequently $\sum_j r_j = 0$.

A membership score for any point $\mathbf{x}$ can then be estimated as a sum of kernel functions $h_j(\cdot)$ centered in the locations of the points $\mathbf{x}_j$ with known membership assessments:

$$r(\mathbf{x}) = \sum_{j \in \mathrm{R}} r_j h_j(\mathbf{x} - \mathbf{x}_j). \tag{2}$$

In the PicSOM system, the kernel functions $h_j(\mathbf{x} - \mathbf{x}_j)$ have been replaced by the use of a function $h(\cdot)$ that can be calculated from the difference between the BMU coordinates on the SOM surfaces. Let $\mathbf{b}(\mathbf{x}) = (b_x(\mathbf{x}), b_y(\mathbf{x}))$ denote the discrete two-dimensional coordinates of the best-matching unit of $\mathbf{x}$. One should note that the values of the BMU function $\mathbf{b}(\mathbf{x}_j)$ can be calculated and tabulated offline for each object $j$ as soon as the SOM has been trained. The membership value estimate for $\mathbf{x}$ can thus be written as

$$r(\mathbf{x}) = \sum_{j \in \mathrm{R}} r_j h \big( b_x(\mathbf{x}) - b_x(\mathbf{x}_j), \, b_y(\mathbf{x}) - b_y(\mathbf{x}_j) \big)$$

$$= \sum_{j \in \mathrm{R}} r_j g \big( b_x(\mathbf{x}) - b_x(\mathbf{x}_j) \big) \, g \big( b_y(\mathbf{x}) - b_y(\mathbf{x}_j) \big) . \tag{3}$$

The latter notation follows from the practice of using separable and symmetric kernels $h(\cdot)$. Now the extent and shape of the scalar function $g(\cdot)$ determines the effect of the SOM surface smoothing. In PicSOM we have used a simple triangular kernel with different widths.

In order to take the BMU smoothing into the formulation, one needs to extend the BMU function $\mathbf{b}(\mathbf{x}_j)$ with the BMU depth index $k$ to be $\mathbf{b}_k(\mathbf{x}_j) = (b_{x,k,j}, \, b_{y,k,j})$, where $k = 1, \ldots, k_{\max}$. Now we have

$$r(\mathbf{x}) = \sum_{j \in \mathrm{R}} r_j \sum_{k=1}^{k_{\max}} f(k) g \big( b_x(\mathbf{x}) - b_{x,k,j} \big) \, g \big( b_y(\mathbf{x}) - b_{y,k,j} \big) . \tag{4}$$

Function $f(k)$ determines the extent of smoothing in the BMU order. Note that the BMUs $b_{x,k,j}$ and $b_{y,k,j}$ of the objects $j$ in the database can be calculated and tabulated offline.

A linear kernel $f(\cdot)$ has been used in our experiments, i.e. the weight decreases linearly with the rank in the ordered list. We have also tried several other shapes of $f(\cdot)$, including Gaussian and one-per-rank, but the linear kernel worked best overall. In our experiments, the most important parameter turned out to be the width of the kernel, not the particular type.

Figure 2 illustrates the smoothing in the two domains separately and combined. The images depict a small neighborhood of a SOM surface trained with Scalable Color features, and a single image of an airplane mapped to its BMU.

The first column shows this single BMU convolved on the map surface with two kernel widths: 3 and 7. This illustrates the traditional approach in PicSOM, where only the map topology is taken into account. The second column shows the same BMU, but now using a BMU depth of 10 or 30, and no map convolution. The values are thus spread to the 10, respectively 30, nearest units in the feature space. It can be readily observed that the two cases on the first row are very similar. Not surprisingly, the nearest units are located closely around the best-matching unit. A map convolution width of 3 encompasses roughly the same amount of units. The difference is that the selection in the first column is done based on the map grid neighborhood, and in the second column on the feature space neighborhood.

On the second row of images in Figure 2 we can see a difference, when the topology of the feature space stretches the BMU depth distribution to the upper right, while the center-symmetric regular map convolution does not take this into account. A similar effect could be achieved with the method of tuning the map convolution to the U-matrix distances [7]. The proposed method, however, is computationally much simpler.

In Figure 2 the last column shows the result of combining the two first columns, i.e. first the values are smoothed in the BMU depth domain, and then the result is smoothed in the SOM surface domain. This combined approach turned out to give the best results in our concept detection experiments.

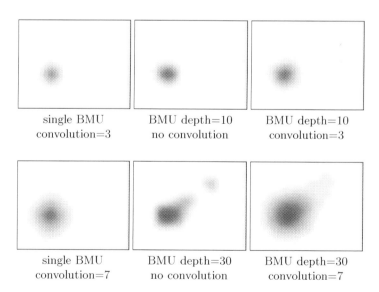

**Fig. 2.** The first column shows a single BMU with SOM convolutions of increasing width. The single impulse is marked with black and decreasing values with shades of gray, with white indicating zero. The second column shows the hits with increasing BMU depth, without SOM convolution. The last row shows the combination of both the BMU depth and the SOM convolution.

# 4    Image Retrieval Experiment

We experimented with SOMs trained on several different features extracted from a set of images from the Pascal Visual Object Class (VOC) 2007 Challenge[1]. The VOC dataset includes several predefined object classes including images annotated according to class membership. The classes are: *aeroplane, bicycle, bird, boat, bottle, bus, car, cat, chair, cow, dining table, dog, horse, motorbike, person, potted plant, sheep, sofa, train* and *TV/monitor*. The full dataset of 9963 images is divided roughly evenly into training and test sets.

## 4.1    Features

The features used were *Color SIFT, Edge Fourier, Edge Histogram, IPLD* and *Scalable Color*. These were selected from a larger set of features because they were the 5 best performing ones (both with and without variable BMU depth). The *Color SIFT* feature is a 256-bin histogram of Opponent-SIFT (opponent color space) features calculated from interest points detected with the Harris-Laplace algorithm [9]. *Edge Fourier* is a $16 \times 16$ FFT of a Sobel edge image, *Edge Histogram* is a histogram of five edge types in $4 \times 4$ subimages. The IPLD feature is based on 256-bin histograms of interest point features. The interest points were detected using a combined Harris-Laplace and Difference-of-Gaussian detector, and SIFT features [10] were calculated for each interest point. The Scalable Color is a Haar transform of the quantized HSV color histogram. Both Edge Histogram and Scalable Color are implemented following the MPEG-7 standard [11].

## 4.2    Performance Measures

Given a training set of example objects belonging to a specific class, one can now calculate the membership score of novel objects from a test set by using Eq. (4) as implemented in the PicSOM algorithm. If the correct answers are known the quality of the SOM model can be measured by standard information retrieval performance measures, such as precision and recall.

In this paper we have opted for the use of non-interpolated average precision (AP) as the performance measure. AP is formed by calculating the precision after each retrieved relevant object. The final measure is obtained by averaging these precisions over the total number of relevant objects, when the precision is defined to be zero for all non-retrieved relevant objects. This measure can be said to incorporate both precision and recall in a single number [12].

## 4.3    Experiment

The convolution width on the map surface was varied from 1 (a single impulse) to 20 units. This was deemed a realistic interval due to the $64 \times 64$ size of the maps. In the feature space, the convolution width, or BMU depth, $k_{\max}$

---

[1] http://pascallin.ecs.soton.ac.uk/challenges/VOC/voc2007/

was varied from 1 to 100. For each run the average precision was calculated. We selected the parameters (convolution width, or both convolution width and BMU depth) with the best average precision results separately for each class and feature combination, with a single BMU and with variable BMU depth. The average precision was calculated as the average over a 6-fold cross validation in the training set, i.e. each of the six subsets were in turn left out and used for validation. Typically, the optimal map convolution was wider when using a single BMU. This is not surprising, since using a greater BMU depth $k_{max}$ in Eq. (4) spreads the values to more units, and a smaller convolution width is needed to reach the same amount of units. The median spatial convolution width changed from 7 to 4 when introducing BMU depth.

The optimal BMU depths found are summarized in Table 1 for each class–feature combination. As can be seen the optimal $k_{max}$ varies quite a lot, and for some cases the optimum is one, i.e. the same as the baseline algorithm which does not use more than one BMU. The feature-wise medians are shown in bold face at the bottom of each column. The class-wise medians are at the end of each row, and 25 in the bottom right corner is the median over the entire table. Some classes vary quite a lot, while some clearly seem to prefer a high (bus, dog, person) or low (aeroplane, horse) BMU depth.

Table 2 summarizes the percentage changes in average precision as measured in the test set when introducing variable BMU depths. Where the optimal BMU depth was one, i.e. there was no improvement above the baseline (in the training set), the table cell has been intentionally left empty. It can be seen that in most cases the result is an improvement in performance, however in some instances there is a small decrease. In some situations, for example the sheep class and

**Table 1.** Optimal BMU depths $k_{max}$ for each class and feature: Color SIFT (cSIFT), Edge Fourier (EF), Edge Histogram (EH), IPLD and Scalable Color (SC). Medians of each row and column are shown in bold face. A priori probabilities of classes are shown in parentheses.

| class | cSIFT | EF | EH | IPLD | SC | |
|---|---|---|---|---|---|---|
| aeroplane (4.47%) | 1 | 5 | 10 | 1 | 10 | **5** |
| bicycle (5.07%) | 20 | 5 | 80 | 1 | 5 | **5** |
| bird (6.24%) | 1 | 10 | 45 | 10 | 10 | **10** |
| boat (3.65%) | 10 | 20 | 30 | 70 | 5 | **20** |
| bottle (5.04%) | 50 | 25 | 50 | 10 | 55 | **50** |
| bus (3.81%) | 20 | 50 | 60 | 55 | 100 | **55** |
| car (15.42%) | 1 | 15 | 75 | 5 | 55 | **15** |
| cat (6.79%) | 30 | 15 | 40 | 10 | 100 | **30** |
| chair (11.21%) | 30 | 1 | 5 | 80 | 65 | **30** |
| cow (2.74%) | 45 | 60 | 1 | 20 | 10 | **20** |
| dining table (5.12%) | 40 | 60 | 40 | 50 | 5 | **40** |
| dog (8.66%) | 35 | 95 | 70 | 75 | 85 | **75** |
| horse (5.75%) | 5 | 20 | 25 | 5 | 5 | **5** |
| motorbike (4.84%) | 30 | 45 | 25 | 5 | 40 | **30** |
| person (42.08%) | 100 | 20 | 45 | 65 | 65 | **65** |
| potted plant (5.29%) | 25 | 5 | 25 | 1 | 70 | **25** |
| sheep (1.96%) | 10 | 50 | 5 | 90 | 5 | **10** |
| sofa (7.30%) | 45 | 10 | 1 | 15 | 30 | **15** |
| train (5.24%) | 35 | 1 | 90 | 20 | 30 | **30** |
| tv/monitor (5.36%) | 5 | 10 | 5 | 75 | 50 | **10** |
| | **27.5** | **17.5** | **35** | **17.5** | **35** | **25** |

**Table 2.** Average precision changes in percent for each class and feature combination, given in percentage. Averages of each row and column are shown in bold face. A priori probabilities of classes are shown in parentheses.

| class | cSIFT | EF | EH | IPLD | SC | |
|---|---|---|---|---|---|---|
| aeroplane (4.47%) | | 2.22 | 3.26 | | -0.85 | **0.93** |
| bicycle (5.07%) | -1.84 | -7.80 | 4.60 | | 5.84 | **0.16** |
| bird (6.24%) | | 0.12 | -7.05 | -5.77 | -0.34 | **-2.61** |
| boat (3.65%) | 7.40 | 2.59 | 4.88 | 3.86 | 3.82 | **4.51** |
| bottle (5.04%) | 4.80 | 0.35 | 1.32 | 0.38 | -10.27 | **-0.69** |
| bus (3.81%) | 25.45 | 4.84 | -3.35 | -0.49 | -0.01 | **5.29** |
| car (15.42%) | | -4.06 | -1.07 | 0.19 | -3.42 | **-1.67** |
| cat (6.79%) | 2.09 | -1.72 | 1.44 | -5.97 | 2.03 | **-0.43** |
| chair (11.21%) | -4.45 | | -0.08 | -5.62 | 4.25 | **-1.18** |
| cow (2.74%) | 22.73 | -4.23 | | 1.21 | -0.50 | **3.84** |
| dining table (5.12%) | -0.70 | -5.38 | -4.95 | 11.37 | 2.97 | **0.66** |
| dog (8.66%) | -7.23 | -0.13 | 5.95 | -1.22 | 2.69 | **0.01** |
| horse (5.75%) | -1.87 | 2.40 | -7.41 | 18.67 | 9.36 | **4.23** |
| motorbike (4.84%) | 8.57 | -2.92 | -7.76 | 0.00 | 5.21 | **0.62** |
| person (42.08%) | 1.03 | 0.52 | 1.08 | 0.46 | 1.02 | **0.82** |
| potted plant (5.29%) | 1.57 | 0.54 | 4.41 | | 4.53 | **2.21** |
| sheep (1.96%) | 9.82 | 10.77 | 26.98 | 5.55 | 3.95 | **11.41** |
| sofa (7.30%) | -2.86 | -0.92 | | -0.86 | 1.84 | **-0.56** |
| train (5.24%) | 4.26 | | -6.53 | 35.48 | 0.00 | **6.64** |
| tv/monitor (5.36%) | -1.06 | -2.97 | -14.29 | 2.44 | -0.56 | **-3.29** |
| | **3.39** | **-0.29** | **0.07** | **2.98** | **1.58** | **1.55** |

the Edge Histogram features, there is a dramatic improvement. The overall improvement is 1.55%. If we select the best single feature for each class the mean average precision increases from 0.2358 to 0.2402, i.e. a 1.86% increase.

It must be emphasized that the parameters of the methods were optimized in the training set, which is separate from the test set. This means that the results should indeed give a realistic indication of the generalization ability of the two different methods. If we optimized the performance directly with the test set, we would get an even more significant performance increase, but this scenario is not realistic as the parameters can easily "overlearn" some features of the dataset and thus not be generally applicable.

## 5    Conclusions

We have proposed a class density estimation method that takes into account the nearest SOM units of projected data vectors both in the feature space and in the SOM grid. In the baseline approach previously used in the PicSOM system the value field on the SOM grid was convolved after projecting an object class to its best-matching units. This is now preceeded by a convolution in the "BMU domain", i.e. in the set of nearest SOM units in the original feature space.

The distribution formed on the SOM surface can be seen as a two-dimensional discrete probability density, and can be used to find unannotated objects which are similar to the modeled class. We have demonstrated that the proposed approach can improve the accuracy when using the PicSOM technique to retrieve objects belonging to the same semantic class in an image database. However, the approach can be more generally applied to any kind of retrieval scenario.

The initial results presented in this paper are promising, however not as conclusive as we had hoped. There is no satisfactory general rule of picking the optimal BMU depths for different class and feature combinations. It thus remains as an open research question what properties of the semantic class and the feature extraction method could explain the optimal value of the $k_{max}$ parameter.

# References

1. Hauptmann, A.G., Christel, M.G., Yan, R.: Video retrieval based on semantic concepts. Proceedings of the IEEE 96(4), 602–622 (2008)
2. Kohonen, T.: Self-Organizing Maps, 3rd edn. Springer Series in Information Sciences, vol. 30. Springer, Berlin (2001)
3. Laaksonen, J., Koskela, M., Oja, E.: Class distributions on SOM surfaces for feature extraction and object retrieval. Neural Networks 17(8-9), 1121–1133 (2004)
4. Laaksonen, J., Koskela, M., Oja, E.: PicSOM—Self-organizing image retrieval with MPEG-7 content descriptions. IEEE Transactions on Neural Networks, Special Issue on Intelligent Multimedia Processing 13(4), 841–853 (2002)
5. Kohonen, T., Kaski, S., Lagus, K., Salojärvi, J., Honkela, J., Paatero, V., Saarela, A.: Self organization of a massive text document collection. IEEE Transactions on Neural Networks 11(3), 574–585 (2000)
6. Pampalk, E., Rauber, A., Merkl, D.: Using smoothed data histograms for cluster visualization in self-organizing maps. In: Dorronsoro, J.R. (ed.) ICANN 2002. LNCS, vol. 2415, pp. 871–876. Springer, Heidelberg (2002)
7. Koskela, M., Laaksonen, J., Oja, E.: Implementing relevance feedback as convolutions of local neighborhoods on self-organizing maps. In: Dorronsoro, J.R. (ed.) ICANN 2002. LNCS, vol. 2415, pp. 981–986. Springer, Heidelberg (2002)
8. Koikkalainen, P.: Progress with the tree-structured self-organizing map. In: 11th European Conference on Artificial Intelligence, European Committee for Artificial Intelligence (ECCAI) (August 1994)
9. van de Sande, K.E.A., Gevers, T., Snoek, C.G.M.: Evaluation of color descriptors for object and scene recognition. In: IEEE Conference on Computer Vision and Pattern Recognition, Anchorage, Alaska, USA (June 2008)
10. Lowe, D.G.: Distinctive image features from scale-invariant keypoints. International Journal of Computer Vision 60(2), 91–110 (2004)
11. ISO/IEC: Information technology - Multimedia content description interface - Part 3: Visual, 15938-3:2002(E) (2002)
12. Manning, C., Schütze, H.: Foundations of Statistical Natural Language Processing. MIT Press, Cambridge (1999)

# Sparse Linear Combination of SOMs for Data Imputation: Application to Financial Database

Antti Sorjamaa[1], Francesco Corona[1], Yoan Miche[1], Paul Merlin[2],
Bertrand Maillet[2], Eric Séverin[3], and Amaury Lendasse[2]

[1] Department of Information and Computer Science
Helsinki University of Technology, Finland
[2] A.A. Advisors-QCG (ABN AMRO) – Variances, CES/CNRS and EIF,
University of Paris-1, France
[3] Department GEA,
University of Lille 1, France

**Abstract.** This paper presents a new methodology for missing value imputation in a database. The methodology combines the outputs of several Self-Organizing Maps in order to obtain an accurate filling for the missing values. The maps are combined using MultiResponse Sparse Regression and the Hannan-Quinn Information Criterion. The new combination methodology removes the need for any lengthy cross-validation procedure, thus speeding up the computation significantly. Furthermore, the accuracy of the filling is improved, as demonstrated in the experiments.

## 1 Introduction

The presence of missing values in the underlying time series is a recurrent problem when dealing with databases [1]. Number of methods have been developed to solve the problem and fill the missing values.

Self-Organizing Maps [2] (SOM) aim to ideally group homogeneous individuals, highlighting a neighborhood structure between classes in a chosen lattice. The SOM algorithm is based on unsupervised learning principle where the training is entirely stochastic, data-driven. No information about the input data is required. Recent approaches propose to take advantage of the homogeneity of the underlying classes for data completion purposes [3]. Furthermore, the SOM algorithm allows projection of high-dimensional data to a low-dimensional grid. Through this projection and focusing on its property of topology preservation, SOM allows nonlinear interpolation for missing values.

This paper describes a new method, which combines several SOMs in order to enhance the accuracy of the nonlinear interpolation. The combination is achieved with a simple linear regression performed on an extracted sample from the data. The maps to be combined are selected first using a ranking of the maps by Multiresponse Sparse Regression (MRSR) and then choosing the best SOMs using the Hannan-Quinn Information Criterion. The combination improves the accuracy of the imputation as well as speeds up the process by removing the cross-validation scheme [4].

J.C. Príncipe and R. Miikkulainen (Eds.): WSOM 2009, LNCS 5629, pp. 290–297, 2009.
© Springer-Verlag Berlin Heidelberg 2009

The global methodology is presented in the next section, including all the methods combined in the global methodology. The Section 3 demonstrates the accuracy of the methodology.

## 2    Global Methodology

The global methodology is summarized in Figure 1.

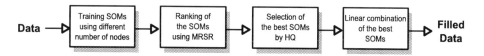

**Fig. 1.** Global methodology summarized

The core of the methodology is the Self-Organizing Map (SOM). Several SOMs are trained using different number of nodes and the imputation results of the best SOMs are linearly combined.

In order to create the linear system, we have to remove a calibration set from the data before any processing. Then, the SOM estimations of the removed calibration data are used as the variables of the linear equations and the removed data itself as the outputs of the equations. The linear system is summarized in the following formula:

$$
\begin{bmatrix}
\hat{s}_{1,1} & \hat{s}_{1,2} & \cdots & \hat{s}_{1,Q} \\
\hat{s}_{2,1} & \hat{s}_{2,2} & \cdots & \hat{s}_{2,Q} \\
\vdots & \vdots & \ddots & \vdots \\
\hat{s}_{L,1} & \hat{s}_{L,2} & \cdots & \hat{s}_{L,Q}
\end{bmatrix}
\times
\begin{bmatrix}
\alpha_1 \\ \alpha_2 \\ \vdots \\ \alpha_Q
\end{bmatrix}
=
\begin{bmatrix}
s_1 \\ s_2 \\ \vdots \\ s_L
\end{bmatrix},
\tag{1}
$$

where $s_i$ denotes the *ith* removed calibration sample, $\hat{s}_{i,j}$ denotes the *ith* calibration data sample estimated by *jth* SOM, $L$ denotes the number of calibration data points, $Q$ the number of the best SOMs used and, finally, the vector $\alpha$ denotes the linear system parameters. The number of SOMs $Q$ is determined by the MultiResponse Sparse Regression and the Hannan-Quinn Information Criterion.

When the $\alpha$ is solved, it can be used to estimate the originally missing values of the dataset from the best SOM estimations selected.

In the following subsections, each of the methods is explained more deeply.

### 2.1    Imputation Using SOM

The SOM algorithm is based on an unsupervised learning principle, where training is entirely data-driven and no information about the input data is required [2]. Here we use a 2-dimensional network, composed of $c$ units (or code vectors) shaped as a square *lattice*. Each unit of a network has as many weights as the length $T$ of the learning data samples, $\mathbf{x}_n$, $n = 1, 2, ..., N$. All units of a network can be collected to a weight matrix $\mathbf{m}(t) = [\mathbf{m}_1(t), \mathbf{m}_2(t), ..., \mathbf{m}_c(t)]$ where $\mathbf{m}_i(t)$ is the $T$-dimensional weight vector of the unit $i$ at time $t$ and $t$ represents the steps of the learning process. Each unit is connected to its neighboring units through a neighborhood function $\lambda(\mathbf{m}_i, \mathbf{m}_j, t)$, which defines the shape and the size of the neighborhood at time $t$. The neighborhood can be constant through the entire learning process or it can change in the course of learning.

The learning starts by initializing the network node weights randomly. Then, for a randomly selected sample $\mathbf{x}_{t+1}$, we calculate the Best Matching Unit (BMU), which is the neuron whose weights are closest to the sample. The BMU calculation is defined as

$$\mathbf{m}_{BMU(\mathbf{x}_{t+1})} = \arg \min_{\mathbf{m}_i, i \in I} \{\|\mathbf{x}_{t+1} - \mathbf{m}_i(t)\|\}, \tag{2}$$

where $I = [1, 2, ..., c]$ is the set of network node indices, the $BMU$ denotes the index of the best matching node and $\|.\|$ is a standard Euclidean norm.

If the randomly selected sample includes missing values, the BMU cannot be solved outright. Instead, an adapted SOM algorithm, proposed by Cottrell and Letrémy [5], is used. The randomly drawn sample $\mathbf{x}_{t+1}$ having missing value(s) is split into two subsets $\mathbf{x}_{t+1}^T = NM_{\mathbf{x}_{t+1}} \cup M_{\mathbf{x}_{t+1}}$, where $NM_{\mathbf{x}_{t+1}}$ is the subset where the values of $\mathbf{x}_{t+1}$ are not missing and $M_{\mathbf{x}_{t+1}}$ is the subset, where the values of $\mathbf{x}_{t+1}$ are missing. We define a norm on the subset $NM_{\mathbf{x}_{t+1}}$ as

$$\|\mathbf{x}_{t+1} - \mathbf{m}_i(t)\|_{NM_{\mathbf{x}_{t+1}}} = \sum_{k \in NM_{\mathbf{x}_{t+1}}} (\mathbf{x}_{t+1,k} - \mathbf{m}_{i,k}(t))^2, \tag{3}$$

where $\mathbf{x}_{t+1,k}$ for $k = [1, ..., T]$ denotes the $k^{th}$ value of the chosen vector and $\mathbf{m}_{i,k}(t)$ for $k = [1, ..., T]$ and for $i = [1, ..., c]$ is the $k^{th}$ value of the $i^{th}$ code vector.

Then the BMU is calculated with

$$\mathbf{m}_{BMU(\mathbf{x}_{t+1})} = \arg \min_{\mathbf{m}_i, i \in I} \left\{\|\mathbf{x}_{t+1} - \mathbf{m}_i(t)\|_{NM_{\mathbf{x}_{t+1}}}\right\}. \tag{4}$$

When the BMU is found the network weights corresponding to the non-missing values of $\mathbf{x}_{t+1}$ are updated as

$$\mathbf{m}_i(t+1) = \mathbf{m}_i(t) - \varepsilon(t)\lambda\left(\mathbf{m}_{BMU(\mathbf{x}_{t+1})}, \mathbf{m}_i, t\right)[\mathbf{m}_i(t) - \mathbf{x}_{t+1}], \forall i \in I, \tag{5}$$

where $\varepsilon(t)$ is the adaptation gain parameter, which is $]0, 1[$-valued, decreasing gradually with time. The number of neurons taken into account during the weight update depends on the neighborhood function $\lambda(\mathbf{m}_i, \mathbf{m}_j, t)$. The number of neurons, which need the weight update, usually decreases with time.

After the weight update the next sample is randomly drawn from the data matrix and the procedure is started again by finding the BMU of the sample. The learning procedure is stopped when the SOM algorithm has converged.

Once the SOM algorithm has converged, we obtain some clusters containing our data. Cottrell and Letrémy proposed to fill the missing values of the dataset by the coordinates of the code vectors of each BMU as natural first candidates for the missing value completion:

$$\pi_{(M_\mathbf{x})}(\mathbf{x}) = \pi_{(M_\mathbf{x})}\left(\mathbf{m}_{BMU(\mathbf{x})}\right), \tag{6}$$

where $\pi_{(M_\mathbf{x})}(.)$ replaces the missing values $M_\mathbf{x}$ of sample $\mathbf{x}$ with the corresponding values of the BMU of the sample. The replacement is done for every data sample and then the SOM has finished filling the missing values in the data.

The procedure is summarized in Table 1. There is a toolbox available for performing the SOM algorithm in [6].

**Table 1.** Summary of the SOM algorithm for finding the missing values

1. SOM node weights are initialized randomly
2. SOM learning process begins
   (a) Input $\mathbf{x}$ is drawn from the learning data set $\mathbf{X}$
      i. If $\mathbf{x}$ does not contain missing values, BMU is found according to Equation 2
      ii. If $\mathbf{x}$ contains missing values, BMU is found according to Equation 4
   (b) Neuron weights are updated according to Equation 6
3. Once the learning process is done, for each observation containing missing values, the weights of the BMU of the observation are substituted for the missing values

## 2.2 MultiResponse Sparse Regression

Multiresponse Sparse Regression, proposed by Timo Similä and Jarkko Tikka in [7] is a variable ranking technique and an extension of the Least Angle Regression (LARS) algorithm [8].

The main idea of the algorithm is the following: Denote by $\mathbf{X} = [\mathbf{x}_1 \dots \mathbf{x}_m]$ the $n \times m$ regressor matrix. MRSR adds each column of the regressor matrix one by one to the model $\hat{\mathbf{Y}}^k = \mathbf{XW}^k$, where $\hat{\mathbf{Y}}^k = [\hat{\mathbf{y}}_1^k \dots \hat{\mathbf{y}}_p^k]$ is the target approximation of the model. The $\mathbf{W}^k$ weight matrix has $k$ nonzero rows at $k$th step of the MRSR. With each new step a new nonzero row, and a new column of the regressor matrix is added to the model.

More specific details of the MRSR algorithm can be found from the original paper [7].

An important detail shared by the MRSR and the LARS is that the ranking obtained is exact, if the problem is linear. Here, in this paper, we linearly combine the SOM estimations of the missing values and, therefore, we have an exact ranking of the estimations.

## 2.3   Hannan-Quinn Information Criterion

Because the MRSR only ranks the SOM estimations, we need a method to actually select the optimal number of input variables. This kind of selection can be considered as a complexity selection or input variable selection.

There are many possible criteria for complexity selection used in machine learning. Typical examples are Akaike's information criterion (AIC) [9] or the Bayesian Information Criterion (BIC) [10]. Their expression is usually based on the residual sum of squares ($Res$) of the considered model (first term of the criterion) plus a penalty term (second term of the criterion). Differences between criteria mostly occur on the penalty term. The AIC penalizes only according to the number of parameters $p$ of the model, shown in Equation 7, whereas the BIC takes into account also the number of samples $N$ used for the model training, Equation 8.

$$BIC = N \times \log\left(\frac{Res}{N}\right) + p \times \log N, \qquad (7)$$

$$AIC = N \times \log\left(\frac{Res}{N}\right) + 2 \times p. \qquad (8)$$

The AIC is known to have consistency problems: while minimizing the AIC, it is not guaranteed that the complexity selection will converge toward an optima, if the number of samples goes to infinity [11]. The main idea raised by this observation is about trying to balance the underfitting and the overfitting when using such a criterion. This is achieved through the penalty term, for example, by having a $\log N$ based term in the penalty, which the BIC has. Unfortunately, in our previous experiments, the BIC criterion failed to give proper results in terms of complexity.

The Hannan-Quinn Information Criterion (HQ) [12] is very close to the other two criteria. The HQ is defined as

$$HQ = N \times \log\left(\frac{Res}{N}\right) + 2 \times p \times \log(\log N). \qquad (9)$$

The idea behind the design of this criterion is to provide a consistent criterion, unlike the AIC, and in which the penalty term $2 \times p \times \log(\log N)$ grows with a very slow rate regarding the number of samples.

In this paper, the HQ criterion is used to select an optimal number of already ranked SOM estimations to be combined. The number of samples corresponds to the number of selected training points from the training dataset and the number of parameters to the number of SOM estimations to be combined.

## 3   Experiments

In the following experiments, we use a financial fund dataset. The dataset is classified and, therefore, our possibilities to mention any specifics are very limited. The dataset can be downloaded from [13].

The dataset contains 120 time series of funds from a total of 121 months each. The data has been normalized and rescaled. The series are correlated in time and between series and there are no missing values originally present in the dataset. Figure 2 shows 15 example series of the original 120 rescaled fund values.

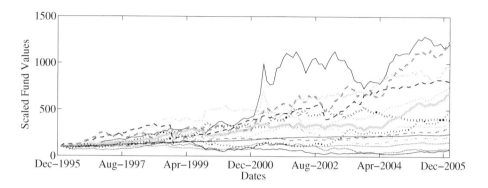

**Fig. 2.** Rescaled and normalized fund values of 15 funds present in the database

Before running any experiments, we randomly remove 20 percent of the data as a test set. The test set contains roughly 2900 values. In our methodology, there is no need for actual validation set, but in order to calculate the linear model parameters for the SOMs, we have to remove a set of data that will be used as output of the linear model. For that purpose, 20 percent of the remaining data are removed, which corresponds to roughly 2300 values, and the set is called calibration set.

According to the methodology, several SOMs are trained using different amount of nodes. Figure 3 shows the training evaluation error with respect to the SOM size. In this paper, the SOM size is actually the length of the dimension of the square lattice. So, for example, size 10 means a square SOM grid of size 10×10, a total of 100 nodes.

From Figure 3 we can see that the best SOM size, according to this simple calibration evaluation, is 6. It means that the som with only 36 nodes is the most optimal to fill in the missing training evaluation values.

Of course, if we would use a standard SOM for the filling, we should use a lengthy Cross-Validation scheme to validate the SOM size. But even that lengthy process does not guarantee that the SOM to be used to fill the test set values is properly validated.

Figure 4 shows the Hannan-Quinn Information Criterion values with respect to the number of SOMs in the combination.

From Figure 4 we can see that the most optimal value is reached with 12 SOMs. The selected SOM sizes are 7, 9, 12, 16, 18, 20, 21, 22, 23, 24, 25 and 26. Here the maximum SOM grid size was 26. From the previous list we can clearly see that the small SOM grids are not accurate enough to be included in

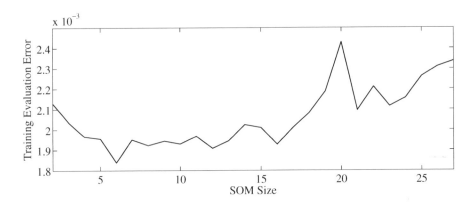

**Fig. 3.** SOM training evaluation errors with respect to the SOM size

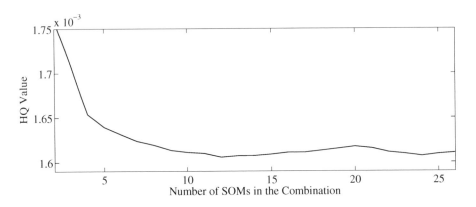

**Fig. 4.** Hanna-Quinn Information Criterion values for the selection of SOMs in the combination

the combination, but several larger sizes are. Comparing this to Figure 4 it is also clear that after the 12 selected SOMs the HQ value starts to increase, which means that the rest of the SOMs do not improve the results.

After the calibration, the obtained models are used to fill in the test set. In Table 2 the errors are summarized.

From Table 2 we can see that the Combination of the SOMs clearly outperforms the single SOM decreasing the test error by 18 percent.

**Table 2.** Test Errors for the SOM and the Combined SOMs

| $10^{-3}$ | Training Evaluation Error | Test Error |
|---|---|---|
| SOM | 1.8 | 1.6 |
| Combined SOMs | | 1.3 |

# 4  Conclusions

As the experiments demonstrate, the new methodology combining several Self-Organizing Maps is at least as accurate in filling of the missing values than single SOM alone. At the same time, the calculation time is reduced significantly (almost divided by 10), because of the removal of the cross-validation phase from the SOM.

Further work consists of finding other ways to combine the SOMs and compare the achieved performance to other popular imputation methods.

**Acknowledgment.** Part of the work of Antti Sorjamaa is supported by a grant from Nokia Foundation, Finland. Part of the work of Amaury Lendasse is supported by Pattern Analysis, Statistical Modelling and Computational Learning (PASCAL2) Network of Excellence funded by the European Union.

# References

1. Sorjamaa, A., Lendasse, A., Cornet, Y., Deleersnijder, E.: An improved methodology for filling missing values in spatiotemporal climate data set. Computational Geosciences (February 2009) (online publication), doi:10.1007/s10596-009-9132-3
2. Kohonen, T.: Self-Organizing Maps. Springer, Berlin (1995)
3. Wang, S.: Application of self-organising maps for data mining with incomplete data sets. Neural Computing and Applications 12(1), 42–48 (2003)
4. Sorjamaa, A., Liitiäinen, E., Lendasse, A.: Time series prediction as a problem of missing values: Application to estsp2007 and nn3 competition benchmarks. In: IJCNN, International Joint Conference on Neural Networks, Documation LLC, Eau Claire, Wisconsin, USA, August 12-17, pp. 1770–1775 (2007), doi:10.1109/IJCNN.2007.4371429
5. Cottrell, M., Letrémy, P.: Missing values: Processing with the kohonen algorithm. In: Applied Stochastic Models and Data Analysis, Brest, France, May 17-20, pp. 489–496 (2005)
6. SOM Toolbox, http://www.cis.hut.fi/projects/somtoolbox/
7. Similä, T., Tikka, J.: Multiresponse sparse regression with application to multidimensional scaling. In: Duch, W., Kacprzyk, J., Oja, E., Zadrożny, S. (eds.) ICANN 2005. LNCS, vol. 3697, pp. 97–102. Springer, Heidelberg (2005)
8. Efron, B., Hastie, T., Johnstone, I., Tibshirani, R.: Least angle regression. In: Annals of Statistics, vol. 32, pp. 407–499 (2004)
9. Akaike, H.: A new look at the statistical model identification. IEEE Transactions on Automatic Control 19(6), 716–723 (1974)
10. Schwarz, G.: Estimating the dimension of a model. Annals of Statistics 6, 461–464 (1978)
11. Bhansali, R.J., Downham, D.Y.: Some properties of the order of an autoregressive model selected by a generalization of akaike's epf criterion. Biometrika 64(3), 547–551 (1977)
12. Hannan, E.J., Quinn, B.G.: The determination of the order of an autoregression. Journal of the Royal Statistical Society, B 41, 190–195 (1979)
13. TSPCi Group Downloads, http://www.cis.hut.fi/projects/tsp/?page=Downloads

# Towards Semi-supervised Manifold Learning: UKR with Structural Hints

Jan Steffen[1], Stefan Klanke[2], Sethu Vijayakumar[2], and Helge Ritter[1]

[1] Neuroinformatics Group, Bielefeld University, Germany
[2] Institute of Perception, Action and Behaviour, University of Edinburgh, UK

**Abstract.** We explore generic mechanisms to introduce *structural hints* into the method of Unsupervised Kernel Regression (UKR) in order to learn representations of data sequences in a semi-supervised way. These new extensions are targeted at representing a dextrous manipulation task. We thus evaluate the effectiveness of the proposed mechanisms on appropriate toy data that mimic the characteristics of the aimed manipulation task and thereby provide means for a systematic evaluation.

## 1   Introduction

Learning of control manifolds is emerging as one of the key challenges in unsupervised learning. Here, the Self-organising Map (SOM) has been influential in various pertinent approaches (cp. e.g.[1]). One more recent method, Unsupervised Kernel Regression (UKR, [6,4]), can be seen as a successor bridging between earlier "Parametrised SOM" (PSOM, [11]) and kernel methods (e.g.[8]).

In previous work [9], we have shown that UKR is well suited for representing human manipulation data. However, due to UKR being unable to incorporate prior knowledge about the data structure, generating *Manipulation Manifolds* (cp. [9]) from training sequences of hand posture data had been realised as supervised *construction* instead of automatic *learning*. In this paper, we present extensions to UKR for learning (periodic) sequences of chronologically ordered data and regularising intra-sequence characteristics which are aimed at learning *Manipulation Manifolds* in a *semi-supervised* manner. As basis for several error measures and thus a systematic evaluation of the new extensions, we perform an analysis on appropriate toy data which mimic the intrinsic characteristics of the targeted manipulation data. Whereas toy data always bare the risk of lacking transferability to the real data case, we here present promising first real data results in our targeted domain of dextrous manipulation.

We briefly recall UKR in Section 2 and present the new extensions in Section 3. Section 4 briefly summarises the original manipulation data and Section 5 addresses the corresponding toy data generation. Section 6 then uses this data for the evaluation of the new UKR extensions. Section 7 concludes the work.

## 2   Unsupervised Kernel Regression (UKR)

UKR is a recent approach to learning non-linear continuous manifolds, that is, finding a lower dimensional (latent) representation $\mathbf{X} = (\mathbf{x}_1, \ldots, \mathbf{x}_N) \in \mathbb{R}^{q \times N}$ of

J.C. Príncipe and R. Miikkulainen (Eds.): WSOM 2009, LNCS 5629, pp. 298–306, 2009.

a set of observed data $\mathbf{Y} = (\mathbf{y}_1, \ldots, \mathbf{y}_N) \in \mathbb{R}^{d \times N}$ and a corresponding functional relationship $\mathbf{y} = \mathbf{f}(\mathbf{x})$. UKR has been introduced as the unsupervised counterpart of the Nadaraya-Watson kernel regression estimator by Meinecke et al. in [6]. Further development has lead to the inclusion of general loss functions, a landmark variant, and the generalisation to local polynomial regression [4]. In its basic form, UKR uses the Nadaraya-Watson estimator [7,12]:

$$\mathbf{f}(\mathbf{x}) = \sum_{i=1}^{N} \mathbf{y}_i \frac{K_{\mathbf{H}}(\mathbf{x} - \mathbf{x}_i)}{\sum_j K_{\mathbf{H}}(\mathbf{x} - \mathbf{x}_j)} \qquad (1)$$

as smooth mapping $\mathbf{f} : \mathbf{x} \in \mathbb{R}^q \to \mathbf{y} \in \mathbb{R}^d$ from latent to observed data space ($K_{\mathbf{H}}$: Kernel with bandwidth $\mathbf{H}$). $\mathbf{X} = \{\mathbf{x}_i\}, i = 1..N$ now plays the role of input data to the regression function (1) and is treated as set of *latent parameters* corresponding to $\mathbf{Y}$. As the scaling and positioning of the $\mathbf{x}_i$'s are free, the formerly crucial bandwidths $\mathbf{H}$ become irrelevant and can be set to 1.

UKR training, that is finding optimal latent variables $\mathbf{X}$, involves gradient-based minimisation of the reconstruction error

$$R(\mathbf{X}) = \frac{1}{N} \sum_i \| \mathbf{y}_i - \mathbf{f}(\mathbf{x}_i; \mathbf{X}) \|^2 = \frac{1}{N} \| \mathbf{Y} - \mathbf{Y}\mathbf{B}(\mathbf{X}) \|_F^2 . \qquad (2)$$

Here, $\mathbf{B}(\mathbf{X})$ with $(\mathbf{B}(\mathbf{X}))_{ij} = \frac{K(\mathbf{x}_i - \mathbf{x}_j)}{\sum_k K(\mathbf{x}_k - \mathbf{x}_j)}$ is an $N \times N$ *basis function matrix*.

To avoid poor local minima, i.e. PCA [3] or Isomap [10] can be used for initialisation. These eigenvector-based methods are quite powerful in uncovering low-dimensional structures by themselves. Contrary to UKR, however, PCA is restricted to linear structures and Isomap provides no continuous mapping.

To avoid a trivial solution by moving the $\mathbf{x}_i$ infinitively apart from each other ($\mathbf{B}(\mathbf{X})$ becoming the identity matrix), several regularisation methods are possible [4]. Most notably, leave-one-out cross-validation (LOO-CV: reconstructing each $\mathbf{y}_i$ without using itself) is efficiently realised by zeroing the diagonal of $\mathbf{B}(\mathbf{X})$ before normalising its column sums to 1. The inverse mapping $\mathbf{x} = \mathbf{f}^{-1}(\mathbf{y}; \mathbf{X})$ can be approximated by $\mathbf{x}^\star = \mathbf{g}(\mathbf{y}; \mathbf{X}) = \arg\min_{\mathbf{x}} \|\mathbf{y} - \mathbf{f}(\mathbf{x}; \mathbf{X})\|^2$.

## 3   UKR for Data Sequences

To enable the originally purely unsupervised UKR training to benefit from prior knowledge about the data structure, we introduce extensions which a) especially consider ordered data *sequences*, b) explicitly allow for *periodic* sequences, c) propagate the original intra-sequence order to their latent representations and d) propagate stability of non-temporal sequence parameters within the sequences.

*a)* We consider given affiliations to sequences which enables us to influence the latent parameter adaptation such that sequence-specific mechanisms can be involved in the training. To this end, we distinguish between one latent *temporal intra*-sequence dimension and the other *inter*-sequence *parameter* dimensions.

*b)* Periodic sequences consist of one periodic temporal and one/several (usually) non-periodic dimensions. To allow for such structure, we provide different univariate kernels $K_l$ for different latent dimensions $l$. The basis functions $(\mathbf{B}(\mathbf{X}))_{\mathbf{ij}}$ (cp. Sec.2) then consist of their normalised products (parametrised by $\Theta_l$):

$$(\mathbf{B}(\mathbf{X}))_{\mathbf{ij}} = \frac{\prod_{l=1}^{q} K_l(x_{i,l} - x_{j,l}; \Theta_l)}{\sum_{k}^{N} \prod_{l=1}^{q} K_l(x_{k,l} - x_{j,l}; \Theta_l)}. \tag{3}$$

In the non-periodic case, the univariate versions of the kernels used in original UKR can be applied (e.g. Gaussian: $K_g(x_i - x_j; \Theta) = \exp\left[-\frac{1}{2}\Theta^2(x_i - x_j)^2\right]$). In analogy to original UKR, we assume no need for bandwidth control. However, to analyse potential cross-effects with the following new extensions, we also investigate different bandwidths for this case. For the periodic case, we propose the following cyclic kernel with bandwidth parameter $\Theta$, periodic in $[0; \pi]$:

$$K_{\circlearrowleft}(x_i - x_j; \Theta) = \exp\left[-\frac{1}{2}\Theta^2 \sin^2(x_i - x_j)\right]. \tag{4}$$

Up to normalisation and scaling, the kernel is equivalent to the von Mises distribution [5] which has been already used by Bishop et al. [2] to represent periodic data characteristics. We chose the presented form for convenience reasons.

In the periodic case, kernel bandwidth regulation is needed since the effective space in corresponding dimensions is constrained due to its periodic nature and fixed bandwidths cannot be compensated by scaling the latent parameters.

*c) "cyclic data order"*: To propagate the original chronological order of $N_S$ data sequences $S_\sigma = (\mathbf{y}_1^\sigma, .., \mathbf{y}_{N_\sigma}^\sigma)$, $\sigma = 1..N_S$ to the corresponding latent parameters $(\mathbf{x}_1^\sigma, .., \mathbf{x}_{N_\sigma}^\sigma)$, the values $x_{i,d_t}^\sigma$, $i = 1..N_\sigma$ in the *temporal* latent dimension $d_t$ need to reflect the order of the original data sequence. In the periodic case, such condition is difficult to induce without any assumptions about the underlying sequences. However, by providing sequences of complete cycles, we can consider the first data point in the sequence as successor of the last one: $\mathbf{x}_0^\sigma = \mathbf{x}_{N_\sigma}^\sigma$. If so, a penalty term in the loss function can preserve the cyclic data order:

$$E_{cseq}(\mathbf{X}) = \sum_{\sigma=1}^{N_S} \sum_{i=1}^{N_\sigma} \sin^2(x_{i,d_t}^\sigma - x_{(i-1),d_t}^\sigma). \tag{5}$$

*d)* One strong assumption which we want to be reflected in the latent space is, that the values of the non-temporal dimensions are approximately constant within single sequences. This consideration stems from the generation of our manipulation data (see next Section for a short description). The basic idea is that the underlying movement parameters usually do not change during single sequences – e.g., for cap turning, the radius of the cap does not change during the turning. We realise this regularisation of intra-sequence parameter variations as penalty term to the loss function which penalises high variances in the non-temporal dimensions $k = 1..q$, $k \neq d_t$:

$$E_{pvar}(\mathbf{X}) = \sum_{\sigma=1}^{N_S} \sum_{k \neq d_t} \frac{1}{N_\sigma} \sum_{i=1}^{N_\sigma} \left(x_{i,k}^\sigma - \langle x_{\cdot,k}^\sigma \rangle\right)^2 \tag{6}$$

**Fig. 1.** Example of a hand posture sequence corresponding to a training manipulation of a bottle cap ($r = 2.0cm$). Note the periodic nature of the movement.

The overall loss function then can be denoted as $E(\mathbf{X}) = R(\mathbf{X}) + \lambda_{cseq} E_{cseq}(\mathbf{X}) + \lambda_{pvar} E_{pvar}(\mathbf{X})$. The new parameters are $(\Theta_1, \ldots, \Theta_q, \lambda_{cseq}, \lambda_{pvar})$.

## 4  Manipulation Data

As described in Sec.1, the presented extensions are aimed at learning the representation of a manipulation task (i.e. turning a bottle cap). The set of training data, which has been used already for the initial manifold *construction* in [9], consists of sequences of hand postures (each a 24D joint angle vectors) recorded during the turning movement for different cap radii ($r = 1.5cm$, $2.0cm$, $2.5cm$, $3.0cm$ and $3.5cm$). The movement itself is periodic in the sense that the beginning and end postures are (in principle) the same. For each radius, we produced five to nine sequences of about 30 to 45 hand postures each – in total 1204 for all sequences and all radii. Figure 1 exemplary visualises one of such sequences.

## 5  Toy Data for Evaluation

To evaluate the new UKR extensions, we generate toy data with similar intrinsic characteristics as the manipulation data in [9] briefly described in the last section. The utilisation of toy data provides us with knowledge about underlying true structures and enables us to compute a variety of error measures not accessible otherwise (cp. Sec. 6 for details). As basis for an adequate toy data generation, we thoroughly investigate the real data. Here, we especially try to uncover the intrinsic data structures reflecting our prior knowledge of the generated manipulation data. From the generation process, we assume the existence of a periodic structure reflecting the periodic nature of the cap turning movement and an additional non-periodic expansion reflecting the different cap radii used for the sequence generation. By using Isomap [10] – a powerful method for uncovering hidden intrinsic structures in large data sets – we are able to reinforce these assumptions: a three-dimensional Isomap embedding of our manipulation data (see Fig.2a) reveals a cylinder-like structure describing a periodicity living in the $x/y$ dimensions and a non-periodic extension in $z$ direction.

To unfold the 2D representation of the periodicity, we can apply *atan2* on the $x/y$-part of the embedding data yielding the basis for the corresponding 1D "angle" $\in [0; \pi]$. In combination with the original $z$ component, we receive a 2D representation of the formerly 3D Isomap embedding and of the 24D original hand posture data, respectively. This data can be used as latent initialisation of

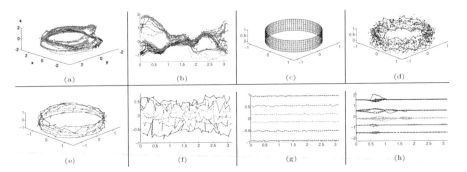

**Fig. 2. (a)** 3D Isomap embedding of 24D hand posture data recorded during the turning movement of a bottle cap. Different colours encode different cap radii. **(b)** atan2-mapping of (a). **(c)** noise-free training data (red, connected); test data (black, single points). **(d)** noisy training/test data. **(e)** Toy data Isomap embedding (cp. (a)). **(f)** atan2-mapping of (e). **(g-h)** Results for toy (g) and real (h) data.

the UKR model[1] as visualised in Fig.2b. Here, it turns out that the different sequences (connected) are not clearly separated and even sequences corresponding to different cap radii (encoded by different colours) partly overlap.

To reflect similar characteristics in our toy data and to provide an informative basis for the later evaluation, we aim at a simple low-dimensional toy data structure that produces Isomap embeddings of a similar form as the real data. To this end, we generate ordered (connected) data samples from the surface of a cylinder geometry (height=1, radius=1, Fig.2c) living in 3D together with noisy versions (Gaussian noise, $\sigma = 0.1$, e.g. Fig.2d). Such data then yield Isomap embeddings which a) provide a periodicity b) a non-periodic *parameter* expansion and c) are organised in *chronologically* ordered sequences ("trials") and thus are quantitatively similar to the Isomap embedding of the real data (Fig.2a/e) and its 2D mapping (Fig.2b/f). Within this cylinder structure, cross sectional rings of different height levels model sequences for different cap radii in the real data. As basis for the evaluation, we generated six training data rings and six overlapping together with five intermediate test data rings (cf. Fig.2c).

In anticipation of the following, Fig.2(g-h) depict the resulting latent parameters from training with toy and real data, respectively, having considered the results from the next section. The similarity of both latent structures supports the appropriateness of the toy data for the use with our real manipulation data.

## 6    Evaluation and Results

We evaluate the new extensions to UKR with the training/test data described in the last section. We incorporate our prior knowledge about the data – periodic sequences and non-periodic height levels corresponding to the periodic movement and the non-periodic radii variation in the manipulation data – in form of the

---

[1] The 2D latent space with one periodic kernel has the topology of a cylinder surface.

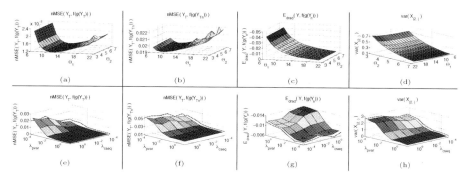

**Fig. 3.** Evaluation results **(a-d)** for $\lambda_{cseq} = \lambda_{pvar} = 1$ and **(e-h)**$(\Theta_1, \Theta_2) = (12, 4)$, red lines:$\lambda_{cseq}/\lambda_{pvar} = 0$. Please refer to text for further explanation.

specification of two associated latent dimensions: one periodic $(K_1(\cdot; \Theta_1) = K_\circlearrowleft)$ *temporal* and one non-periodic $(K_2(\cdot; \Theta_2) = K_g)$ *parameter* dimension. The loss function then consists of the reconstruction error and the penalty terms introduced in Section 3: $E(\mathbf{X}) = R(\mathbf{X}) + \lambda_{cseq} \cdot E_{cseq}(\mathbf{X}) + \lambda_{pvar} \cdot E_{pvar}(\mathbf{X})$.

As proposed in the last section, we compute 3D Isomap embeddings (for this data very robust in the choice of the neighbourhood parameter; here $K = 10$) of the noisy training data $\mathbf{Y}$ (cf. Fig.2e), and again use atan2 to retrieve a 2D latent initialisation for the UKR model (Fig.2f).

The evaluation focusses on the effect of different combinations of the inverse bandwidths $\Theta_1, \Theta_2$, and the penalty weightings $\lambda_{cseq}, \lambda_{pvar}$. From our toy data structure, we derive initial guesses for good bandwidth parameters $(\Theta_1 = 14, \Theta_2 = 5$ based on average inter-point distances) and evaluate correspondingly $\Theta_1$ for values $\{7, 8, .., 14, .., 21\}$ and $\Theta_2$ for $\{3, 3.5, .., 5, .., 7\}$. As for $\lambda_{cseq}$ and $\lambda_{pvar}$, no assumptions could be made, we choose $\lambda_{cseq}, \lambda_{pvar} \in \{0, 10^{-4}, 10^{-3}, 10^{-2}, 10^{-1}, 10^0, 10^1\}$. For each tuple $(\Theta_1, \Theta_2, \lambda_{cseq}, \lambda_{pvar})$, 10 training runs with 10 noisy versions of the training data are conducted. Each run consists of 500 optimisation steps including LOO-CV (exemplary result: Fig.2(g)). Initial tests yielded the most promising results for $\lambda_{cseq} = \lambda_{pvar} = 1$ which thus provides a good starting point for the evaluation of $\Theta_1$ and $\Theta_2$.

Fig.3(a-d) depict the corresponding reconstruction errors for varying bandwidth parameters $\Theta_1, \Theta_2$. Fig.3a shows the normalised mean square error (nMSE) between noise-free test data $\mathbf{Y}_T$ (the underlying true cylinder geometry) and its UKR reconstructions $\mathbf{f}(\mathbf{g}(\mathbf{Y}_T))$, visualising UKR's ability to generalise to unseen data from the underlying structure. Fig.3b shows the nMSE between $\mathbf{Y}_T$ and the reconstruction of its noisy versions $\mathbf{f}(\mathbf{g}(\mathbf{Y}_{\mathbf{Tn}}))$, visualising UKR's robustness in representing the underlying structure and its ability to correct noisy input data. The bias of $\mathbf{f}(\mathbf{g}(\cdot))$ towards the inner of the underlying structure (a known problem in original UKR) is depicted in Fig.3c for noisy training data $\mathbf{Y}_n$.

Fig.3(a-b) show a clear error dependency on $\Theta_1$ and minimal errors for $\Theta_1 = 12$ (Fig.3a) or $\Theta_1 = 10$ (Fig.3b). However, as the bias significantly increases with decreasing $\Theta_1$ (Fig.3c), we use $\Theta_1 = 12$ for further evaluation. As assumed before, there is no significant dependency on $\Theta_2$ due to the free positioning of

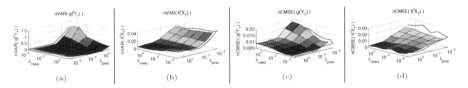

**Fig. 4.** Application err's $(\Theta_1, \Theta_2) = (12, 4)$, Red line: $\lambda_{cseq}/\lambda_{pvar} = 0$ (details: see text)

the latent parameters in the non-periodic dimensions. This is shown in Fig.3: whereas the errors stay approximately constant (Fig.3(a-c)), the variance in the latent *parameter* dimension varies strongly for changing $\Theta_2$ (Fig.3(d)).

Fig.3(e-h) depict errors for fixed bandwidth parameters $(\Theta_1, \Theta_2) = (12, 4)$ and different combinations of $\lambda_{cseq}, \lambda_{pvar}$. Fig.3(e-f) reveal that high values of $\lambda_{cseq}$ – which stronger force correctly ordered latent parameters – negatively influence the reconstruction error. However, high values of $\lambda_{pvar}$ damp the reconstruction error in general and are able to overrule the negative effect of the sequence order penalty. Indeed, as depicted in Fig.3g, both high weightings of $E_{cseq}$ and $E_{pvar}$ yield high radius errors. Logically consistent, high values of $\lambda_{pvar}$ strongly damp the variance in the latent data dimension (cp. Fig.3h).

For applications exploiting the aimed sequence-reflecting latent structure, not only the pointwise nMSE, but also structure-related errors are of interest. Fig.4a shows a normalised variance in the latent parameter dimension ("nVAR") of observed fix-parameter sequences (lines in observed space) mapped into latent space $(\mathbf{g}(^{\mathbf{r}}\mathbf{Y_T}))$ providing a measure for the distortion of the line projection and thus for the distortion in the parameter dimension. The plot uncovers that high weightings of $E_{pvar}$ (reducing general reconstruction errors; cp.Fig.3(e-h)) only result in stable sequence projections for strongly weighted $E_{cseq}$. Fig.4b shows the inverse projection direction, corresponding to reproducing/synthesising sequences in original data space with fixed sequence parameters. Again, only combined high $E_{pvar}$- and $E_{cseq}$-weightings produce stable sequences. Fig.4(c-d) investigate the corresponding inverse situations. Fig.4c visualises temporal synchronisation distortions of the latent space projections of sequence parameter modulations in observed space for fixed points in time. To take account for the periodic nature of the latent temporal dimension, we calculate nMSEs on the angular deviations from the mean ("nCMSE") of the analysed line and take the underlying kernel period into account. Like this, the nCMSE has similar characteristics as the nVAR for non-periodic dimensions. Here, high $\lambda_{cseq}$-weightings result in higher distortions of the projections. However, for the targeted high weightings of $E_{pvar}$, the negative effect of higher values for $\lambda_{cseq}$ still is in a reasonable region. Fig.4d visualises the inverse mapping, measuring the distortions of projections of lines in latent space with zero-variance in the temporal dimension back into observed space. Again, for high $\lambda_{pvar}$, the effect of the sequence penalty $E_{cseq}$ is strongly dominated by the effect of $E_{pvar}$.

To sum up: whereas the choice of $\Theta_2$ is less important, the inverse bandwidth $\Theta_1$ should be set to a value (slightly) smaller than the inverse of the average point distance in the corresponding dimension (here: $\Theta_1 = 12$). The results for varying

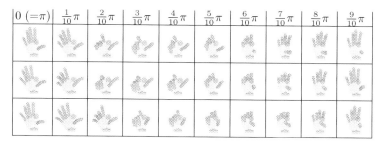

| $0 \ (=\pi)$ | $\frac{1}{10}\pi$ | $\frac{2}{10}\pi$ | $\frac{3}{10}\pi$ | $\frac{4}{10}\pi$ | $\frac{5}{10}\pi$ | $\frac{6}{10}\pi$ | $\frac{7}{10}\pi$ | $\frac{8}{10}\pi$ | $\frac{9}{10}\pi$ |
|---|---|---|---|---|---|---|---|---|---|
| | | | | | | | | | |
| | | | | | | | | | |
| | | | | | | | | | |

**Fig. 5.** Promising results for the targeted manipulation task. Horizontal dim.: time; vertical dim.: cap radius. Depicted are reprojections $\mathbf{f}(\mathbf{x}; \mathbf{X})$ of regularly sampled latent positions $\mathbf{x}$ of the trained UKR. Please consider also the video available under http://www.techfak.uni-bielefeld.de/~jsteffen/mov/wsom2009/.

$\lambda_{cseq}/\lambda_{pvar}$ are relatively robust. Here, they optimally effect the generation of the desired latent structures for $\lambda_{cseq}/\lambda_{pvar} = 1$.

## 7 Conclusion

We presented extensions to the unsupervised manifold learning method UKR, which now allow for semi-supervised learning of structured manifolds. We evaluated the new extensions on toy data in a general and manipulation relevant context as basis for future work on real manipulation data. First promising results using our new insights are visualised in Fig. 5: the targeted task of representing the periodic movement of turning a bottle cap has been successfully achieved.

**Acknowledgement.** This work has been carried out with support from the German Collaborative Research Centre "SFB 673 - Alignment in Communication" granted by the DFG, from the EU FP6 SENSOPAC project funded by the European Commission and from the German Cluster of Excellence 277 CITEC.

## References

1. Barreto, G., Araújo, A., Ritter, H.: Self-Organizing Feature Maps for Modeling and Control of Robotic Manipulators. Intelligent and Robotic Systems 30(4) (2003)
2. Bishop, C., Legleye, C.: Estimating conditional probability densities for periodic variables. In: Advances in Neural Information Processing Systems, vol. 7, pp. 641–648 (1995)
3. Jolliffe, I.T.: Principal Component Analysis, 2nd edn. Springer, New York (2002)
4. Klanke, S.: Learning Manifolds with the Parametrized Self-Organizing Map and Unsupervised Kernel Regression. PhD thesis, Bielefeld University (2007)
5. Mardia, K.: Statistics of Directional Data. Academic Press, London (1972)
6. Meinicke, P., Klanke, S., Memisevic, R., Ritter, H.: Principal Surfaces from Unsupervised Kernel Regression. IEEE Trans. on PAMI 27(9) (2005)
7. Nadaraya, E.A.: On Estimating Regression. Theory of Probability and Its Application 9 (1964)

8. Schölkopf, B., Smola, A., Müller, K.: Nonlinear Component Analysis as a Kernel Eigenvalue Problem. Neural Computation 10(5), 1299–1319 (1998)
9. Steffen, J., Haschke, R., Ritter, H.: Towards Dextrous Manipulation Using Manifolds. In: Proc. IROS (2008)
10. Tenenbaum, J.B., de Silva, V., Langford, J.C.: A global geometric framework for nonlinear dimensionality reduction. Science 290(5500), 2319–2323 (2000)
11. Walter, J., Ritter, H.: Rapid Learning with Parametrized Self-organizing Maps. Neurocomputing 12, 131–153 (1996)
12. Watson, G.S.: Smooth Regression Analysis. Sankhya, Ser. A 26 (1964)

# Construction of a General Physical Condition Judgment System Using Acceleration Plethysmogram Pulse-Wave Analysis

Heizo Tokutaka[1], Yoshio Maniwa[2], Eikou Gonda[3], Masashi Yamamoto[4],
Toshiyuki Kakihara[5], Masahumi Kurata[5], Kikuo Fujimura[5],
Li Shigang[5], and Masaaki Ohkita[1]

[1] SOM Japan Inc.,
4-637, Koyamacho-kita Tottori, Japan
[2] Yabu City Minamidani Clinic
[3] Yonago National College of Technology
[4] Shikano-Onsen Hospital
[5] Tottori University
tokuhema@hal.ne.jp
http://www.somj.com

**Abstract.** Among the popular lifestyle-related diseases are smoking, overweight and stress. A daily health check is important because there is no clear objective symptom for these diseases. We developed diagnostic software which shows the state of the blood vessels using a Basic SOM model, and performs synthetic plethysmogram analysis of 4 components using the map location (the state of the blood vessel, vascularity), looseness, pulse/minute, and pulse stability.

**Keywords:** plethysmogram, SOM, clinical case study.

## 1  Introduction

To carry oxygen and nutrition all over the body is a blood vessel's most important task. When healthy, the blood vessel has the elasticity and flexibility corresponding to the pressure of the blood. When dangerous factors such as smoking, overweight, stress and other lifestyle-related diseases influence a blood vessel, arteriosclerosis gets worse. It causes cerebral infarction and cerebral hemorrhages due to a high blood pressure. These accidents can happen suddenly without clear symptoms, so that a daily check is becoming increasingly important for prevention.

From the fingertip, the acceleration plethysmogram is obtained without subjective stress in a short period of time. We developed diagnostic software which graphically shows the state of the blood vessel by using a Basic SOM model.

Much information is included in the plethysmogram about the blood movement in the vessel, which goes from the center (heart) to the end (fingertips). The bloodstream travels through the blood vessels from the heart to the capillary vessels in a wave-like motion. The plethysmograph is affected by the ictus

J.C. Príncipe and R. Miikkulainen (Eds.): WSOM 2009, LNCS 5629, pp. 307–315, 2009.

 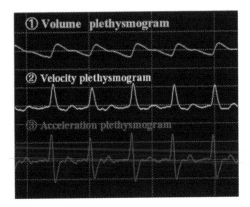

**Fig. 1.** The left figure shows the set-up for measuring the plethysmogram. The panels in the right figure are ①Volume plethysmogram, ②Velocity plethysmogram and ③Acceleration plethysmogram.

cordis, the haemodynamics and the physiological condition caused by the change in the properties of an arteriole. The effects can be observed as distortions in the wave profiles. The inside diameter of the blood vessel changes due to a swelling of the blood vessel. The wave motion that occurs at that time is called the volume plethysmogram (refering Fig. 1①.). The volume plethysmogram has the problem that the base line never becomes stable. Therefore, it is difficult to estimate the inflection point because the wave becomes sparse when it raises. Hence, techniques to differentiate the waveform have been proposed. The acceleration plethysmogram (2nd derivative) is one recent example that has considered and evaluated [1]. The volume plethysmogram by a volumetric change of the blood vessel, the velocity plethysmogram (1st derivative) and the acceleration plethysmogram (2nd) are shown in Fig. 1. A departure from flatness is more visible in the acceleration plethymogram than in the volume plethysmogram, and it becomes, thus, easier to evaluate the waveform. The plethysmogram used for the diagnosis at present is the acceleration plethysmogram (referring Fig. 1③), and a doctor is evaluating the plethysmogram by watching the location of the inflection point or by calculating a blood vessel age formula.

The waveforms of the acceleration plethysmogram and the presently used characteristics (or features) are shown in Fig. 2. Waveforms are typically categorized into seven classes, as illustrated in Fig. 2. The figure shows the gradual changes from the waveform of a healthy signal (Group A) to a possibly unhealthy waveform (Group G), which could be caused by an incomplete blood circulation [1]. When the labels "a" to "e" are put on the wave extremal points, "b" is smallest for a healthy subject and "d" smallest when an incomplete circulation is possible (refer to Fig. 2). Doctors and researchers can assess the state of the blood vessel from the waveforms, but a non-specifically trained person. The plethysmogram analysis software described below has been developed using a Self Organizing Maps (SOM) [2,3] so that a non-trained person may interpret the waveform. The conventional tool is calculating the vein age by Eq. (1),

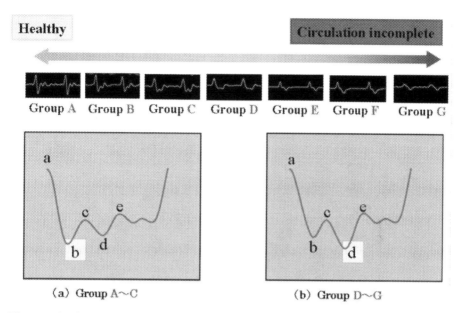

**Fig. 2.** The features of the acceleration plethysmogram. The group (a) in the right hand side show healthier waves than the one (b).

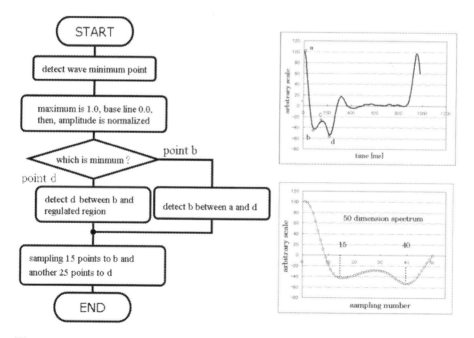

**Fig. 3.** The pre-processing of the acceleration Plethysmogram pulse wave before developing the SOM

which the medical doctor [4] has developed using the peak values of a , b, c, d, e shown in Fig. 2 (almost all of the commercial tools work this way). Our tool has cut down 50 points from the sampling points of the acceleration plethysmogram wave using the algorithm which is described below (Fig. 3) and which tried to classify the wave forms using the Self-Organizing Maps (SOM) [2,3]. It is developed in such a way that our tool Pulsar-SOM [5,6] becomes more easy to use, even by the general public. Thus, our tool Pulsar-SOM is developed for the general public for easily easily assessing the condition of their capillary vessels [5,6].

$$[Vein\ age] = 43.5 \times \frac{(b - c - d - e)}{a} + 65.9 \tag{1}$$

## 2  Synthetic Plethysmogram Diagnosis

Pulser_SOM is an application which obtains the state of the blood stream from the plethysmogram. The waveform of the acceleration plethysmogram, to differentiate the volume plethysmogram, obtained from the fingertip, has the characteristic of predicting the state of the blood stream. This characteristic cannot be

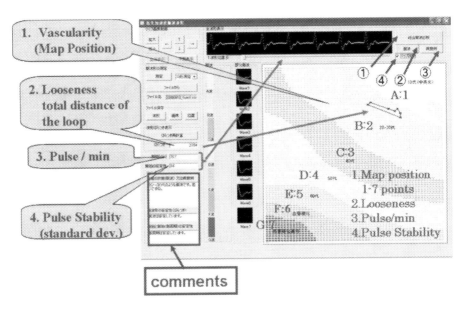

**Fig. 4.** The general pulse wave analysis tool in which the pulse wave analysis and the clinical example can be seen on the same screen. At present, an acceleration Plethysmogram pulse wave is analyzed, by pushing the button "the pulse wave" of ②. Then, the analysis can be moved to the general pulse wave analysis by pushing the button ① and using the data of 1 - 4 in the figure. Putting the mark in the button ④, the Japanese explanation of each color region on the map of an acceleration Plethysmogram pulse wave analysis mode appears.

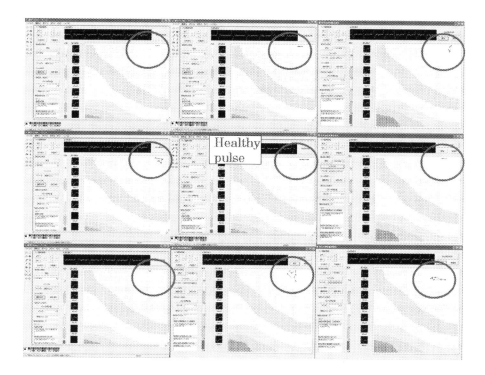

**Fig. 5.** Examples of a classification as healthy

understood by the non-trained individual, only by doctors and researchers. However, a non-trained person, who doesn't have the special knowledge, can predict the state of the blood vessel using Pulser_SOM. Pulser_SOM uses the plethysmogram sensor of U-Medica Inc. First, we explain what role each part of this software has. This software is developed based on the algorithm [5,6] in Fig. 3.

We show the example of the measurement in Fig. 4. As shown in Fig. 4, the plethysmograms classify from "Group A" in the upper right to "Group G" in the lower left. The upper right section in each sub-figure of Fig. 5 shows an example of a healthy person. Dr. Maniwa, who is one of the authors, tried to classify the clinical examples of the plethysmogram [7]. Each region of the map in Fig. 6 shows a clinical example. As shown in Fig. 6, there are 6 clinically explained/relevant regions: "healthy", "healthy, but pulse frequent tendency", "metabo recoverable group: acylglycerol and/or blood sugar high level", "fatigue, stress, shortage of sleep", "climacteric disorder", "arteriosclerosis". Although the plethysmogram of the examinee in Fig. 4 is in the healthy region of "2", its clinical example in Fig. 6 is in the "metabo recoverable group".

As shown in Fig. 4, the position of the partial wave is shown by a point (The point can be expanded, so that the partial wave number becomes readable.) on the map. To the center of gravity of this point group corresponds the colors of A-G in the "legend", and a comment which corresponds to its color is shown in an editbox of "vascularity". In this research, the A-G colors are converted

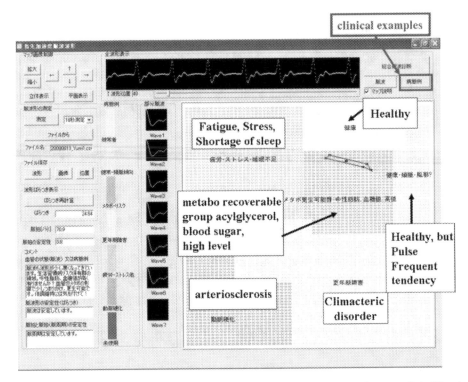

**Fig. 6.** "The clinical example" of the examinee is shown by pushing button ③ in Fig. 4 which is the button of "the clinical examples". This examinee corresponds to the area of "metabo recoverable group" by the color of "the clinical examples" on the left of the figure. Also, the black mark surrounded by the line which is linked by five points of Wave1-Wave5 in the figure is the center of gravity (or the representative point ), where Wave6 is hidden among Wave1- Wave5. Putting the mark in the button ④ which is already shown in Fig. 4, the Japanese explanation of each color region on the map of "the clinical example" analysis mode of the examinee appears.

into numbers 1-7. The transition of the partial wave shown on the map is tied with the shortest distance as defined by TSP (Traveling Salesman Problem), and its distance is shown by "looseness". Next "pulse" and "pulse stability" are also described by numbers. Using four items, and pushing the button ① of "Synthetic plethysmogram diagnosis" in Fig. 4, a new map is created in Figs. 7 and 8. Figures 4, 5 and 6 are obtained from the Basic SOM model. However, Figures 7 and 8 are obtained from Torus-type SOM. For the basic SOM model, the number of input data points is 1500, the number of neurons 60 80, the number of iterations 10,000, the learning coefficient 0.4, and the neighborhood radius 60, gaussian. For the torus-type SOM, the number of input data points is 400, the number of neurons 20 30, the number of iterations 100,000 times, the learning coefficient 0.01, and the neighborhood radius 10, gaussian.

The data in the upper right section of Fig. 7 shows "vascularity" 2.0, "looseness" 27.5, "pulse" 70.7, "pulse stability" 3.8. The position of the center of

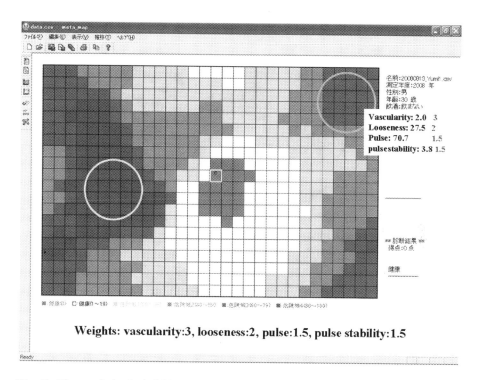

**Weights: vascularity:3, looseness:2, pulse:1.5, pulse stability:1.5**

**Fig. 7.** The mark (point) SOM map which was analyzed using the 4 components obtained in the pulse wave analysis shown in Fig. 4 (the values of the weight in the following Eq. (2) are described in the upper right in red letters as Vascularity: 3, Looseness: 2, Pulse: 1.5, and Pulse stability: 1.5).

gravity in the map of Fig. 4 is classified as "Group B", therefore, it is 2.0 if the group region exchanges the number. The others are shown as the number of data on the left hand side in Fig. 4. The number of data is calculated using Eq. (2).

$$\text{MHP}_i = \frac{\sqrt{\sum_{n=1}^{n} W_n (WV_n - NV)^2} - \sqrt{\sum_{n=1}^{n} W_n (X_{ni} - NV)^2}}{\sqrt{\sum_{n=1}^{n} W_n (WV_n - NV)^2}} \times 100 \quad (2)$$

with $WV_n$ the worst value of each component, and $NV$ the normal value 1, $X_{ni}$ the value of examinee $i$ of the $n$th component, $W_n$ the value of the weight of the $n$th component. The weights $W_n$ are given by Dr. Y. Maniwa (one of the authors) from his medical points. The values of the weights in Figs. 7 and 8 are described as Vascularity: 3, Looseness: 2, Pulse: 1.5, and Pulse stability: 1.5.

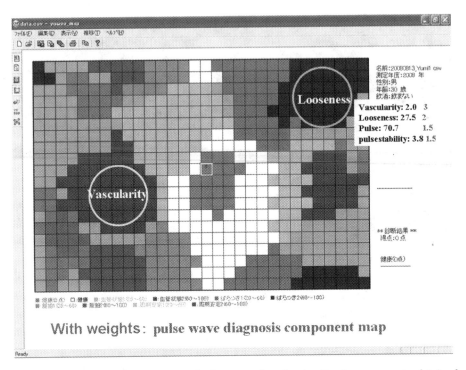

**Fig. 8.** The component SOM map which was analyzed using the 4 components obtained in the pulse wave analysis shown in Fig. 4 (the values of the weight in the following eq. 1 are described in the upper right corner in red letters as Vascularity: 3, Looseness: 2, Pulse: 1.5, and Pulse stability: 1.5).

The $MHP$ (Myakuha Health Mark Point) is converted into the plethysmogram score calculated as $100 - MHP$. When the plethysmogram score becomes high, all four components go to a bad situation. When it becomes low, they go to a good situation. As shown in Fig. 7, the color has 5 grades. If the examinee gets the worst value for each component, the $MHP$ is 0 points, therefore $100 - MHP$ is 100 points. When the plethysmogram score becomes high, the plethysmogram's healthy condition goes to a bad situation. The worst region is shown by the "red" color. The component SOM mapping in Fig. 8 shows the region of the worst value for each of the components. "Vascularity" has a "green" color, "Looseness" a "blue" color, "Pulse" a "red" color, "Pulse stability" a "purple" color. Because the examinee has the normal value for all components, he is in the position of the normal value "0", and its color is "light blue". As a result, we can evaluate the examinee's data quantitatively by developing a SOM map of four components.

## 3    Conclusion

We developed the diagnosis help software which shows the state of the blood vessel (vascularity) graphically using the Basic SOM model, and performs a synthetic plethysmogram diagnosis of 4 components using the map location (the state of the blood vessel, vascularity), looseness, pulse/minute, and stability of a pulse. In addition, we developed the classification of the clinical examples using the waveform of the plethysmogram. As a result, the examinee can assess the present health state quantitatively because the health state is given by a numerical value. We are convinced that the 4 components and the clinical example classification provide suggestive evidence to medical doctors, researchers and non-experts.

# References

1. Sano, Y., Kataoka, Y., Ikuyama, T., Wada, M., Konno, H., Kawamura, K., Watanabe, T., Nishida, A., Koyamauchi, H.: The evaluation of the blood circulation and its application by the acceleration plethysmogram. The Journal of Science of Labour 61(3), 129–143 (1985) (in Japanese)
2. Kohonen, T.: Self-Organaizing Maps. Springer, Heidelberg (2001)
3. Tokutaka, H., Ohkita, M., Fujimura, K.: Self-Organizing Maps and its Application. Springer, Japan (2007) (in Japanese)
4. Takazawa, K., Fujita, M., Yabe, K., Sasaki, T., Kobayashi, T., Maeda, K.: Clinical usefulness of the second derivative of a plethysmogram. J. Cardiol 23(suppl. 37), 207–217 (1993)
5. Noso, N., Fujimura, K., Ohkita, M.: Automatic Clustering on a Self-Organization Maps -Acceleration plethysmogram map. In: Proceedings of Fuzzy System Symposium 2007, pp. 407–410 (2007) (in Japanese)
6. Noso, N.: Automatic Classification and Analysis of Acceleration Plethysmogram on a Self-Organization Maps. In: Proceedings of Acceleration Plethysmogram Complex Systems Workshop of Japan, p. 3 (2007) (in Japanese)
7. Tokutaka, H., et al.: Construction of Acceleration Plethysmogram Analysis System with SOM and the Visible Analysis of each Clinical Example. In: The 81st Japan Society for Occupational Health in Sapporo Convention Center, E109, June 24-27 (2008) (in Japanese)

# Top-Down Control of Learning in Biological Self-Organizing Maps

Thomas Trappenberg[1,2], Pitoyo Hartono[3], and Douglas Rasmusson[1,4]

[1] Dalhousie Neuroscience Institute, Halifax, NS, Canada
[2] Dalhousie University, Faculty of Computer Science, Halifax, Canada
[3] Future University, Hakodate, Japan
[4] Dalhousie University, Department of Physiology and Biophysics, Halifax, Canada

**Abstract.** This paper discusses biological aspects of self-organising maps (SOMs) which includes a brief review of neurophysiological findings and classical models of neurophysiological SOMs. We then discuss some simulation studies on the role of topographic map representation for training mapping networks and on top-down control of map plasticity.

## 1 Introduction

Experience driven development and learning is of central importance in the brain, and self-organizing maps (SOMs) have long been a principle model for such systems (Kohonen, 1982; Willshaw and von der Malsburg, 1976). Work has continued in refining models for better matches with experiments and enhanced theoretical treatments (e.g. Tanaka, 1991). While map plasticity is generally celebrated as physical evidence of learning, learning-theoretic studies of the role of SOMs for learning on a cognitive level have been sparse.

This paper provides some discussions on modelling biological SOMs. We briefly point to some of the basic biological findings and literature on cortical maps, with some concentration on issues of top-down control of learning. We then review recent results by Zhou and Merzenich (2007) which demonstrate enormous changes in tonotopic maps of adult animals and the importance of behavioral relevant learning for map formation. The following brief review of biological models discusses the differences, and the relations, between some classical papers (Willshaw and von der Malsburg, 1976; Amari, 1977; Kohonen, 1982). Finally, we show that topographic map representations can assist learning of feed forward networks and address some questions of top-down control of map formation.

## 2 Topographic Maps in the Brain

Topographic maps are a common feature of cerebral cortical areas that process sensory information. The features that are mapped in the various sensory cortices differ somewhat, being mostly spatial in visual cortex (with visual space corresponding to position on the retina) and somatosensory cortex (position on the body surface) as compared to sound frequency in auditory cortex which results from a transformation of frequency to

J.C. Príncipe and R. Miikkulainen (Eds.): WSOM 2009, LNCS 5629, pp. 316–324, 2009.

position on the cochlea. An important characteristic of all maps and in all species studied is that there is significant distortion in the maps such that the parts of the sensory sheet that are biologically most important to the animal have a larger representational area. This is often called a magnification factor and results in the well-known distorted picture of the body representation in somatosensory cortex termed the homunculus in which the finger and lip representations are proportionally much larger than on the body itself. This enlargement of specific subregions in some animals has also provided important model systems that have facilitated the study of these maps, for example the high-frequency regions of auditory cortex of the bat that are involved in echolocation or the whisker region of rodent somatosensory cortex in which anatomically discrete barrels have a one-to-one correspondence to the large whiskers on the animals snout.

The original formation of topographic maps is largely a developmental problem involving chemical signaling between neurons in addition to network activity that results from sensory experience. While we can study how the framework of topographic maps can be laid down in the absence of neural activity (by applying drugs that block the electrical signals), it is impossible to do the reverse. Consequently, the role of network activity can best be studied in the more mature animal, after the developmental processes are complete, by perturbing some aspect of the map and observing how the mapping functions react. One approach is to remove one part of the input, say by denervating one part of the body or producing a small lesion in the cochlea or retina. A common finding of such experiments is that the deprived regions of cortex do not remain quiescent, but gradually begins to respond to stimulation of adjacent parts of the sensory sheet (Rasmusson, 1982; Robertson and Irvine, 1989; Kaas et al., 1990); it is often said that the map reorganizes itself. A related technique is to directly lesion part of the cortical map; again a consistent finding is that the surrounding regions gradually begin to respond to adjacent sensory inputs (Winship and Murphy, 2008). A second approach is to ask whether the map can be changed by increased neural activity, either by enhanced sensory input or by direct stimulation of a small part of the cortex. Here the usual finding is that the representational area enlarges with progressive stimulation, whether in the somatosensory cortex with finger stimulation (Jenkins et al., 1990), or the auditory system with repeated presentations of specific sound frequencies (Recanzone et al., 1993).

Before the precise mechanisms underlying these reorganizations can be uncovered, it is necessary to ask where exactly does plasticity occur, that is, where are the synaptic connections that are being changed during reorganization. The cortical responses are the result of a chain of three or more neurons proceeding from the sensory sheet to the cortex; at each level of processing there is a similar topographic map. It is possible, therefore, that plasticity occurs at one of these preceding levels and that the cortical response is simply a reflection of reorganization at these earlier stages. The cortex was the original focus of most early studies largely because it is more easily accessible. In fact, recent studies have demonstrated some degree of reorganization in the thalamus, which contains the final projection neurons that go to the cortex. The problem that arises here is that each of these sensory pathways is not only a bottom-up system with projections only going upstream. Feedback from the cortex to the thalamus has long been known and in fact outnumbers thalamocortical fibers by 10 to 1 (Steriade et al., 1997). A new thalamic response might then be due to reorganization within the cortex relayed back to the thalamus or to plasticity within the thalamus itself.

## 3    Goal-Directed Learning Controls Topographic Map Plasticity

While most of the work on the formation of topographic maps is targeted towards early development, we concentrate on the ongoing refinements of cortical maps during adult plasticity. A nice demonstration of representational plasticity in adult mammals was recently given by Zhou and Merzenich (2007) and is shown in Fig. 1. They raised rat pups in a noisy environment that severely impaired the development of tonotopicity in the primary auditory cortex (A1), which lasted into adulthood. An example of such a map in an adult rat is shown in Fig. 1A. The hatched areas represent areas with neurons that showed abnormal, poor frequency tuning. These rats did not recover a normal tonotopic representation in A1 even though they were stimulated in adulthood with sounds of different frequencies. However, when the same sound pattern were used to train the rats in a discrimination task in order to get food reward, the developmentally degraded rats were able to recover a normal tonotopic map as shown in Fig. 1B. This example demonstrates that goal directed learning can influence map plasticity. A purely bottom-up stimulation of networks, which is the common focus in SOM modelling, is certainly not enough to explain the results by Zhou and Merzenich (2007). Before returning to questions on modelling these results, we first review classical SOM models in the next section.

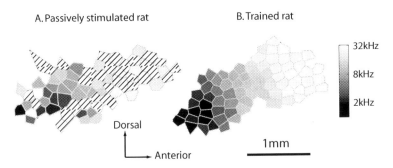

**Fig. 1.** (A) Frequency map in A1 of a rats that was developmentally degraded by being raised in noisy environments. The hatched areas contain neurons with poor frequency tuning. (B) Tonotopic map of an adult rat that recovered with training on a frequency discrimination task. [Data courtesy of Xiaoming Zhou and Micheal Merzenich.]

## 4    Basic Cortical Map Models

A model for activity-dependent, self-organized development of topographic cortical maps was studied over 30 years ago by Willshaw and von der Malsburg (1976). They considered a two dimensional cortical sheet as illustrated in Figure 2A. The states

A. Willshaw-von der Malsburg model          B. Kohonen model

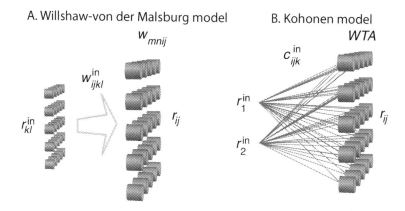

**Fig. 2.** Common architecture of self-organizing maps

of the nodes in this cortical sheet represent activation rates in localized neuronal populations with a leaky integrator dynamics. Population models were derived by Wilson and Cowan (1972) with separate excitatory and inhibitory populations, and subsequently studied with centre-surround architectures (Wilson and Cowan, 1973). Amari (1977) abstracted these models further in two important ways, by combining the excitatory and inhibitory populations, and by using a continuous descriptions of the neural sheet. The dynamics of the neural sheet described by neural field equations,

$$\tau \frac{\partial \mathbf{u}(\mathbf{x}, t)}{\partial t} = -\mathbf{u}(\mathbf{x}, t) + \int_{\mathbf{y}} \mathbf{w}(\mathbf{x}, \mathbf{y}) \mathbf{r}(\mathbf{y}, t) \mathrm{d}\mathbf{y} + I^{\mathrm{ext}}(\mathbf{x}, t) \tag{1}$$

$$\mathbf{r}(\mathbf{x}, t) = g(\mathbf{u}(\mathbf{x}, t)), \tag{2}$$

of which the equations by Willshaw and von der Malsburg are an discretized example. The general activation function $g$, which often has the form of a sigmoid, relates the internal state variable, $u$, to an externally observable rate variable, $r$. The integration kernel $w$ describes the interaction within the cortical sheet (recurrent network), and has typically the characteristics of short-distance excitation and long-distance inhibition. For example, with a cortical sheet on a torus to minimize boundary effects, a common interaction kernel is a shifted Gaussian,

$$\mathbf{w}(|x - y|) = A_{\mathrm{w}} \left( e^{-(x-y)^2 / 4\sigma^2} - C \right), \tag{3}$$

where $A_{\mathrm{w}}$, $\sigma$, and $C$ are constants, and $x$ and $y$ are the locations of the interacting neural populations. This kernel can be learned through Hebbian learning (Stringer et al., 2002), but most studies of self-organizing maps consider this interaction kernel as fixed when learning the afferent (input) weights to this neural sheet. Neural field models are widely used to model reaction times in cognitive neuroscience (e.g. Trappenberg, 2009).

The activity-dependent self-organization of the inputs to the neural sheet is also based on Hebbian learning as outlined in (Willshaw and von der Malsburg, 1976). Briefly, the neural sheet implements a maximum-likelihood estimator (Wu et al., 2001), where the estimate is given by a localized activity packet that develops through competitive neural field dynamics (Amari, 1977). This process is approximated by a winner-takes-all mechanism and a neighborhood function in Kohonen's formulation (Kohonen, 1982) . The neural activity in the activity packet is then correlated with the input pattern through Hebbian learning, with most active nodes in the centre of the activity packet strengthened more than nodes in the periphery of the activity packet.

A final approximation in Kohonen's formulation of SOMs is the replacement of distributed activity patterns for feature values in the input sheet with a direct representation of feature values in an input vector (Figure 2B). Note that we discuss here the case where the dimensionality of the feature space is the same as the dimensionality of the feature map, although SOMs can be used to map high dimensional feature spaces into lower dimensional maps (Obermeyer et al., 1990). The activity packet that develops though the neural dynamics determines the tuning curves in neural sheets (see Trappenberg, 2009 for an illustration). Therefore, the radial basis function nodes in Kohonen's network can be seen as a model of unimodal tuning curves, and the input weights in Kohnonen's formulation correspond to the centre of the tuning curve. This centre of the tuning curve is often called the preferred response (orientation in V1) of a neuron.

Kohonen's formulation, while more abstract, has several advantages to the model by Willshaw and von der Malsburg. Integrating the neural field dynamics is usually very time consuming, and the replacement of neural field representations with tuning curves reduces the computational burden dramatically. Furthermore, topographic organizations are easily observable when plotting centers of tuning curves, which are directly modelled in Kohonen's model. This information has first to be decoded in the neural field formulation. However, the neural field dynamics need to be considered when studying the dynamics of responses in neural tissue. Also, it is not clear how multiple concurrent objects in a sensory field can be simulated in Kohonen's model.

## 5  The Importance of Topographic Representations in Learning Mapping Functions

To demonstrated how topographic representations can help train mapping networks, we provide here a basic example of a system that must learn to solve a decoding task. The decoding task is to convert an analog signal to a digital representation (binary output vector). The system includes SOM in Kohonen's formulation, which receives the analog input, and a simple perceptron to map SOM states to the desired local, binary representation as shown in Figure 3A. The SOM was then trained on random examples of analog input signals for a specific number of training steps before using this representation to train the perceptron.

Such a decoding task can, in principle, be solved with a perceptron with one hidden layer since such machine learning systems are universal approximators (Hornik et al., 1989). Thus, why should we complicate the system with a SOM layer? To demonstrate this we use a simple perceptron (no hidden layers) as the output module of the network. Avoiding hidden layers makes it harder to solve the mapping problem. Also, a simple

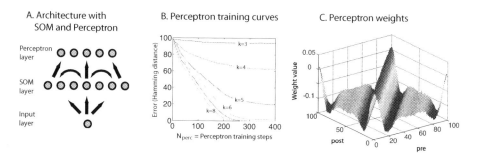

**Fig. 3.** Example of a decoder resulting from training a perceptron on self-organized representations. (A) Illustration of the architecture. (B) Learning curves for the perceptron with different number of learning steps in the SOM. (C) Weight values of the perceptron with well organized SOM and trained perceptron.

perceptron can be viewed as a reinforcement learner, which is the desired architecture to model the results of Zhou and Merzenich (2007).

Training curves for the perceptron are shown Figure 3B. The different curves correspond to different *number of steps* ($= 2^k$) of training the SOM before training the perceptron. We found that some organization was necessary for the perceptron to learn, and that further self-organization helped the perceptron to reduce its training time to reach a desired accuracy. We found similar results in more complicated mapping tasks in which multilayer perceptrons were used as the output layer (not shown here).

Given enough training steps for SOM learning and perceptron learning, the decoder was able to accurately learn the decoding task. The weights of the perceptron after training are shown in Figure 3C. The pattern of weights has a centre-surround organization. This is a consequence of extended activity packets in the SOM layer and the forced, localized output representation. As some of the neighboring nodes are activated in the SOM layer, the output layer learns to suppress these activation through cente-suround inhibition.

## 6    Modelling Top-Down Control of SOM Plasticity

The above example demonstrate the usefulness of topographic organizations in intermediate layers of mapping networks. The next question is how top-down control of map plasticity can be implemented. One possible direction, which we are currently exploring, is using the error signal of the output layer as modulation signal for learning.

Two key components of the behavioral task by Zhou and Merzenich are attention and motivation. In the mammalian brain, both of these functions involve analysis in systems outside the traditional sensory systems and would contribute to sensory plasticity in a top-down manner. One example is the basal forebrain neurons that use acetylcholine as a neurotransmitter to project to many regions in cortex. Acetylcholine has been implicated in many models of plasticity (Rasmusson, 2000) and these neurons are preferentially activated during tasks with high attentional demands (Sarter et al., 2003). The prefrontal cortex is one higher cognitive region that regulates activity of the basal fore-brain neurons and could thus provide modulatory acetylcholine to the sensory cortices in such a top-down manner (Rasmusson et al., 2007).

The following simulations of experiments similar to the ones by Zhou and Merzenich-were were done with a standard Kohonen model. The activation of a node $i$, $\mathbf{r}_i$, was determined by the tuning curve, $t(\mathbf{x}, c_i)$, of the neuron with preferred feature $c_i$

$$\mathbf{r}_i = \eta + bt(\mathbf{x}, c_i), \tag{4}$$

in addition to some uniformly distributed noise, $\eta$. The strength value $b$ was altered in the experiments to simulate attentional processes. To represent the initial noise-reared case, we simulated the map development with an environment in which the response of each node in the SOM layer was noise-dominated by setting $b = 0.1$. With this parameter, no map development was achieved. Similar, even if we start the system with a perfect topographic organization (left graph in Figure 4), the map deteriorates with training (middle graph).

After these initial 1000 training steps, we simulate the effect of top-down control resulting from goal-directed training in the experiments in two ways. While standard procedures in SOM development call for a reduction of the learning rate to stabilize learning (stability over plasticity), and to narrow the neighborhood function to allow better local organizations after an initial global phase, we keep the learning rate and width of the neighborhood function constant throughout the simulations. This can be seen as an increase in the case when the parameters are lowered with time. There are several possible explanations for ongoing adult plasticity. For example, it is possible that the apparent slowdown in adult plasticity might only be a reflection of the dynamic reinforcement of existing representations after the brain has developed a model that can sufficiently model world states. But it is also possible that neural plasticity is actively modulated. The current model does not distinguish between these possible sources of adult plasticity. The second important ingredient is the boosting of the signal relative

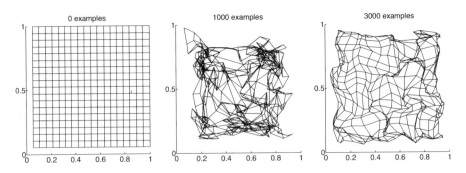

**Fig. 4.** Examples of organization in a Kohonen map. The map is started form a perfect organiza-tion and then trained for 1000 training steps in the presence of dominating noise which deterio-rates the map considerably. After this, the map is further trained with an enhanced input signal relative to noise, which helps to recover the map.

to noise by using $b = 10$. This reflects the outcome of goal-directed learning via motivation and attention that facilitates the enhancement of the relevant signals. With these changes, the maps starts to organize with continuous SOM learning. After 2000 further training steps, the map recovered dramatically, as shown in the right graph of Figure 4.

## 7    Conclusion

We discussed biological SOMs with a focus on ongoing adult plasticity and top-down controle. New experimental findings have clearly demonstrated that there can be considerable changes in cortical maps resulting from goal-directed learning, and that attentional processes are important for map formation.

The above discussions provide some background and simulations with a basic model. In particular, we simulated two factors which seem necessary to explain the results of experiments by Zhou and Merzenich. These two factors are the continuation of sensible plasticity thresholds, and the enhancement of task-relevant activity. The demonstration in this paper leave, of course, many open questions. For example, the differences of the contributing factor should be studied in more detail to guide further experimental work. Also, the experiments by Zhou and Merzenich not only show a deficit in tonotopic map organization, but also in the development of tuning curves with clear preferred frequency tuning. Such details can only be studied with more detailed models such as the Willshaw-von der Malsburg model. A better knowledge of such learning mechanisms could ultimately help to enhance teaching methods or to develop effective pharmacological interventions to promote recovery after brain injuries.

## Acknowledgment

This research was supported by NSERC (Canada). TT would like to thank Takayuki Nakata for inspiring discussions while at Future University, Hakodate.

## References

Amari, S.: Dynamics of pattern formation in lateral-inhibition type neural fields. Biological Cybernetics 27, 77–87 (1977)

Hornik, K., Stinchcombe, M., White, H.: Multilayer feedforward networks are universal approximators. Neural Networks 2, 359–366 (1989)

Jenkins, W., Merzenich, M., Ochs, M., Allard, T., Guic-Robles, E.: Functional reorganization of primary somatosensory cortex in adult owl monkeys after behaviorally controlled tactile stimulation. J. Neurophysiol. 63, 82–104 (1990)

Kaas, J., Krubitzer, L., Chino, Y., Langston, A., Polley, E., Blair, N.: Reorganization of retinotopic cortical maps in adult mammals after lesions of the retina. Science 248, 229–231 (1990)

Kohonen: Self-organized formation of topologically correct feature maps. Biological Cybernetics 43, 59–69 (1982)

Obermeyer, K., Ritter, H., Schulten, K.: A principle for the formation of the spatial structure of cortical feature maps. Proc. Natl. Acad. Sci., USA 87, 8345–8349 (1990)

Rasmusson, D.: Reorganization of raccoon somatosensory cortex following removal of the fifth digit. J. Comp. Neurol. 205, 313–326 (1982)

Rasmusson, D.: The role of acetylcholine in cortical synaptic plasticity. Behav. Brain Res. 115, 205–218 (2000)

Rasmusson, D., Smith, S., Semba, K.: Inactivation of prefrontal cortex abolishes cortical acetylcholine release evoked by sensory or sensory pathway stimulation in the rat. Neuroscience 149, 232–241 (2007)

Recanzone, G., Schreiner, C., Merzenich, M.: Plasticity in the frequency representation of primary auditory cortex following discrimination training in adult owl monkeys. J. Neurosci. 13, 87–103 (1993)

Robertson, D., Irvine, D.: Plasticity of frequency organization in auditory cortex of guinea pigs with partial unilateral deafness. J. Comp. Neurol. 282, 456–471 (1989)

Sarter, M., Bruno, J., Givens, B.: Attentional functions of cortical acetylcholine inputs: what does it mean for learning and memory? Neurobiol. Learning Mem. 80, 245–256 (2003)

Steriade, M., Jones, E., McCormick, D.: Thalamus. Elsevier, Amsterdam (1997)

Stringer, S., Trappenberg, T., Rolls, E., Araujo, I.: Self-organising continuous attractor networks and path integration: One-dimensional models of head direction cells. Network: Computation in Neural Systems 13, 217–242 (2002)

Tanaka, S.: Theory of ocular dominance column formation. Biological Cybernetics 64, 263–272 (1991)

Trappenberg, T.: Decision making and population decoding with strongly inhibitory neural field. In: Mavritsaki, D.H.E. (ed.) Computational modelling in behavioural neuroscience: Closing the gap between neurophysiology and behaviour. Psychology Press, London (2009)

Willshaw, D.J., von der Malsburg, C.: How patterned neural connexions can be set up by self-organisation. Proc. Roy. Soc. B 194, 431–445 (1976)

Wilson, H., Cowan, J.: Excitatory and inhibitory interactions in localized populations of model neurons. Biophys. J. 12, 1–24 (1972)

Wilson, H., Cowan, J.: A mathematical theory of the functional dynamics of cortical and thalamic nervous tissue. Kybernetik 13, 55 (1973)

Winship, I., Murphy, T.: In vivo calcium imaging reveals functional rewiring of single somatosensory neurons after stroke. J. Neurosci. 28, 6592–6606 (2008)

Wu, S., Nakahara, H., Amari, S.-I.: Population coding with correlation and an unfaithful model. Neural Computation 13, 775–797 (2001)

Zhou, X., Merzenich, M.M.: Intensive training in adults refines a1 representations degraded in an early postnatal critical period. Proc. Natl. Acad. Sci. 104, 4423–4428 (2007)

# Functional Principal Component Learning Using Oja's Method and Sobolev Norms

Thomas Villmann[1,*] and Barbara Hammer[2]

[1] University of Applied Sciences Mittweida, Dept. of Mathematics
Technikumplatz 17, 09648 Mittweida, Germany
thomas.villmann@hs-mittweida.de
[2] Clausthal University of Technology, Inst. of Computer Science
Julius-Albert-Str. 4, 38678 Clausthal, Germany

**Abstract.** In this paper we present a method for functional principal component analysis based on the Oja-learning and neural gas vector quantizer. However, instead of the Euclidean inner product the Sobolev counterpart is applied, which takes the derivatives of the functional data into account and, therefore, uses information contained in the functional shape of the data into account. We investigate the theoretical foundations of the algorithm for convergence and stability and give exemplary applications.

**Keywords:** functional PCA, neural gas Sobolev-norms.

## 1 Introduction

Data processing of *functional data* is a challenging topic in machine learning data analysis [11]. There is a broad area of application: biomedicine, chemometrics and chemistry, physics and astrophysics as well as geosciences and remote sensing analysis, to name just a few. The problems to be solved range from time series analysis and prediction, identification of characteristic patterns and classification to spectral data analysis.

The characteristic feature distinguishing usual vectorial data from functional one is that the vector components represent functions $v_i = f(x_i)$. Hence, data processing method should not proceed the vector components independently but taking into account their *spatial* position within the vector. However, there exist only few methods in machine learning, which make use of this property [6],[12]. In this work we investigate the usability of *Sobolev-metrics* for an adequate functional data handling. The main advantage of this metric family in comparison to others is that it can be related to an inner product. Thus, Sobolev-metrics can be used in machine learning methods using derivatives of norms or inner products. In this paper we mainly concentrate on OJA's Hebbian learning of principal components. Applying the respective *Sobolev-inner-product* (SIP) to this model we obtain a method for *functional principal component analysis* (FPCA).

---

* Corresponding author.

J.C. Príncipe and R. Miikkulainen (Eds.): WSOM 2009, LNCS 5629, pp. 325–333, 2009.
© Springer-Verlag Berlin Heidelberg 2009

## 2   Functional Metrics, Norms and Inner Products

There exist only few data processing methods which are specifically designed to process functional data paying attention to the property of inherent spatial dependencies. Most of them deal with the function description in terms of basis functions like Fourier-, Laplace-, wavelet expansions or others, such that methods can be applied to the respective coordinate space [11],[12]. An interesting alternative was proposed by LEE&VERLEYSEN in [6]. The functional norm by LEE&VERLEYSEN motivated by geometrical considerations is defined as

$$\|\mathbf{f}\|_p^{fc} = \left( \sum_{k=1}^{D} \left( A_k\left(\mathbf{f}\right) + B_k\left(\mathbf{f}\right) \right)^p \right)^{\frac{1}{p}} \tag{1}$$

with $\mathbf{f} = (f_1, \ldots, f_D)$ and $f_k = f(x_k)$, $x_k \in X \subseteq \mathbb{R}$ whereby we assume w. l. o. g. that $x_k < x_{k+1}$ for all $k$. Frequently we have equidistant values $x_k = k \cdot h$ with $h$ being a constant. Further, $A_k$ and $B_k$ are defined as

$$A_k\left(\mathbf{v}\right) = \begin{cases} \frac{\tau}{2} |v_k| & \text{if } 0 \le v_k v_{k-1} \\ \frac{\tau}{2} \frac{v_k^2}{|v_k| + |v_{k-1}|} & \text{if } 0 > v_k v_{k-1} \end{cases} \quad \text{and} \quad B_k\left(\mathbf{v}\right) = \begin{cases} \frac{\tau}{2} |v_k| & \text{if } 0 \le v_k v_{k+1} \\ \frac{\tau}{2} \frac{v_k^2}{|v_k| + |v_{k+1}|} & \text{if } 0 > v_k v_{k+1} \end{cases} \tag{2}$$

with the usual choice $\tau = 1$. The $\|\mathbf{f}\|_p^{fc}$-norm is based $Minkowski\text{-}p\text{-}norm$ $\|\mathbf{f}\|_p$ with $\|\mathbf{f}\|_p^{fc} \le \|\mathbf{f}\|_p$. It was successfully applied in functional vector quantization as demonstrated in [6]. However, it does not fulfill the parallelogram equation for norms $\|\cdot\|$:

$$\|\mathbf{f} - \mathbf{g}\|^2 + \|\mathbf{f} + \mathbf{g}\|^2 = 2\left( \|\mathbf{f}\|^2 + \|\mathbf{g}\|^2 \right) \tag{3}$$

and, hence, it can not be related to an inner product [15].

The $\|\mathbf{f}\|_p$-respective function space is $\mathcal{L}_p(X)$, which forms a Hilbert-space for $p = 2$ with Euclidean inner product

$$\langle f, g \rangle_{\mathrm{E}} = \int_X f(x)\, g(x)\, dx \tag{4}$$

generating the norm $\|\mathbf{f}\|_2$ [14].

We now introduce the $p-Sobolev\text{-}norm$ ($p$–SIP) of degree $K$. Let $f \in \mathcal{C}^K(X)$ be a $K$–times continuous-differentiable integrable function (in the Lebesgue sense) over $X$. Then the norm is defined as

$$\|f\|_{p,K}^{\mathcal{S}} = \|f\|_p + \sum_{1 \le j \le K} \left\| f^{(j)} \right\|_p \asymp \|f\|_p + \left\| f^{(K)} \right\|_p \tag{5}$$

which defines a distance $s_{p,K}^{\mathcal{S}}(f, g) = \|f - g\|_p^{\mathcal{S}}$. The space $\mathcal{C}^K(X)$ together with the norm (5) forms a $Banach\text{-}space$ $\mathcal{S}_{p,K}$.[1] Further, in case of $p = 2$ the space $\mathcal{S}_{2,K}$ becomes a $Hilbert\text{-}space$ with the inner product given by

---

[1] Yet, there are more general definitions possible. We here restrict ourself to this simplification which are sufficient for the most applications of machine learning problems. For a further reading we refer to [1] or [2].

$$\langle f, g \rangle_{S,K} = \langle f, g \rangle_{\mathrm{E}} + \sum_{1 \le j \le K} \left\langle D^{(j)} f, D^{(j)} g \right\rangle_{\mathrm{E}} \asymp \langle f, g \rangle_{\mathrm{E}} + \left\langle D^{(K)} f, D^{(K)} g \right\rangle_{\mathrm{E}} \quad (6)$$

and $D^{(k)}$ being the $k$th differential operator [2]. Moreover, for this case an interesting connection to the Fourier-analysis can be made using the *Parseval-equation*: The Sobolev-norm can equivalently determined by

$$\|f\|_K^S = \sqrt{\int_{-\infty}^{\infty} (1 + \omega^K)^2 \left| \hat{f}(\omega) \right|^2 d\omega}. \quad (7)$$

whereby $\hat{f}$ is the Fourier-transform of $f$.

Clearly, all definitions can be transferred to vectorial representations of functions replacing the integrals by sums and taking the differential operators $D^{(k)}$ as difference operators. Further, the Sobolev-norm can also be defined in a parametrized form

$$\|f\|_{p,K,\alpha}^S = \|f\|_p + \alpha \left\| f^{(K)} \right\|_p \quad (8)$$

with the respective inner product for $p = 2$

$$\langle f, g \rangle_{S,\alpha,K} = \langle f, g \rangle_{\mathrm{E}} + \alpha \sum_{1 \le j \le K} \left\langle D^{(k)} f, D^{(k)} g \right\rangle_{\mathrm{E}}. \quad (9)$$

## 3   Functional Principal Component Analysis (FPCA)

In this chapter we will give two approaches for FPCA. The first method uses the function representation in terms of orthogonal basis functions, whereas the second approach utilizes the Sobolev-inner-product $2-SIP$.

### 3.1   FPCA Based on Orthogonal Basis Functions

In this section we assume that the real function $f, g$ over $X \subseteq \mathbb{R}$ can be represented by orthogonal basis functions $\phi_k$ which form a basis of the functional space containing $f$ and $g$. Thereby, orthogonality is defined by $\langle \phi_k, \phi_j \rangle_{\mathrm{E}} = \delta_{k,j}$. The basis may contain a infinite number of basis functions. Prominent examples for basis systems are the the set of monomials $1, x, x^2, \ldots, x^k, \ldots$ or the Fourier-system of $\sin(k\omega x), \cos(k\omega x)$ with $k = 0, 1, 2, \ldots$.

Using a basis system of $K$ linear independent functions an arbitrary function $f$ can be approximated by $f(x) = \sum_{k=1}^{K} \alpha_k \phi_k(x)$, which can be seen as a discrete Euclidean inner product $\langle \boldsymbol{\alpha}, \boldsymbol{\phi}(x) \rangle_{\mathrm{E}}$ of the coordinate vector $\boldsymbol{\alpha} = (\alpha_1, \ldots, \alpha_k)^{\mathrm{T}}$ with the function vector $\boldsymbol{\phi} = (\phi_1(x), \ldots, \phi_k(x))^{\mathrm{T}}$. If the basis functions are the Fourier functions and $f$ given as functional vector $\mathbf{f}$, then the Sobolev-norm $\|\mathbf{f}\|_k^S$ can be immediately computed via (7).

We denote by $\mathcal{A}$ the function space spanned by all basis functions $\phi_k$: $\mathcal{A} = span(\phi_1, \ldots, \phi_k)$. Following the suggestions in [11] and [12] to transfer the ideas of usual multivariate PCA to FPCA. We consider the Euclidean inner product $\langle f, g \rangle_E$ from (4), which can be rewritten in terms of the basis functions as

$$\langle f, g \rangle_E = \sum_{k=1}^{K} \sum_{j=1}^{K} \alpha_k \beta_j \langle \phi_k, \phi_j \rangle_E \tag{10}$$

whereby in the second line the Fubini-lemma was used to exchange the integral and the sums. Let $\mathbf{\Phi}$ be the symmetric matrix spanned by $\Phi_{k,j} = \langle \phi_k, \phi_j \rangle_E$ using the symmetry of an inner product. Using this definition, the last equation can be rewritten as $\langle f, g \rangle_E = \langle f, g \rangle_{\mathbf{\Phi}}$ with the new inner product $\langle f, g \rangle_{\mathbf{\Phi}} = \boldsymbol{\alpha}^T \mathbf{\Phi} \boldsymbol{\beta}$. We remark that $\mathbf{\Phi}$ is independent of both $f$ and $g$. If the basis is orthogonal, $\mathbf{\Phi}$ is diagonal with entries $\Phi_{k,k} = 1$. Thus, the inner product of functions is reduced to the inner product (10) of the coordinate vectors

$$\langle f, g \rangle_E = \langle \boldsymbol{\alpha}, \boldsymbol{\beta} \rangle_E \tag{11}$$

For handling non-orthogonal basis systems we refer to [12].

Looking at (11) we see that performing elementary vector operations on the coordinate vectors in the Euclidean space $\mathbb{R}^K$ equipped with the (discrete) Euclidean inner product (4) is equivalent to the respective operations in the inner product space $\mathcal{A}$ with the Euclidean inner product (4). This statement allows a straightforward application to FPCA: FPCA can performed on a set $F = \{f_k\}_{k=1...N}$ of functions $f_k$ by usual vectorial PCA analysis of the respective set of coordinate vectors $\boldsymbol{\alpha}_k$ as explained in [11].

### 3.2   Oja's PCA-Learning for Functional Data

E. OJA developed an online-learning algorithm to determine the first principal component of data vectors $\mathbf{v} \in V \subseteq \mathbb{R}^n$ adaptively [8],[9]. The first principal component $\mathbf{w}$ related to the maximum eigen value for the data set $V$ is obtained by the stochastic adaptation. For a given input $\mathbf{v}$ the learning rule is

$$\triangle \mathbf{w} = \varepsilon_t O (\mathbf{v} - O \mathbf{w}) \tag{12}$$

with $O$ being the output

$$O = \mathbf{v}^T \cdot \mathbf{w} = \mathbf{w}^T \cdot \mathbf{v}.$$

Formally, the output $O$ can be written as an inner product $O = \langle \mathbf{v}, \mathbf{w} \rangle_E$. Then, the update yields

$$\triangle \mathbf{w}(t) = \varepsilon_t \langle \mathbf{w}(t), \mathbf{v} \rangle_E (\mathbf{v} - \langle \mathbf{w}(t), \mathbf{v} \rangle_E \mathbf{w}(t)). \tag{13}$$

with $\varepsilon_t > 0$, $\varepsilon_t \underset{t \to \infty}{\to} 0$, $\sum_t \varepsilon_t = \infty$ and $\sum_t \varepsilon_t^2 < \infty$, which is a converging stochastic process [3].

Obviously, this variant can be immediately applied to the above outlined approach of FPCA based on function representations using orthogonal basis functions. However, if the functional data are given in vectorial form, there exist an interesting alternative. Instead of the Euclidean inner product we plug the (parametrized) $2-$SIP of degree $K$ (9) into the learning rule (13) and get

$$\triangle \mathbf{w} = \varepsilon_t \left[ \langle \mathbf{w}, \mathbf{v} \rangle_{\mathcal{S}, \alpha, K} \left( \mathbf{v} - \langle \mathbf{w}, \mathbf{v} \rangle_{\mathcal{S}, \alpha, K} \mathbf{w} \right) \right] \qquad (14)$$

as new update rule for a functional data vector. We denote the equilibrium vector $\mathbf{w}^*$ of this dynamic as *functional (first) principal component*.

In the following, we will analyze the dynamic in more detail. Obviously, for small (vanishing) values of $\alpha$, the original Oja-learning rule is preserved. For non-vanishing $\alpha$ we consider the first term $T_1 = \langle \mathbf{v}, \mathbf{w} \rangle_{\mathcal{S}, \alpha, K} \mathbf{v}$ in (14), which can be approximately written as

$$T_1 = \mathbf{v} \langle \mathbf{v}, \mathbf{w} \rangle_{\mathrm{E}} \mathbf{v} + \alpha \sum_{k=1}^{K} \mathbf{v} \left\langle \mathbf{D}^{(k)} \mathbf{v}, \mathbf{D}^{(k)} \mathbf{w} \right\rangle_{\mathrm{E}} \qquad (15)$$

with $\mathbf{D}^{(k)}$ being the matrix approximating the $k$th differential operator. Using $\left\langle \mathbf{D}^{(k)} \mathbf{v}, \mathbf{D}^{(k)} \mathbf{w} \right\rangle = \left( \mathbf{D}^{(k)} \mathbf{v} \right)^{\mathrm{T}} \cdot \mathbf{D}^{(k)} \mathbf{w}$ we obtain

$$T_1 = \mathbf{v} \cdot \mathbf{v}^{\mathrm{T}} \left( \mathbf{1} + \alpha \mathbf{D} \right) \mathbf{w}$$

with $\mathbf{D} = \sum_{k=1}^{K} \left( \mathbf{D}^{(k)} \right)^{\mathrm{T}} \mathbf{D}^{(k)}$. Hence, the equation (14) describes the usual PCA-learning according the OJA-learning rule (12) *modified by the distortion* controlled by the $\alpha$-strength and the maximum degree $K$ of the differential operators.

As in the original PCA-learning, the second term $T_2 = \langle \mathbf{w}, \mathbf{v} \rangle_{\mathcal{S}, \alpha, K}^2 \mathbf{w}$ in (14) ensures the stability of the dynamic. We consider for its analysis the vectors

$$\hat{\mathbf{v}} \in V^{(K)} = V \otimes D^{(1)} V \otimes \ldots \otimes D^{(K)} V$$

taken as concenation of $\mathbf{v}$ and its derivatives $\alpha D^{(k)} \mathbf{v}$. The new prototypes $\hat{\mathbf{w}}$ are defined analogously. Using the linearity of differential operators, the functional PCA-learning can be reformulated in terms of the original Oja-learning in the now extended data space $V^{(K)}$:

$$\triangle \hat{\mathbf{w}} = \epsilon \left( \langle \hat{\mathbf{w}}, \hat{\mathbf{v}} \rangle_{\mathrm{E}} \left( \hat{\mathbf{v}} - \langle \hat{\mathbf{w}}, \hat{\mathbf{v}} \rangle_{\mathrm{E}} \hat{\mathbf{w}} \right) \right). \qquad (16)$$

Then, the equilibrium solution $\mathbf{w}^*$ of the functional PCA (14) is obtained simply by the projection of the equilibrium of (16): $\mathbf{P}_0 \hat{\mathbf{w}}^* = \mathbf{w}^*$. Therefore, it follows immediately from the stability analysis of the original Oja-learning given in [8] that the equilibrium $\hat{\mathbf{w}}^*$ of (16) is the first principal component in $V^{(K)}$ and, hence, the update (14) converges to its projection $\mathbf{P}_0 \hat{\mathbf{w}}^*$.

Finally, we remark that the approach can obviously applied in complete analogy to the full adaptive PCA-approach following the Oja-Sanger-update [13], which we do not explain here because the lack of the space.

### 3.3  Sparse Coding for Functional Data

*Sparse coding* (SC) of data is biologically motivated coding approach [10]. SC by means of Oja-learning was recently proposed – the sparse coding neural gas (SCNG) [4]. It takes into account the accelerated convergence by neighborhood cooperativeness provided by the involved neural gas algorithm [7]. Hence, there is a natural way to transfer the new FPCA approach to SCNG.

For sparse coding it is supposed that $N$ data $\mathbf{f}_k$ are available according to a distribution $P$, each containing $D$ values, i.e. $\mathbf{f}_k \in \mathcal{F} \subseteq \mathbb{R}^D$ and $\|\mathbf{f}_k\| = 1$. We here assume that the $\mathbf{f}_k$ are functional data vectors. A set of $M$, may be overcomplete but constrained or a may be non-orthogonal basis function vectors $\phi_j \in \mathbb{R}^D$ should be used for representation of the data in form of linear combination:

$$\mathbf{f}_k = \sum_j \alpha_{j,k} \cdot \phi_j + \boldsymbol{\xi}_k \tag{17}$$

with $\boldsymbol{\xi}_k \in \mathbb{R}^D$ being the reconstruction error vector and $\alpha_{j,k}$ are the weighting coefficients with $\alpha_{j,k} \in [0,1]$, $\sum_j \alpha_{j,k} = 1$ and $\boldsymbol{\alpha}_k = (\alpha_{1,k}, \ldots, \alpha_{M,k})$. The cost function

$$E_k = \|\boldsymbol{\xi}_k\|^2 - \lambda \cdot S_k$$

for $\mathbf{f}_k$ has to be minimized. Thereby, the regularization term $S_k$ serves as constraint judging the *sparseness* of the representation and can be taken as the entropy $S_k = H(\boldsymbol{\alpha}_k)$ of the vector $\boldsymbol{\alpha}_k$. Then, minimum sparseness is achieved iff $\alpha_{j,k} = 1$ for one arbitrary $j$ and zero elsewhere. Using this scenario, optimization is reduced to minimization of the description errors $\|\boldsymbol{\xi}_k\|^2$ or equivalently to the optimization of the basis vectors $\phi_j$. Optimum vectors for a set of data vectors are such $\phi_j$, which are chosen as principal components of subsets of $\mathcal{F}$. Minimum principal component analysis requires at least the determination of first principal component. Taking into account higher components improves the approximation. However, if the data space is nonlinear, principal component analysis (PCA) may be suboptimal. One possible way to overcome this problem is to split the data space into subsets and to carry out a PCA on each but with only a few PCA-components. SCNG takes only the first principal component but automatically detects the partition of the data for a predefined number $N$ of subsets $\Omega_i$ according to usual neural gas. In *functional* SCNG $N$ prototypes $\mathbf{W} = \{\mathbf{w}_i\}$ approximate the FPCA $\phi_i$ of the subsets $\Omega_i$. A functional data vector $\mathbf{f}_k$ belongs to $\Omega_i$ iff its correlation to $\phi_i$ defined by the inner product (9) is maximum:

$$\Omega_i = \left\{ \mathbf{f}_k | i = \underset{j}{\mathrm{argmax}} \langle \phi_j, \mathbf{f}_k \rangle_{S,\alpha,K} \right\} \tag{18}$$

The approximations $\mathbf{w}_i$ of $\phi_i$ are obtained via the FPCA according to (14) combined with neural gas following the idea of SCNG. During learning, in each time step a data vector $\mathbf{f}_j \in \mathcal{F}$ is selected according to $P$ and the prototypes $\mathbf{w}$ are updated according to

$$\triangle \hat{\mathbf{w}}_i = \varepsilon h_\sigma \left( \hat{\mathbf{f}}_j, \mathbf{W}, i \right) \left\langle \hat{\mathbf{w}}, \hat{\mathbf{f}}_j \right\rangle_{S,\alpha,K} \left( \hat{\mathbf{f}}_j - \left\langle \mathbf{w}, \hat{\mathbf{f}}_j \right\rangle_{S,\alpha,K} \mathbf{w} \right) \tag{19}$$

using the previously introduced embedding procedure.

$$h_{\sigma_t}\left(\mathbf{f}_k, \mathbf{W}, i\right) = \exp\left(\frac{-r_i\left(\hat{\mathbf{f}}_k, \mathbf{W}\right)}{\sigma_t}\right)$$

is the neural gas neighborhood function with neighborhood range $\sigma_t > 0$. For $t \to \infty$ the range is decreased as $\sigma_t \to 0$ and, hence, only the best matching prototype is updated in (19) in the limit. The function $r_i\left(\hat{\mathbf{f}}_k, \mathbf{W}\right)$ is the rank function

$$r_i\left(\hat{\mathbf{f}}_k, \mathbf{W}\right) = N - \sum_{j=1}^{N} \theta\left(\left\langle \hat{\mathbf{w}}_i, \hat{\mathbf{f}}_k \right\rangle_{\mathrm{E}} - \left\langle \hat{\mathbf{w}}_j, \hat{\mathbf{f}}_k \right\rangle_{\mathrm{E}}\right) \tag{20}$$

counting the number of pointers $\hat{\mathbf{w}}_j$ for which the relation $\left\langle \hat{\mathbf{w}}_i, \hat{\mathbf{f}}_k \right\rangle_{\mathrm{E}} > \left\langle \hat{\mathbf{w}}_j, \hat{\mathbf{f}}_k \right\rangle_{\mathrm{E}}$ is valid [4]. $\theta(x)$ is the Heaviside-function. Then, in the equilibrium of the stochastic process (19) one has $\Omega_i(t) \to \Omega_i$ for a certain subset configuration, which is related to the data space shape and the density $P$ [16]. Further, one gets $\phi_i = \mathbf{P}_0 \hat{\mathbf{w}}_i^*$ in the limit for each $\Omega_i$. We denote this *functional* SCNG by FSCNG.

## 4    Example Application

We applied the FSCNG for to data sets. The first consists of the monthly averaged temperature of 35 Canadian weather stations and the second one are the respective precipitation profiles both taken from [5]. The settings were identical for both experiments: We used 2 prototypes and performed two runs of FSCNG using the Sobolev inner product $\langle f, g \rangle_{\mathcal{S},\alpha,K}$ for $\alpha = 0.5$ with $K = 0$ and $K = 1$, respectively. Accordingly, for $K = 0$ the FSCNG is reduced to standard SCNG

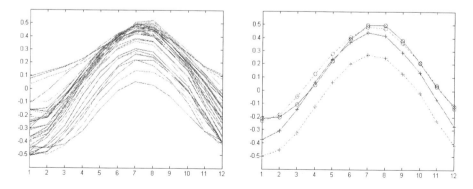

**Fig. 1.** FSCNG applied to the temperature profiles of 35 Canadian weather stations. The (normalized) profiles are depicted left. Right, the prototype results from FSCNG are shown (dashed, $K = 0$; solid, $K = 1$; ○ - prototype 1, +- prototype 2).

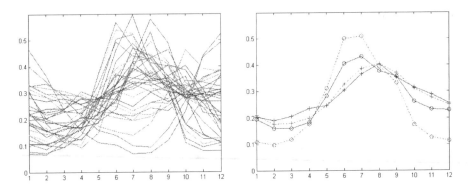

**Fig. 2.** FSCNG applied to the precipitation profiles of 35 Canadian weather stations. The (normalized) profiles are depicted left. Right, the prototype results from FSCNG are shown (dashed, $K = 0$; solid, $K = 1$; ∘ - prototype 1, +- prototype 2).

and for $K = 1$ the first derivative is taken into account. The results are depicted in Fig.1 for the temperature values and in Fig.2 for the precipitation profiles. We clearly state for both experiments the influence of the derivative for prototype learning. In particular, one observes for the first temperature data set that blue solid lines, which correspond to the $K = 1$ value, show a stronger curved shape as consequence of taking the first derivative into account.

## 5   Conclusion

In contribution we propose the application of Sobolev inner products for PCA-learning in case of functional data based on Oja's Hebbian online learning, which, however, takes the derivatives into account. We provide the theoretical background for this methodology. In particular, on the one hand side, we have shown that this variant can be seen as pertubation of the standard Oja-learning. In this view the controlling parameter $\alpha$ determines the strength of the distortion. On the other hand, the stability analysis can be carried out in complete analogy to the standard rule if the respective embedding space is considered there instead of the usual data space and the solution prototypes are the projection of the prototypes in the embedding space onto the original data space. Hence, the Sobolev inner product application in Oja-learning converges to the maximum eigenfilter solution in the embedding space and such that to its respective projection in the original data space. We plugged this learning rule into the recently introduced sparse coding neural gas and demonstrate the abilities of the resulting functional eigen analysis for two exemplary real world data sets. Further, it should be mention here that the corresponding Sobolev norms can be applied for each machine learning algorithm which takes the norm or its derivatives into account. This particularly concerns algorithms like all variants of learning vector quantization, self-organizing maps, neural gas etc. Doing so a functional vector quantizer can be obtained as alternative to plugging the functional norm provided by LEE&VERLEYSEN into it.

# References

1. Kantorowitsch, I., Akilow, G.: Funktionalanalysis in normierten Räumen, 2nd revised edn. Akademie-Verlag, Berlin (1978)
2. Kolmogorov, A., Fomin, S.: Reelle Funktionen und Funktionalanalysis. VEB Deutscher Verlag der Wissenschaften, Berlin (1975)
3. Kushner, H., Clark, D.: Stochastic Appproximation Methods for Constrained and Unconstrained Systems. Springer, New York (1978)
4. Labusch, K., Barth, E., Martinetz, T.: Learning data representations with sparse coding neural gas. In: Verleysen, M. (ed.) Proceedings of the European Symposium on Artificial Neural Networks ESANN, pp. 233–238. D-Side Publications (2008)
5. Landgrebe, D.: Signal Theory Methods in Multispectral Remote Sensing. Wiley, Hoboken (2003)
6. Lee, J., Verleysen, M.: Generalization of the $l_p$ norm for time series and its application to self-organizing maps. In: Cottrell, M. (ed.) Proc. of Workshop on Self-Organizing Maps (WSOM) 2005, Paris, Sorbonne, pp. 733–740 (2005)
7. Martinetz, T.M., Berkovich, S.G., Schulten, K.J.: 'Neural-gas' network for vector quantization and its application to time-series prediction. IEEE Trans. on Neural Networks 4(4), 558–569 (1993)
8. Oja, E.: Neural networks, principle components and suspaces. International Journal of Neural Systems 1, 61–68 (1989)
9. Oja, E.: Nonlinear pca: Algorithms and applications. In: Proc. of the World Congress on Neural Networks Portland, Portland, pp. 396–400 (1993)
10. Olshausen, B., Finch, D.: Emergnece of simple-cell receptive field properties by learning a sparse code for natural images. Nature 381, 607–609 (1996)
11. Ramsay, J., Silverman, B.: Functional Data Analysis, 2nd edn. Springer Science+Media, New York (2006)
12. Rossi, F., Delannay, N., Conan-Gueza, B., Verleysen, M.: Representation of functional data in neural networks. Neurocomputing 64, 183–210 (2005)
13. Sanger, T.: Optimal unsupervised learning in a single-layer linear feedforward neural network. Neural Networks 12, 459–473 (1989)
14. Triebel, H.: Analysis und mathematische Physik. BSB B.G., 3rd revised edn., Teubner Verlagsgesellschaft, Leipzig (1989)
15. Villmann, T.: Sobolev metrics for learning of functional data - mathematical and theoretical aspects. Machine Learning Reports, 1(MLR-03-2007), 1–15 (2007) ISSN:1865-3960, http://www.uni-leipzig.de/~compint/mlr/mlr_01_2007.pdf
16. Villmann, T., Claussen, J.-C.: Magnification control in self-organizing maps and neural gas. Neural Computation 18(2), 446–469 (2006)

# A Computational Framework for Nonlinear Dimensionality Reduction and Clustering

Axel Wismüller

Depts. of Radiology and Biomedical Engineering, University of Rochester, New York
601 Elmwood Avenue, Rochester, NY 14642-8648, U.S.A.
axel_wismueller@urmc.rochester.edu

**Abstract.** We introduce the Exploration Machine (Exploratory Observation Machine, XOM) as a novel versatile instrument for scientific data analysis and knowledge discovery. XOM systematically inverts structural and functional components of topology-preserving mappings. In contrast to conventional approaches known from the literature, this novel computational framework for self-organization does not require to incorporate additional graphical display or coloring techniques, or to modify topology-preserving mapping algorithms by additional regularization in order to recover the underlying cluster structure of inhomogeneously distributed input data. Thus, XOM can be seen as an approach to bridge the gap between nonlinear embedding and classical topology-preserving feature mapping. At the same time, XOM results in tremendous computational savings when compared to conventional topology-preserving mapping, thus allowing for direct structure-preserving visualization of large data collections without prior data reduction.

## 1 Introduction

A classical approach to exploratory data analysis has been contributed by so-called 'topology-preserving mappings' which have been pioneered by Kohonen's discovery of the Self-Organizing Map (SOM) almost three decades ago. In the meantime, several thousands of scientific publications have documented the importance of topology-preserving mappings as a useful instrument for pattern recognition and machine learning, for bibliographical data see e.g. [9]. The conventional way of looking at topology-preserving mappings can be characterized as follows: Data are presented to an input space, often referred to as 'feature space'. These data are then processed by means of a topological structure defined in a *different* space often referred to as 'grid' or 'index space'. In this context, the general convention is that input data are not used directly to define the topology of the 'grid space'. The key issue of this paper is to raise the question whether one might disregard this convention, i.e. use the input data *directly* in order to define the topology of the 'grid space'. The systematic pursuit of this simple idea leads to the concept of the Exploration Machine (Exploratory Observation Machine, XOM) presented in this contribution. Although the resulting modified perspective of topology-preserving mappings may at first appear unconventional,

J.C. Príncipe and R. Miikkulainen (Eds.): WSOM 2009, LNCS 5629, pp. 334–343, 2009.
© Springer-Verlag Berlin Heidelberg 2009

it leads to a novel computational framework that can simultaneously contribute to different domains of advanced machine learning and scientific visualization.

In the remainder of this paper, we first motivate the Exploration Machine algorithm by a didactic pictorial application to incremental optimization in order to illustrate its relation to topology-preserving mappings. Second, we formally summarize its algorithmic steps. Third, we demonstrate its applicability to different tasks of advanced machine learning, as exemplified by structure-preserving dimensionality reduction and data clustering. Fourth, we quantitatively evaluate the performance of XOM in comparison to classical and advanced recent embedding algorithms on a real-world data set. Finally, we discuss how its capability to perform *both* structure-preserving visualization *and* data partitioning qualifies XOM as a novel flexible workflow framework for exploratory data analysis. Here, it is a specific virtue of XOM that it *combines* originally separated domains of machine learning, namely structure-preserving dimensionality reduction and data clustering from a *unified* viewpoint within a *single* computational approach.

## 2    A Pictorial Introduction to XOM

In order to specify the fundamental differences of the Exploration Machine approach in comparison to the Self-Organizing Map (SOM) or other topology-preserving mappings, a simple application should be discussed as a pictorial illustration of how the XOM algorithm systematically inverts the data processing workflow in topology-preserving mappings. For this application, both the XOM and the SOM approach can be used, thus allowing for a conceptual comparison. We emphasize that, for this special application, XOM does not provide any specific advantage when compared to SOM. Instead, it is only presented here in order to motivate the XOM algorithm and to clarify its relation to SOM and other topology-preserving mappings as known from the literature.

Finding approximate solutions to the Travelling Salesman Problem (TSP) is a classical subject of 'connectionist' computing techniques. Figs. 1A–B visualize

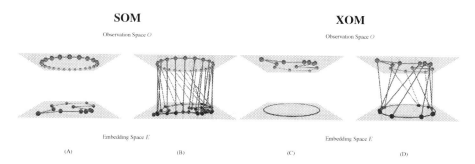

**Fig. 1.** Didactic comparison of the Exploration Machine (XOM) and the Self-Organizing Map (SOM) for finding approximate solutions of a Travelling Salesman Problem (TSP). (A), (B) SOM; (C), (D) XOM. For explanation, see text.

the concept of finding TSP solutions by topology-preserving mappings as described in the literature, e.g. [11,3,1], whereas Figs. 1C–D illustrate the XOM approach for comparison. – The classical strategy is explained in Figs. 1A–B: In order to find the shortest closed-loop path through a set of cities (red line), the city coordinates representing the input data (red dots) are defined within the 'embedding space' $E$ which is often referred to as 'data space' or 'feature space' in the literature, see Fig. 1A. In a second step, a 'structure hypothesis' is defined by a set of local information-processing units, so-called 'neurons' (green dots) arranged in a ring-shaped topology within a different space $O$ which is often referred to as 'grid space', 'index space', or 'model cortex'. In the following, we will call this space $O$ the 'observation space'. – Subsequently, to each neuron in $O$ a so-called 'image vector' is attributed in space $E$, e.g. by random initialization (blue dots, Fig. 1B). Note that the number of neurons may exceed the number of cities. By using a topology-preserving mapping $T$, the image vectors are arranged to match the city coordinates in $E$. Once the training of the mapping is completed, the ring topology of the corresponding neurons in the space $O$ can be used to define a path through the image vectors in $E$ (blue line). By some appropriate heuristics, this path through the image vectors can be transformed into a path through the cities (red line in Fig. 1B).

For comparison, Figs. 1C–D illustrate the XOM approach: For XOM, in contrast to Figs. 1A–B, the city coordinates, i.e. the input data, are *directly* used to define a topological structure in the observation space $O$. Thus, each city is represented by a single neuron (green dots). In a second step, a ring-shaped uniform distribution is defined in the embedding space $E$ representing a structure hypothesis (red line in Fig. 1C). During the training phase of $T$, data points are randomly sampled from this distribution. Subsequently, to each input data item, i.e. each city in $O$, an image vector is attributed in the embedding space $E$, e.g. by random initialization on the ring (blue dots in Fig. 1D). During training of $T$, data points are randomly sampled from the ring-shaped distribution in $E$, and the positions of the image vectors are incrementally updated, where the image vectors can move about the ring in order to represent the topology induced by the cities. Once the training of $T$ is completed, the final arrangement of the image vectors on the ring in $E$ can be used to define a path through the input vectors in $O$, which is equivalent to a path through the cities (green line in Fig. 1D). For completeness, it should be mentioned that it is not necessary to constrain the migration of the image vectors onto the ring, this restriction is only introduced to keep the illustration as simple as possible. – Again, it should be noted that in the TSP application of Fig. 1, the use of XOM instead of SOM does not provide any specific advantage, and that it is only introduced for didactic reasons. Here, neither SOM nor XOM can compete with specialized methods described in the literature for finding solutions of the TSP, in particular, if larger TSPs are considered.

In summary, Figs. 1A–B illustrate the data processing workflow according to the literature on topology-preserving mappings, where input data are presented

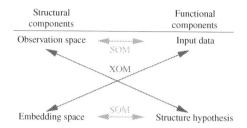

**Fig. 2.** Functional and structural components of topology-preserving mappings (such as SOM) and their systematic inversion introduced by XOM. For explanation, see text.

in the embedding space $E$, and a structure hypothesis is defined in the observation space $O$. In contrast, the XOM approach in Figs. 1C–D introduces a fundamental change w.r.t. these conventions by *completely inverting the role of input data and structure hypotheses:* The input data are *directly* used to define a topological structure in the observation space $O$, whereas the structure hypotheses are represented by sampling distributions in the embedding space $E$. Thus, XOM systematically reverses the relation between functional and structural components of topology-preserving mappings, as summarized in Fig. 2.

In the XOM algorithm, it is the input data that *directly* determines the topological structure represented by the mapping. As an immediate consequence, each input data item is mapped directly, i.e. assigned to its own specific image vector. Hence, no implicit approximation of input data items by image vectors is involved. Instead, sampling and adaptation of the image vectors is entirely restricted to the embedding space. Moreover, this reduces the number of free parameters in topology-preserving mappings by one, namely the number of 'neurons', which can be set equal to the number of input data items.

When used for nonlinear embedding, the systematic inversion of input data and structure hypotheses in XOM provides important advantages: In particular, the formulation of the dynamics in the embedding space entails a substantial reduction of computational costs in comparison to topology-preserving mappings, as the best-match search in each iteration step does not require computational operations in the high-dimensional input data space, but now occurs in the usually low-dimensional embedding space. This leads to tremendous savings in computation time in the case of high-dimensional input data. Finally, based on the inversion of input data and structure hypotheses, the Exploration Machine can analyze non-metric input data directly, i.e. without the need to transform such input data into metric representations prior to processing.

Based on the pictorial motivation presented in the previous section, we can now formally summarize the Exploration Machine algorithm as follows: For simplicity, let us first consider $N$ real-valued input vectors $r_j$ in the observation space $O$, each of dimensionality $D$. The XOM algorithm can then be resolved into three simple steps:

**The Exploration Machine (XOM) Algorithm:**

1. Define topology of input data in the observation space $O$ by computing distances $d(\boldsymbol{r}_i, \boldsymbol{r}_j)$ between data vectors $\boldsymbol{r}_i, i \in \{1, \ldots, N\}$. This step is omitted, if input data is already given as a set of distances between input data items.
2. Define a 'hypothesis' on the structure of the data in the embedding space $E$, represented by 'sampling' vectors $\boldsymbol{x}_k \in E, k \in \{1, \ldots, K\}, K \in \mathbb{N}$, and initialize an 'image' vector $\boldsymbol{w}_i \in E, i \in \{1, \ldots, N\}$ for each input vector $\boldsymbol{r}_i$.
3. Reconstruct the topology induced by the input data in $O$ by moving the image vectors in the embedding space $E$ using the computational scheme of a topology-preserving mapping $T$. The final positions of the image vectors $\boldsymbol{w}_i$ represent the output of the algorithm.

A simple choice for $T$ is Kohonen's self-organizing map algorithm [9], e.g. in its basic incremental version, which is used for all computer simulations in this contribution. However, any other topology-preserving mapping described in the literature may be selected for $T$. Besides numerous variants of Kohonen's self-organizing map algorithm, such as 'batch' versions, various kinds of topographic vector quantizers, e.g. [7] as well as modifications of the quoted methods should be mentioned. In essence, the algorithmic concept of XOM is independent of the specific choice of $T$. However, the respective properties of $T$ determine theoretical convergence properties, the number of free parameters, mapping performance, and computational complexity.

# 3   Properties and Applications of XOM

**Structure-Preserving Visualization:** The XOM algorithm specified above can be used for structure-preserving visualization. Here, the sampling distribution in step 2 of the algorithm may typically be chosen as a uniform distribution on a low-dimensional mainfold which not necessarily has to be a subset of a linear space. Thus, XOM can be used for dimensionality reduction of high-dimensional observations, or even for non-metric data, given as distances, 'similarities', or 'proximities' between input data items. In this context, it should be mentioned that conventional topology-preserving feature mapping has substantially contributed to many domains of real-world data analysis and found numerous applications ranging from self-organizing semantic maps to web interfaces. For an overview, we refer to [9]. However, a major drawback of these feature maps is that they cannot accurately recover the underlying structure of inhomogeneously distributed data, in particular, boundaries between eventual clusters cannot be detected and visualized directly. Numerous heuristic approaches have been proposed to alleviate this problem, i.e. to integrate the advantages of structure-preserving visualization into the field of topology-preserving data representation. These approaches include specific graphical display and coloring techniques, e.g. [12,5,8], or generalizations of the SOM algorithm motivated on mathematical [6] grounds or from modeling processes of biological structure formation [13]. – Here, the Exploration Machine can be seen as an approach to bridge the gap between structure-preserving visualization and topology-preserving data representation *without* the necessity to incorporate additional graphical display or coloring techniques, or to modify topology-preserving mapping algorithms by additional regularization. Unlike conventional topology-preserving feature maps described in the literature, XOM can consistently recover and directly visualize the underlying cluster structure of inhomogeneously distributed data, as shown for the real-world data set in Fig. 3 below.

**Data Clustering by XOM:** It is important to realize that there is no principal restriction to the choice of the structure hypothesis in the second step of the XOM algorithm specified above. This flexibility to choose arbitrary sampling distributions in the embedding space leads to the discovery that XOM can be used for *data partitioning*, i.e. *clustering* as well. The key idea is to simply select the sampling vectors from 'clustered', i.e. non-uniform distributions, e.g. from a mixture of several distributions centered at different positions in the embedding space. A typical choice could be a mixture of Gaussian distributions centered at different locations in the image space. In detail, we propose to proceed as follows: (i) Use the input data to define the observation space directly, as has been discussed above in the context of the Travelling-Salesman Problem example. (ii) Choose an appropriate structure hypothesis in the embedding space consisting of a mixture of 'clustered' distributions. As a typical example, a sampling distribution may be selected that consists of several Gaussian distributions with arbitrary parameters. The centers of the Gaussian distributions may define an arbitrary topological structure, e.g. they may be located on a regular grid or on the vertices of a regular simplex. However, there are no restrictions w.r.t. the specific choice of the sampling distributions used for this purpose. (iii) Run the XOM algorithm as explained in section 2. (iv) Assign the resulting image vectors to the specific sampling distributions of the embedding space in a hard or fuzzy way, e.g. by computing and comparing the distances of the image vectors to the centers of the respective sampling distributions. By definition of appropriate distance measures, e.g. the assignment likelihood, this can be performed in a fuzzy manner as well. Note that XOM can also support non-metric clustering.

## 4   Experiments on a Real-World Data Set

To prove the applicability of our approach to the machine learning domains specified above, we present results on real-world data, see Fig. 3. The data consists of 147 feature vectors in a 79-dimensional space encoding gene expression profiles obtained from microarray experiments.

**Visualization by Dimensionality Reduction:** Fig. 3A shows a XOM visualization result obtained from a structure-preserving dimensionality reduction of gene expression profiles related to ribosomal metabolism, as a detailed visualization of a subset included in the genome-wide expression data taken from Eisen et al. [4]. The figure illustrates the exploratory analysis of the 147 genes labeled as '5' (22 genes) and '8' (125 genes) according to the cluster assignment

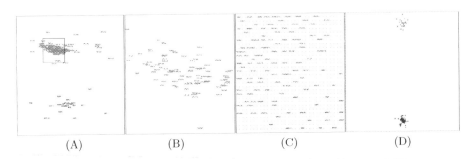

(A)                    (B)                    (C)                    (D)

**Fig. 3.** Visualization of genome expression profiles related to ribosomal metabolism using the Exploration Machine. (A) Genome map obtained by structure-preserving dimensionality reduction using the Exploration Machine. (B) Enlarged section of (A). (C) Result obtained by SOM. (D) XOM clustering result. For explanation, see text.

by Eisen et al. [4]. Besides several genes involved in respiration, cluster '5' (blue) contains genes related to mitochondrial ribosomal metabolism, whereas cluster '8' (orange) is dominated by genes encoding ribosomal proteins and other proteins involved in translation, such as initiation and elongation factors, and a tRNA synthetase. – In the XOM genome map of Fig. 3A, it is clearly visible at first glance that the data consists of two distinct clusters. Comparison with the functional annotation known for these genes [2] reveals that the map overtly separates expression profiles related to mitochondrial and to extramitochondrial ribosomal metabolism. Fig. 3B shows an enlarged section of the map indicated by the small frame in Fig. 3A. Fig. 3C shows a data representation obtained by a SOM trained on the same data using a regular grid of 30 × 30 'neurons'. As can be clearly seen in the figure, the SOM *cannot* achieve a structure-preserving mapping result as provided by the Exploration Machine in Fig. 3A: Although the genes related to mitochondrial and to extramitochondrial ribosomal metabolism are collocated on the map, the distinct cluster structure underlying the data remains invisible, if the color coding is omitted.

In other words, the Exploration Machine, in contrast to the Self-Organizing Map, makes the underlying input data cluster structure visible in this example. At the same time, in comparison to the SOM result in Fig. 3C, the computational expense for XOM embedding in Fig. 3A is reduced by a factor of more than 1400 (!) for this data set.[1] The reason for this tremendous computational saving is that for XOM the nearest neighbor search required to identify the winner neuron has only to be performed in the two-dimensional embedding space instead of the 79-dimensional input data space. In addition, as each input data item is attributed directly to its own image vector, only 147 instead of 900 'neurons' have to be updated in each iteration. Finally, if it is argued according to [9] that the number of iterations should be some multiple of the number of neurons,

---

[1] The observed computational benefit of XOM vs. SOM in Tab. 1 is lower than the theoretically predicted value, which is due to implementation using completely different software platforms.

**Table 1.** Comparative evaluation of computation times and structure preservation for nonlinear embedding of the ribosome gene expression data of Fig. 3. The table lists average computation times using an ordinary PC (Intel Pentium 4 CPU, 1.6 GHz, 512 MB RAM). A scale-invariant version of the Sammon error $E'$ was computed as a measure of structure preservation, absolute and relative values compared to the XOM result are given. The free parameters of all the methods examined in the comparison (except PCA) were optimized to obtain the best results, i.e. to minimize $E'$. Note that XOM yields competitive structure preservation at acceptable computation times.

| Method | Comp. Time (s) | $E'$ | Rel. $E'$ |
|---|---|---|---|
| XOM | 0.72 | $2.21 \cdot 10^3$ | 1.00 |
| Sammon | 8.25 | $2.45 \cdot 10^3$ | 1.11 |
| LLE | 1.36 | $2.77 \cdot 10^3$ | 1.25 |
| PCA | 0.03 | $2.82 \cdot 10^3$ | 1.28 |
| Isomap | 0.27 | $3.36 \cdot 10^3$ | 1.52 |
| SOM | 21.6 | $10.19 \cdot 10^3$ | 4.61 |

i.e. image vectors, the number of iterations in the SOM has to be increased by a factor of $900/147$ when compared to XOM in this example. In summary, we end up with an overall computational saving as quoted above. – We conjecture that these properties of the Exploration Machine, i.e. the combination of superior structure preservation with tremendous savings in computational expense, can serve as important assets for its applications to 'data mining' and 'knowledge discovery' in large data bases, such as text mining in large document collections. A quantitative comparison of the embedding result shown in Fig. 3A with several other embedding algorithms is summarized in Tab. 1.

**Clustering Based on Structure Hypothesis from Visualization:** Fig. 3D shows the XOM clustering result of the ribosome gene expression data set using the method explained in section 3. As can be seen in Fig. 3D, the two clusters are separated completely. Note that the decision to use *two* clusters – instead of any different number of clusters – can be conveniently based on the results of structure-preserving XOM visualization in Fig. 3A, which can, thus, serve as a useful preprocessing step to clustering. The visualization obtained by the Self-Organizing Map in Fig. 3C, however, is not clearly indicative for the presence of exactly two clusters. Note that to perform XOM clustering in Fig. 3D, only the structure hypothesis is changed from a uniform distribution on a unit square as used in Fig. 3A to a set of two Gaussians centered at different locations of the exploration space, here at the top and the bottom of a unit square.

# 5   Discussion

In this paper, we have shown that the Exploration Machine is capable of creating low-dimensional representations of high-dimensional inputs for structure-preserving visualization. In this context, we have shown that the Exploration

Machine, when compared to the Self-Organizing Map, better preserves an underlying cluster structure of high-dimensional input data. At the same time, XOM provides a tremendous reduction of computational expense when compared to SOMs. Finally, we have shown that XOM can be used to serve as a method for data clustering as well. Further extensions, variants, and applications of the Exploration Machine, such as related to supervised learning, analysis of non-metric data, out-of-sample extension, and constrained incremental learning, have been thoroughly investigated in [14]. – In summary, the ribosome gene expression data set example shows how XOM can serve as an integrative framework for exploratory analysis of complex high-dimensional data sets: (i) In a first step, nonlinear embedding by XOM can be performed as explained in section 3 in order to detect an eventual cluster structure and to create an estimate of how many clusters may be appropriate for data partitioning. This problem is often referred to as 'cluster validity' estimation[2] (ii) In a second step, clustering by XOM as explained in section 3 can be performed by simply choosing an adequate sampling distribution in the embedding space, i.e. a distribution which corresponds to the estimated structure assumptions that have been developed based on visual inspection of the embedding results obtained in the first step.

We recommend the combination of both steps as a pragmatic unified workflow approach for exploratory data analysis. Specifically, this method can approach the 'cluster validity' issue which is still an unsolved problem of pattern recognition[3]. To the best of our knowledge, the Exploration Machine is the first method in the literature that can be used for *both* clustering *and* cluster validity estimation by structure-preserving data visualization. – Numerous methods for solving one of either steps above have been described in the literature, i.e. to perform *either* nonlinear embedding *or* data clustering, i.e. these tasks have often been treated as independent problems of pattern recognition. As shown in this contribution, the Exploration Machine can approach *both* domains from a unified viewpoint within a *single* computational framework.

# References

1. Angeniol, B., De La Croix Vaubois, G., Le Texier, J.Y.: Self-organizing feature maps and the travelling salesman problem. Neural Networks 1, 269–293 (1988)
2. Cherry, J.M., Ball, C., Weng, S., Juvik, G., Schmidt, R., Adler, C., Dunn, B., Dwight, S., Riles, L., Mortimer, R.K.: Nature 387, 67–73 (1997)
3. Durbin, R., Willshaw, D.: An analogue approach to the travelling salesman problem using an elastic net method. Nature 326, 689–691 (1987)
4. Eisen, M.B., Spellman, P.T., Brown, P.O., Botstein, D.: Cluster analysis and display of genome-wide expression patterns. Proc. Natl. Acad. Sci. USA 95, 14863–14868 (1998)

---

[2] Besides the number of clusters, other criteria, such as their statistical weights or their 'topological' relations may be inferred in this step as well, in order to facilitate an adequate definition of a structure hypothesis for XOM clustering in step (ii).

[3] For the questionable results obtained from theoretically motivated so-called 'cluster validity indices' see e.g. the classical paper by Milligan and Cooper [10].

5. Flexer, A.: On the use of self-organizing maps for clustering and visualization. In: Żytkow, J.M., Rauch, J. (eds.) PKDD 1999. LNCS, vol. 1704, pp. 80–88. Springer, Heidelberg (1999)
6. Goodhill, G.J., Sejnowski, T.: A unifying objective function for topographic mappings. Neural Comp. 9, 1291–1303 (1997)
7. Graepel, T., Burger, M., Obermayer, K.: Phase transitions in stochastic self-organizing maps. Physical Review E 56(4), 3876–3890 (1997)
8. Kaski, S.: Data exploration using self-organizing maps. Act Polytech Scand, Mathematics, Computing and Management in Engineering Series No. 82 (1997)
9. Kohonen, T.: Self-Organizing Maps, 3rd edn. Springer, Heidelberg (2001)
10. Milligan, G.W., Cooper, M.C.: An examination of procedures for determining the number of clusters in a data set. Psychometrika 50, 159–179 (1985)
11. Ritter, H., Martinetz, T., Schulten, K.: Neural Networks and Self-Organizing Maps. Addison-Wesley, New York (1992)
12. Vesanto, J.: SOM-based data visualization methods. Intelligent Data Analysis 3, 111–126 (1999)
13. Wiskott, L., Sejnowski, T.: Constrained optimization for neural map formation: a unifying framework for weight growth and normalization. Neural Comp. 10, 671–716 (1998)
14. Wismüller, A.: Exploratory Morphogenesis (XOM): A Novel Computational Framework for Self-Organization. Ph.D. thesis, Technical University of Munich, Department of Electrical and Computer Engineering (2006)

# The Exploration Machine – A Novel Method for Data Visualization

Axel Wismüller

Depts. of Radiology and Biomedical Engineering, University of Rochester, New York
601 Elmwood Avenue, Rochester, NY 14642-8648, U.S.A.
axel_wismueller@urmc.rochester.edu

**Abstract.** We present a novel method for structure-preserving dimensionality reduction. The Exploration Machine (Exploratory Observation Machine, XOM) computes graphical representations of high-dimensional observations by a strategy of self-organized model adaptation. Although simple and computationally efficient, XOM enjoys a surprising flexibility to simultaneously contribute to several different domains of advanced machine learning, scientific data analysis, and visualization, such as structure-preserving dimensionality reduction and data clustering.

## 1 Motivation

The exceedingly growing amount of available data obtained by retrieval in computer-based data collections and web resources raises the question of how to organize and extract useful knowledge, given this abundance of information. Structure-preserving data reduction can frequently be accomplished by two alternative approaches, namely data partitioning in the sense of 'clustering' and dimensionality reduction often referred to as 'embedding'.

We introduce a novel algorithm called 'Exploration Machine' (Exploratory Observation Machine – XOM) that can approach *both* domains from a unified viewpoint within a *single* computational framework. After introducing the XOM algorithm, we perform a quantitative evaluation by computer simulations on benchmark data sets. Finally, we present results on real-world data.

## 2 The Exploration Machine Algorithm

The Exploration Machine (XOM) algorithm can be resolved into simple, geometrically intuitive, steps. For simplicity, let us first consider $N$ real-valued input vectors $\boldsymbol{r}_i$ in the 'observation space' $O$, each of dimensionality $D$.

1. Define the topology of the input data in the observation space $O$ by computing distances $d(\boldsymbol{r}_i, \boldsymbol{r}_j)$ between the data vectors $\boldsymbol{r}_i, i \in \{1, \ldots, N\}$. This step is omitted, if the input data is already given as a set of distances between input data items.

J.C. Príncipe and R. Miikkulainen (Eds.): WSOM 2009, LNCS 5629, pp. 344–352, 2009.

2. Define a 'hypothesis' on the structure of the data in the embedding space $E$, represented by 'sampling' vectors $\boldsymbol{x}_k \in E$, $k \in \{1, \ldots, K\}$, $K \in \mathbb{N}$, and initialize an 'image' vector $\boldsymbol{w}_i \in E$, $i \in \{1, \ldots, N\}$ for each input vector $\boldsymbol{r}_i$.

3. Reconstruct the topology induced by the input data in $O$ by moving the image vectors in the embedding space $E$ using the computational scheme of a topology-preserving mapping $T$. The final positions of the image vectors $\boldsymbol{w}_i$ represent the output of the algorithm.

It should be noted that there is no restriction whatsoever to the distance measure used in the first step of the XOM algorithm. In particular, geodesic distances, an ordinal rank metric, or even nonmetric dissimilarities may be used as well. In addition, there is no need to compute a complete distance matrix in this step, i.e. distances between specific pairs of input data items may not be defined. The choice of the distance measure used for defining the topology of the input data may depend on the specific focus of the XOM application. For completeness, we emphasize that wherever the terms 'vector' and 'distance' are used throughout this paper (e.g. for data, sampling, or image 'vectors'), generalization to arbitrary non-vectorial data structures and arbitrary distance measures between these structures is straightforward. For simplicity and clarity, however, this is not explicitly mentioned every time.

In the second step of the XOM algorithm, there is no principal restriction to the choice of the sampling distribution serving as structure hypothesis on the data: It may be selected from arbitrary underlying probability distributions, or just simply represent a list of given sampling items $\boldsymbol{x}_k \in E$. As shown later, it is this inherent flexibility that allows XOM to contribute to different domains of scientific data analysis. Typical choices for sampling distributions are: for structure-preserving visualization, use uniform distribution (e.g. in a 2D square, as used in figs. 2 and 3); for data clustering use several Gaussian distributions with different centers (e.g. located on the nodes of a regular simplex, as used in fig. 4.)

In the third step of the XOM algorithm, the topology-preserving mapping $T$ can be considered as a free variable. A simple choice for $T$ is Kohonen's self-organizing map algorithm [4], e.g. in its basic incremental version. Here, the image vectors $\boldsymbol{w}_i$ are incrementally updated by a sequential learning procedure. For this purpose, the neighborhood couplings between the input data items are represented by a so-called cooperativity function $\psi$. Typically, $\psi$ is chosen as a Gaussian

$$\psi(\boldsymbol{r}, \boldsymbol{r}'(\boldsymbol{x}(t)), \sigma(t)) := \exp\left(-\frac{(\boldsymbol{r} - \boldsymbol{r}'(\boldsymbol{x}(t)))^2}{2\sigma(t)^2}\right) \tag{1}$$

or a characteristic function on a $D$-dimensional hypercube around $\boldsymbol{r}'(\boldsymbol{x}(t))$. In the XOM context, $\boldsymbol{r}'(\boldsymbol{x}(t))$ represents the 'best-match' input data vector. For a randomly selected sampling vector $\boldsymbol{x}(t) \in E$, the best-match input data vector is identified by: $\|\boldsymbol{x} - \boldsymbol{w}_{\boldsymbol{r}'}\| = \min_{\boldsymbol{r}} \|\boldsymbol{x} - \boldsymbol{w}_{\boldsymbol{r}}\|$. Once the best-match input data vector has been identified, the image vector $\boldsymbol{w}_{\boldsymbol{r}}$ is updated by the sequential adaptation step according to the learning rule

$$\boldsymbol{w}_{\boldsymbol{r}}(t+1) = \boldsymbol{w}_{\boldsymbol{r}}(t) + \epsilon(t)\,\psi(\boldsymbol{r}, \boldsymbol{r}'(\boldsymbol{x}(t)), \sigma(t))\,(\boldsymbol{x}(t) - \boldsymbol{w}_{\boldsymbol{r}}(t)), \tag{2}$$

where $t$ represents the iteration step, $\epsilon(t)$ a learning parameter, and $\sigma(t)$ a measure for the width of the neighborhood taken into account by the cooperativity function $\psi$. In general, $\sigma(t)$ as well as $\epsilon(t)$ are changed in a systematic manner depending on the number of iterations $t$ by some appropriate annealing scheme. A typical choice is an exponential decay, e.g.

$$\kappa(t) = \kappa(0) \left( \frac{\kappa(t_{\max})}{\kappa(0)} \right)^{\frac{t}{t_{\max}}} \quad t \in [0, t_{\max}], \tag{3}$$

where $\kappa := \epsilon$ or $\kappa := \sigma$, respectively. The algorithm is terminated, once a problem-specific cost criterion is satisfied, or a maximum number of iterations has been completed.

Although the above computational scheme is formally identical to Kohonen's self-organizing map algorithm, the meaning of the variables $r$, $w$, and $x$ *completely differs* in the Exploration Machine: Whereas in Kohonen's algorithm the sampling vectors $x$ represent the input data, this role is attributed to the vectors $r$ in XOM. As an important consequence, in contrast to Kohonen's self-organizing map algorithm, each image vector $w$ is attributed to its own specific input data vector. Hence, no implicit approximation of input data items by image vectors is involved. Instead, sampling and adaptation of the image vectors is entirely restricted to the embedding space. In other words, XOM completely inverts the role of input data and structure hypotheses, given the conventions of topology-preserving mappings as known from the literature. Note that, besides the afore mentioned basic incremental self-organizing map algorithm, any other topology-preserving mapping described in the literature may be selected for $T$. Besides numerous variants of Kohonen's self-organizing map algorithm, such as 'batch' versions, various kinds of topographic vector quantizers, e.g. [3] as well as modifications of the quoted methods should be mentioned. In essence, the algorithmic concept of XOM is independent of the specific choice of $T$. However, the characteristics of $T$ determine theoretical convergence properties, the number of free parameters, mapping performance, and computational complexity.

**Algorithmic Properties of XOM:** There are deep differences between the Exploration Machine approach and the use of topology-preserving mappings as known from the literature. Specifically, these differences induce that (i) the dynamics of self-organization is formulated *directly* in the embedding space $E$ in which structure formation occurs, and not indirectly via movements in the space $O$ of the high-dimensional input data. As shown later, this can lead to tremendous computational savings. Second (ii), the coupling of the movements of the image vectors is now governed by the actual distance topology of the input data and *not* by the possibly inaccurate structure hypothesis as in existing approaches[1].

As in topology-preserving mappings, we still need a structure hypothesis. But now it is succinctly spelled out in the choice of the sampling distribution

---

[1] A typical choice for defining such structure hypotheses in topology-preserving mappings is to use two-dimensional, discrete, periodic (e.g. quadratic or hexagonal) grids.

and its underlying space $E$ that govern the exploration movements of the image vectors *without* affecting their interactions. Therefore, when compared to topology-preserving mappings, the Exploration Machine algorithm can be interpreted compactly as an inversion of the roles of structure hypotheses and input data w.r.t. observation and embedding spaces.

The new scheme endows XOM with a surprising flexibility to contribute to many domains of scientific data analysis and visualization beyond structure-preserving dimensionality reduction, as exemplified below. It also induces favorable algorithmic properties: In particular, the formulation of the dynamics in the embedding space entails a substantial reduction of computational complexity in comparison to topology-preserving mappings, as the best-match search in each iteration step does not require computational operations in the high-dimensional input data space, but now occurs in the usually low-dimensional embedding space. This leads to tremendous savings in computation time in the case of very high-dimensional real-world data, such as for the embedding example of the whole-genome gene expression data in fig. 3.

**Clustering by XOM:** Although the Exploration Machine has originally been invented as a novel method for structure-preserving dimensionality reduction, it is essential to realize that XOM can be applied to other domains of data analysis as well. Data clustering, for example, can be performed by exploiting the flexibility to design arbitrary structure hypotheses in the embedding space. Here, the key idea is to simply select the sampling vectors from non-uniform distributions, e.g. from a mixture of several distributions centered at different positions in the embedding space. A typical choice could be a mixture of Gaussian distributions centered at different locations in the image space[2]. After running the XOM algorithm, the image vectors can be assigned to these distributions, e.g. by computing and comparing the distances of the image vectors to the centers of the respective distributions. By definition of appropriate distance measures, e.g. the assignment likelihood, this can be performed in a fuzzy manner as well.

## 3   Experiments

**'Hepta' Data Sets.** In order to quantitatively evaluate the quality of dimensionality reduction by XOM and to relate its results to other methods known from the literature, we have performed extensive computer simulations. To this end, we have investigated the degree of structure preservation which can be achieved by classical and advanced recent nonlinear embedding methods, namely Principal Component Analysis (PCA), Locally Linear Embedding [5], and Isomap [7]. For this purpose, we used 40 data sets similar to a synthetic benchmark data set called 'Hepta' proposed in [8] for the evaluation of structure preservation. A single realization of a 'Hepta' data set is depicted in Fig. 1. It

---

[2] An extreme choice for a structure hypothesis suitable for clustering in step 2 of the algorithm would be to repetitively sample from a set of $K$ isolated points, if one wishes to perform clustering using $K$ clusters.

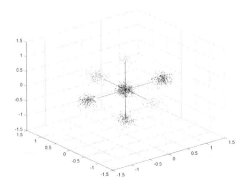

**Fig. 1.** 'Hepta' Data Set

(A)                                          (B)

**Fig. 2.** XOM embedding of the 'Hepta' data sets. The figure depicts the best (A) and the worst (B) embedding result obtained by XOM for the 40 data sets constructed according to the specifications explained in Fig. 1.

consists of 2300 points randomly sampled from seven Gaussian distributions, thus forming 'clusters' in $\mathbb{R}^3$. The centroids of the six non-central Gaussian distributions span the coordinate axes of the $\mathbb{R}^3$, the respective clusters consist of 300 data points each. The central Gaussian distribution consists of 500 data points, i.e. the respective point selection can be interpreted as a cluster whose density is higher than the density of the surrounding six clusters. For our investigation, we created 40 'Hepta' data sets according to these specifications. – To quantitatively evaluate structure preservation, we used a scale-invariant version of Sammon's error [6].

Our results are summarized in Tab. 1. On average, XOM outperformed LLE and Isomap with regard to structure preservation, although Isomap yielded better results in a few data sets. Interestingly, we frequently obtained poor results for PCA. This is caused by the spatial symmetry of the data set which makes the projection axis in PCA very sensitive to noise, i.e. to the random choice of data points sampled from the Gaussian distributions specified in the 'Hepta' data set construction. Thus, different clusters are frequently projected onto each other in the embedding result, i.e. cannot be separated, which leads to impaired

**Table 1.** Comparative evaluation of computation times and structure preservation for nonlinear embedding of 40 'Hepta' data sets as specified in Fig. 1. Average and minimum values as well as the standard deviation of Sammon's error $E'$ were computed as a scale-invariant measure of structure preservation. The free parameters of all the methods examined in the comparison (except PCA) were optimized to obtain the best results, i.e. to minimize $E'$. Note that XOM yields competitive structure preservation at acceptable computation times.

| Method | Comp. Time (s) | $E'$ | $\min(E')$ | $\sigma(E')$ |
|--------|----------------|------|------------|--------------|
| Isomap | 468 | $1.851 \cdot 10^5$ | $1.319 \cdot 10^5$ | $0.384 \cdot 10^5$ |
| LLE | 11600 | $3.681 \cdot 10^5$ | $1.776 \cdot 10^5$ | $1.267 \cdot 10^5$ |
| PCA | 0.3 | $2.216 \cdot 10^5$ | $1.279 \cdot 10^5$ | $0.608 \cdot 10^5$ |
| XOM | 4.6 | $1.732 \cdot 10^5$ | $1.426 \cdot 10^5$ | $0.247 \cdot 10^5$ |

structure preservation. For illustration, the best and the worst of the 40 embedding results obtained by XOM are shown in Fig. 2. Tab. 1 also lists computation times using an ordinary PC (Intel Pentium 4 CPU, 1.6 GHz, 512 MB RAM). PCA outperformed all other methods with regrad to computation time. XOM required considerably smaller computation times than LLE and Isomap. – We emphasize that the results depicted in Tab. 1 depend on the structure of the data set, and do not allow to draw final conclusions on the overall performance of the nonlinear embedding algorithms with regard to the general degree of structure preservation or their computational expense. In addition, the choice of other measures for structure preservation may also result in different ranking scenarios. For example, we conjecture that PCA will be superior in situations where the data is approximately located in a linear subspace of the observation space. LLE and Isomap will perform better in situations where the data is not distributed inhomogeneously in the observation space, i.e. does not exhibit an underlying distinct cluster structure of almost isolated data patches, but rather consists of a 'connected' single cluster. In such data sets, both LLE and Isomap can accurately reconstruct the data with a smaller number of nearest neighbors, which will also reduce their computational expense considerably. – However, even taking all these limitations into account, our investigation at least shows that there exist classes of data sets where XOM yields competitive results in comparison to the methods known from the literature.

**Visualization of Genome-Wide Expression Patterns:** We used the Exploration Machine to visualize genome-wide expression patterns by structure-preserving dimensionality reduction. Fig. 3 (A) shows a genome map created by nonlinear XOM embedding of gene expression profiles in the yeast *Saccharomyces cerevisiae* obtained from DNA microarray hybridization experiments. The data is taken from [2], where it is described in detail. It includes 79-dimensional vectors representing concatenated time courses obtained for 2467 genes functionally annotated in the *Saccharomyces* Genome Database [1]. The expression profiles were average-corrected and scaled to unit variance. – Each of the 2467 points on the map represents the 79-dimensional expression profile of a single gene. Specific

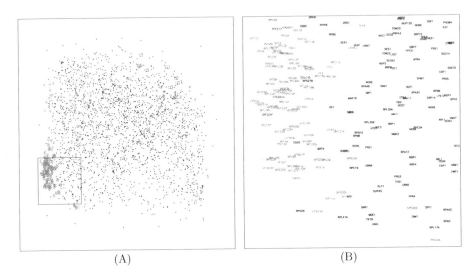

**Fig. 3.** Visualization of genome-wide expression profiles by the Exploration Machine. (A) Genome map created by nonlinear XOM embedding of 2467 79-dimensional gene expression profiles in the yeast *Saccharomyces cerevisiae* obtained from DNA microarray hybridization experiments, data published in [2]. (B) Enlarged section of the genome map depicted by the inset in (A). For further explanation, see text.

groups of genes related to each other with respect to their biological function according to a cluster annotation by Eisen et al. [2] are color-coded and labeled by numbers. In detail: '1': spindle pole body assembly and function, '2': the proteasome, '3': mRNA splicing, '4': glycolysis, '5': the mitochondrial ribosome, '6': ATP synthesis, '7': chromatin structure, '8': the ribosome and translation, '9': DNA replication, and '10': the tricarboxylic acid cycle and respiration.

Fig. 3 (B) shows the enlarged section of the genome map depicted by the inset in (A). For the practical use of XOM genome maps as presented in Fig. 3, appropriate graphical user interfaces can easily be implemented that supply annotation information on the map when needed. Correspondingly, plots of the individual expression profiles can easily be projected onto the map. A notable result from the visualization in fig. 3 is that genes of similar biological function are collocated on the map. Our computation time for the XOM genome map was 72 seconds on an ordinary PC (Intel Pentium 4 CPU, 1.6 GHz, 512 MB RAM). We quantitatively compared our result of fig. 3 with the results obtained by several other embedding algorithms using Sammon's error function [6] as a criterion for embedding quality: We obtained error values of $5.91 \cdot 10^5$ (1.00) for XOM, $6.50 \cdot 10^5$ (1.10) for Sammon's mapping, $6.56 \cdot 10^5$ (1.11) for Principal Component Analysis (PCA), and $7.24 \cdot 10^5$ (1.22) for a SOM with a regular grid of $125 \times 125$ neurons, where numbers in brackets indicate relative values compared to the XOM result. Computation times were 72 s for XOM, 216 s for Sammon's mapping, 2 s for PCA, and 881 s for SOM. – The example shows that XOM is particularly suited to contribute to knowledge discovery in large

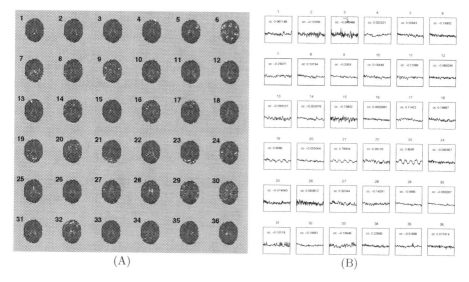

**Fig. 4.** Exploration Machine cluster analysis in human brain mapping, based on a functional MRI visual stimulation experiment. (A) Cluster assignment maps. White regions denote pixels assigned to a respective cluster. (B) Cluster-specific prototypical time-series, i.e. 'codebook vectors' and cluster-specific correlation coefficients between codebook vectors and stimulus function. Cluster numbers correspond to assignment maps in (A).

high-dimensional data collections, such as databases from biology, by succeeding in fast and concise structure-preserving visualization.

**Functional MRI for Human Brain Mapping:** In the previous section, the Exploration Machine has been successfully applied to structure-preserving *dimensionality reduction*. To demonstrate its applicability to *data clustering* as well, we performed exploratory functional MRI analysis for human brain mapping in a visual stimulation experiment. Here, the basic idea is to group pixels according to their similarity of pixel-specific signal dynamics time-series. Experimental protocols, data acquisition and pre-processing have been published previously [10]. Each functional MRI slice includes approximately $5 - 10 \cdot 10^3$ pixels, with a number of 98 acquisitions over a time of 300 s. Thus, the task is to cluster several thousand time-series vectors in $\mathbb{R}^{98}$. Figures 4 (A) and (B) show an example of cluster assignment maps and corresponding cluster-specific prototypical signal-time series, so-called 'codebook vectors', that can be interpreted as the average time-series of all the pixels belonging to a specific cluster. As can be seen from the figures, clusters 21 and 23 clearly identify task-related activity in the visual cortex, reflected by the high correlation between codebook vectors and the box-car shaped stimulus function used in this experiment. Cluster 24 includes pixels representing cerebrospinal fluid of internal ventricles, whereas cluster 4 is indicative for a through-plane motion artifact. A quantitative ROC analysis revealed areas under ROC curves of $0.984 \pm 0.03$ for XOM, $0.983 \pm$

0.02 for Minimal-Free-Energy VQ [10], and $0.979 \pm 0.02$ for SOM for the detection of task-related activation. We conclude that our method is well-suited to perform sophisticated high-dimensional cluster analysis tasks in functional MRI real-world data yielding competitive results comparable to those obtained by established algorithms published in the literature of this domain [10].

## 4     Conclusion

In this paper, we have introduced the Exploration Machine as a novel method for nonlinear embedding based on a systematic inversion of the data processing workflow in topology-preserving mappings. We have shown that XOM is capable of structure-preserving dimensionality reduction as demonstrated for real-world applications to data visualization and data clustering. – Further analyses, extensions, variants, and applications of the Exploration Machine, such as related to computational complexity, supervised learning, analysis of non-metric data, out-of-sample extension, and constrained incremental learning, have been thoroughly investigated in [9]. – In the current literature, clustering and structure-preserving dimensionality reduction are frequently treated as independent problems of information processing. As shown in this paper, XOM creates a link between these subjects by simultaneously contributing to different fundamental domains of machine learning within a single computational framework for exploratory data analysis.

## References

1. Cherry, J.M., Ball, C., Weng, S., Juvik, G., Schmidt, R., Adler, C., Dunn, B., Dwight, S., Riles, L., Mortimer, R.K.: Nature 387, 67–73 (1997)
2. Eisen, M.B., Spellman, P.T., Brown, P.O., Botstein, D.: Cluster analysis and display of genome-wide expression patterns. Proc. Natl. Acad. Sci. USA 95, 14863–14868 (1998)
3. Graepel, T., Burger, M., Obermayer, K.: Self-organizing maps: Generalizations and new optimization techniques. Neurocomputing 21, 173–190 (1998)
4. Kohonen, T.: Self-Organizing Maps, 3rd edn. Springer, Heidelberg (2001)
5. Roweis, S.T., Saul, L.K.: Nonlinear dimensionality reduction by locally linear embedding. Science 290(5500), 2323–2326 (2000)
6. Sammon, J.W.: A nonlinear mapping for data structure analysis. IEEE Transactions on Computers C 18, 401–409 (1969)
7. Tenenbaum, J.B., de Silva, V., Langford, C.: A global geometric framework for nonlinear dimensionality reduction. Science 290(5500), 2319–2323 (2000)
8. Ultsch, A.: Maps for the visualization of high-dimensional data spaces. In: Proceedings of the Workshop on Self-Organizing Maps 2003 (WSOM 2003), Hibikino, Kitakyushu, Japan, pp. 225–230 (2003)
9. Wismüller, A.: Exploratory Morphogenesis (XOM): A Novel Computational Framework for Self-Organization. Ph.D. thesis, Technical University of Munich, Department of Electrical and Computer Engineering (2006)
10. Wismüller, A., Lange, O., Dersch, D.R., Leinsinger, G.L., Hahn, K., Pütz, B., Auer, D.: Cluster analysis of biomedical image time-series. International Journal of Computer Vision 46(2), 103–128 (2002)

# Generalized Self-Organizing Mixture Autoregressive Model

Hujun Yin[1] and He Ni[2]

[1] School of Electrical and Electronic Eng., The University of Manchester, Manchester, UK
h.yin@manchester.ac.uk
[2] School of Finance, Zhejiang Gongshang University, Hangzhou, China
nihe@mail.zjgsu.edu.cn

**Abstract.** The self-organizing mixture autoregressive (SOMAR) model regards a time series as a mixture of regressive processes. A self-organizing algorithm is used with the LMS algorithm to learn the parameters of these regressive models. The self-organizing map is used to simplify the mixture as a winner-take-all selection of local models, combined with an autocorrelation coefficient based measure as the similarity measure for identifying correct local models. The SOMAR has been shown previously being able to uncover underlying autoregressive processes from a mixture. This paper proposes a generalized SOMAR that fully considers the mixing mechanism and individual model variances that make modeling and prediction more accurate for non-stationary time series. Experiments on both benchmark and financial time series are presented. The results demonstrate the superiority of the proposed method over other time-series modeling techniques on a range of performance measures.

## 1 Introduction

Time series modeling and forecasting is an active, challenging and reoccurring topic in statistics and signal processing owing to their wide use in real-world applications such as communications, speech processing, finance, astronomy and neuro-physiology. Linear regression and autoregressive models such as autoregressive (AR), moving average (MA) and autoregressive moving average (ARMA), are commonly used methods. Most linear models assume that the time series being dealt with is stationary and uni-modal [3, 8] and assume a structured linear relationship of constant coefficients between the current value of the time series and its previous values and the error terms. Such conditions are not often met in practice. They are the pitfall of (linear) regressive models when the time series is non-stationary. Developing methods for modeling non-stationary and multimodal time series has become an active area of research. The autoregressive integrated moving average (ARIMA) [3], a generalized ARMA model, can better handle slow changing non-stationary time series by modeling the difference of the consecutive time series values instead of the value itself. The generalized autoregressive conditional heteroscedastic (GARCH) model [2] models the variance of the residual as a linear function of the previous variances, along with the autoregressive model of the time series. It has been a benchmark model for financial data, which exhibits varying volatilities from time to time. The mixture

J.C. Príncipe and R. Miikkulainen (Eds.): WSOM 2009, LNCS 5629, pp. 353–361, 2009.
© Springer-Verlag Berlin Heidelberg 2009

autoregressive (MAR) model [17] represents another approach that considers the process as a mixture of regressive models and is a generalized Gaussian mixture transition distribution. It can handle non-stationary cycles and conditional hetero-scedasticity and is often solved by the expectation and maximization (EM) method.

Various adaptive neural networks have been adopted to extend linear regressive models such as multilayer perceptron (MLP), radial basis functions (RBFs), support vector machines (SVM) and recurrent networks [9]. Nonstationarity implies that the time series change their dynamics in different time regions. It is unreasonable for a single model to capture the dynamics of the entire series. A potential solution is to use a mixture model approach to divide the entire model into several smaller ones. Then regression and prediction are made by the local models. The self-organizing map (SOM) can be used to partition time series. For instance, Dablemont et al. [6] applied SOM-based local models with RBF networks as regressors. Cao [4] used SVM regressors on SOM-clustered local segments. However, these models are two-stage modeling. Both clustering and local modeling may not be jointly optimized.

There were two early approaches to analyzing temporal signals or sequences with the SOM. One is to train a SOM on static states (i.e. time series values), and then temporal patterns or sequences of states can be identified by marking sequential locations of the state on the trained map. Such approaches can be used to monitor dynamic processes or trajectories of a temporal process such as industrial plants [1]. Another approach, which is often found in the literature, is to group consecutive time points into segments (using a sliding window). Then these segments are used as the input vectors to train the SOM. We term this method as vector SOM or simply SOM.

Several variants have since been proposed to extend SOM's ability for temporal modeling such as the recurrent SOM (RSOM) [10] and the recursive SOM (RecSOM) [16,15]. These variants integrate the information of a sequence via recursive operations. As they differ in the notion of context, their efficiency in terms of representing temporal context are different. Neural gas (NG) [12] is another variant of SOM. Instead of having a fixed network topology throughout, NG can dynamically deploy its resources to suit varying topology of the data and has been applied to tasks including temporal modeling [12] and has been enhanced by merge NG (MNG) [15].

Earlier, Lampinen and Oja proposed a self-organizing map of spatial and temporal AR models [11], where each unit represents an AR model with its reference vector as the model parameters. The method in fact is a multiple AR model with the component models forming a spatial topology. However, the model has difficulties to converge to the underlying regressive models due to the simple error-based similarity measure. We have extended it to a mixture regressive model, termed the self-organizing mixture autoregressive (SOMAR) model [13,14], with a different partition mechanism and similarity measure to reflect the characteristics of homogeneous time series. Both the mixture and local models are jointly trained, and thus it offers better modeling performance [13,14]. Here the SOMAR model is further analyzed in light of the MAR model and generalized to a full mixture model.

The remainders of the paper are as follows. Section 2 briefly describes various regressive models. Section 3 presents SOM-based autoregressive models and the proposed generalized SOMAR model, followed by experimental results on both benchmark data and real-world data and comparisons with several methods in Section 4. Finally, conclusions are given in Section 5.

## 2   Regressive Time Series Models

### 2.1   Autoregressive Models: AR, ARMA, GARCH and ARIMA

Linear regressive models have been the primary tool in modeling time series. An autoregressive model of order $p$, denoted as AR($p$), can be described as,

$$x_t = c + \sum_{i=1}^{p} \phi_i x_{t-i} + \varepsilon_t = c + \Phi^T \mathbf{x}_{t-1}^{(p)} + \varepsilon_t \tag{1}$$

where $\Phi = [\phi_1, ..., \phi_p]^T$ are the parameters, $\mathbf{x}_{t-1}^{(p)} = [x_{t-1}, ..., x_{t-p}]^T$ is the concatenated input vector, $c$ is a constant and $\varepsilon$ is white noise of zero mean and variance $\sigma^2$.

An ARMA model with $p$-order AR terms and $q$-order MA terms is called ARMA($p$, $q$) model and can be written as,

$$x_t = c + \sum_{i=1}^{p} \phi_i x_{t-i} + \sum_{i=0}^{q} \mu_i \varepsilon_{t-1} \tag{2}$$

where $\{\mu_0, ..., \mu_q\}$ are the parameters of the moving average. The error terms are assumed to be independent identically-distributed (i.i.d.) random variables sampled from a normal distribution with zero mean and variance $\sigma^2$. When this condition does not hold, the GARCH model provides a generalized alternative, in which the variance of the error terms is modeled by another regressive model.

A standard GARCH($\theta$, $q$) model is characterized by Eq. (1) and the following variance model,

$$\sigma_t^2 = \alpha_0 + \sum_{i=1}^{\theta} \alpha_i \varepsilon_{t-i}^2 + \sum_{i=0}^{p} \beta_i \sigma_{t-i}^2 \tag{3}$$

where $\varepsilon_t$ is the error term with the assumption $\varepsilon_t = \sigma_t v_t$ and $v_t$ is i.i.d. with zero mean and unit variance. $\{\alpha\}$ and $\{\beta\}$ are the model parameters of the variance.

ARIMA model uses lags or differencing of the time series in the ARMA model. ARIMA($p,d,q$) model is characterized by the following equation,

$$\left(1 - \sum_{i=1}^{p} \phi_i L^i\right)(1 - L)^d x_t = \left(1 + \sum_{i=1}^{q} \mu_i L^i\right)\varepsilon_t \tag{4}$$

where $L$ is the lag operator, i.e. $Lx_t = x_{t-1}$ and $p$, $d$, and $q$ are the orders of the autoregressive, integrated, and moving average parts of the model respectively. Note that ARMA($p,q$), i.e. Eq. (1), can be expressed as $\left(1 - \sum_{i=1}^{p} \phi_i L^i\right)x_t = \left(1 + \sum_{i=1}^{q} \mu_i L^i\right)\varepsilon_t$.

As can be seen, the ARIMA model operates on the difference of the lagged time series. Such simple transformation can be effective in dealing with slow changes in non-stationarity. That is, the difference operator transforms a slow drift non-stationary process into a stationary process.

## 2.2 Mixture Autoregressive (MAR) Model

A nonlinear or non-stationary time series can be regarded as a mixture of stationary processes characterized by the standard autoregressive models. The $K$-component MAR model is defined by [17],

$$F(x_t \mid \Gamma_{t-1}) = \sum_{k=1}^{K} \pi_k \varphi(\frac{x_t - \phi_{k0} - \phi_{k1} x_{t-1} - \dots - \phi_{kp_k} x_{t-p_k}}{\sigma_k}) \tag{5}$$

where $F(x_t \mid \Gamma_{t-1})$ is the conditional distribution of $x_t$ given the past information up to $t$-1, $\Gamma_{t-1}$; $\varphi(.)$ is the standard normal distribution; $\{\pi_1,\dots,\pi_K\}$ are the mixing parameters and $\pi_1+\dots+\pi_K=1$, $\pi_k>0$, $k=1,\dots, K$; $p_k$ is the order of the $k$-th AR model; and $\sigma_k^2$ is the variance of the $k$-th distribution. This model is denoted as MAR($K$; $p_1,\dots,p_K$) model. The MAR has the ability to handle cycles and conditional heteroscedasticity in time series and its parameters are estimated via the EM algorithm and model selection by a Bayesian information criterion (BIC) [17].

# 3 Self-Organizing Mixture Autoregressive Models

## 3.1 Self-Organizing AR (SOAR) Model

Lampinen and Oja proposed a self-organizing AR (SOAR) network [11]. It is a map of neurons, each representing an AR model with its parameters as the reference vector $\mathbf{w}_i$. The experiment showed that the SOAR model can learn to distinguish texture images [11]. The method in fact is a multiple AR model. However the model has difficulties in converging to correct AR models. The training procedure is:

1) At each time step $t$, find the best matching unit by measuring the estimation error of each node, $e_{i,t} = x_t - \mathbf{w}_i^T \mathbf{x}_{t-1}^{(p)}$. In order to reduce the effect of the fluctuation or noise in the errors, an exponential average over the recent errors is used,

$$u_{i,t} = \lambda e_{i,t} + (1-\lambda)u_{i,t-1} \tag{6}$$

where $\lambda$ is a smoothing factor, $e_i(t)$ is the current error of node $i$ and $u_i(t$-1) is the past averaged error.

2) Update the best matching unit as well as its neighborhood on the map by the recursive LMS or Widrow-Hoff rule,

$$\mathbf{w}_{i,t} = \mathbf{w}_{i,t-1} + \eta h(v,i)e_{i,t}\mathbf{x}_t^{(p)} \tag{7}$$

where $\eta$ is learning rate and $h(i,v)$ is the neighborhood function of indexes of node $i$ and winner $v$.

However the performance of the SOAR model in finding the underlying AR processes in the mixture is poor [13]. Due to the stochastic nature of AR processes, although the overall MSE decreases, at each input, one can always expect large fluctuation even when the true model parameters are used and further smoothing is applied. In other words, this method has difficulties in converging to the true model parameters of the underlying AR processes. Nevertheless, the SOAR model localizes the time series by local models.

## 3.2  Self-Organizing MAR (SOMAR) Model

Based on the similar principle and the MAR model, the self-organizing MAR (SOMAR) was proposed [13]. It constitutes a simplified MAR model with the winner-take-all for local AR models. To ensure a robust learning, a new winner selection or similarity measure was proposed. A stochastic process is characterized by white noise residuals. As a sufficient condition, the modeling errors or the residuals should be or close to white noise if the modeling is following the correct path. Therefore, the autocorrelation of the error instead of the error itself is used to evaluate the similarity between the input vector and the neurons' weights representing the model parameters. To estimate the autocorrelation, a small batch of the errors is used,

$$R_{i,t}(k) = \frac{1}{m\sigma_e^2} \sum_{l=0}^{m-p-1} (e_{i,t-l} - \mu_e)(e_{i,t+k-l} - \mu_e) \qquad (8)$$

where $m$ is the length of the batch, $\mu_e$ and $\sigma_e^2$ are the mean and the variance of the errors in the batch respectively.

The winner is selected according to the sum of (absolute value of) autocorrelation coefficients (SAC),

$$v = \arg\min_i \left( \sum_{k=-m}^{m} |R_{i,t}(k)| \right) \qquad (9)$$

The use of correlation measure for identifying local models is justified by the fact that a correct model produces white noise residuals. That is, if the model is correct or adequate, the residual is unpredictable or structure-less. Such effective correlation-based tests are often used in statistics and neural networks for checking the fitness of a model, e.g. [5], though there are other whiteness tests in the literature.

## 3.3  Generalized SOMAR (GSOMAR) Model

Both the SOMAR and SOAR models represent a simplified, homescedastic and winner-take-all version of the MAR model. At any time, only one local AR model (the winner) is selected to represent the time series, all models are assumed of equal variance and the mixing factors are either unit for the winner or zero otherwise. Although some empirical use of neighboring nodes has been proposed for forecasting [13], the model is not a full mixture model. To fully employ the mixture model, all components will be required to contribute to the mixture coherently both in training and testing. The mixing factors and model variances have to be learnt as well. The SOM has been extended before to a mixture model. The self-organizing mixture network (SOMN) [18] is such an example, in which each node represents a conditional distribution. The SOMN has also been shown to converge faster and be more robust than the EM algorithm for heteroscedastic mixture distributions. To make the SOMAR a full mixture of AR models, the algorithm of the SOMN can be used to learn the mixing factors and variances. In addition to the weights (or AR model parameters), the mixing factors and model variances are updated in the training (modeling). Further assuming that the component models are uncorrelated, so their covariances are zeros. The variances of local models are scalar. Then the updating rules for the mixing weights and variances have the following simple forms,

$$\pi_{i,t} = \pi_{i,t-1} + \eta(\hat{P}_i - \pi_{i,t-1})$$ (10)

$$\sigma_{i,t}^2 = \sigma_{i,t-1}^2 + \eta(\sigma_{i,e}^2 - \sigma_{i,t-1}^2)$$ (11)

where $\hat{P}_i$ is the winning frequency and $\sigma_{i,e}^2$ is the error variance of node $i$.

The trained mixture model, representing the MAR model, Eq. (5), can be fully used for forecasting the time series as well as model's volatility. In forecasting, the learnt mixing factors are further weighted by the neighborhood function of the SOM, acts as the posterior probability of a component class given an input sample [18].

## 4   Experimental Results and Comparisons

### 4.1   Artificial Data

As an illustrative example, a mixture two AR(2) processes was generated with their model parameters set to [0.2, -0.3] and [0.4, -0.1] and variances to 3 and 5, respectively. The learning process of the GSOMAR is shown in Fig. 1.

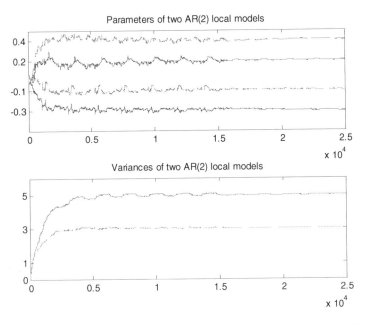

**Fig. 1.** Parameter and variance estimation of a mixture of two AR(2) processes. Fine tuning phase [13] starts at t=15000. Dashed (red) lines represent one process and solid (blue) lines the other.

## 4.2 Mackey-Glass Data

The Mackey-Glass series has been widely used as a benchmark data for testing nonlinear models. The data set was generated by a dynamic system defined by the following differential equation,

$$\frac{dx}{dt} = \beta x_t + \frac{\alpha x_{t-\delta}}{1 + x_{t-\delta}^{10}}$$

with the parameter values set as $\delta=17$, $\alpha=0.2$, and $\beta=-0.1$. In total 2000 points were generated. The Mackey-Glass data is regarded as consisting of a number of unknown AR processes. In the experiment, the series was grouped into 12 consecutive values as the input vectors. The order of the AR processes was chosen by the BIC. The prediction result is shown in Table 1. The performance of statistical benchmark models GARCH and ARIMA are 4.48 and 4.35 respectively. The results are the average over 10 independent runs on the same data set. Different data set may lead to slightly difference performance. However, it can be seen that GSOMAR and SOMAR markedly outperform the others and GSOMAR further improves on SOMAR.

**Table 1.** Forecasting performance on Mackey-Glass data by various adaptive models

|          | GSOMAR | SOMAR | SOAR | SOM | RSOM | RecSOM | NeualGas | MNG | SOM+SVM |
|----------|--------|-------|------|-----|------|--------|----------|-----|---------|
| MSE($^{-2}$) | **3.24** | 3.62 | 4.29 | 4.48 | 4.32 | 4.10 | 4.38 | 4.35 | 4.52 |

## 4.3 Foreign Exchange Rates

The data was obtained from the PACIFIC Exchange Rate Service provided by W. Antwiler at UBCs Sauder School of Business. It consists of 15years' daily exchange rates (British pound vs. US dollar, Euro and HK dollar) excluding weekends and bank holidays when the currency markets were closed. In total 3200 consecutive points were used, in which the first 3,000 points were used as the training set, the next 100 points as the validation set, and the remaining 100 points as the test set. The training, validation and testing sets were windowed with the length of 15 points to form input vectors (again validated by the BIC).

To compare with other regressive models, the following commonly used performance measures have been calculated:

*Predicted return* (%): The percentage of correct prediction of the return ($\ln x_{t+1} / x_t$), which is also used as a criterion to check whether the prediction is made in the right direction. In other words, it shows how many percentages of the predicted returns have the same signs as their corresponding actual returns.

*MSE of predicted rate* ($^{-2}$): The MSE between the actual exchange rates and the predicted ones in the test set.

*Accumulated profit* (P%): The accumulated profit is the percentage gain of the accumulated profits over the testing period, say 100 trading days.

**Table 2.** Performance on FX rate prediction by various adaptive models. The best performances are marked in bold.

| GBP vs | GSOMAR | SOMAR | SOAR | SOM | RSOM | RecSOM | NeualGas | MNG | SOM+ SVM |
|---|---|---|---|---|---|---|---|---|---|
| USD % | 59.70 | **59.73** | 52.84 | 52.63 | 52.26 | 52.58 | 54.08 | 54.16 | 53.43 |
| USD $^{-2}$ | 3.87 | **3.80** | 4.28 | 4.20 | 4.24 | 4.70 | 4.23 | 4.20 | 4.12 |
| USD $P\%$ | **5.53** | 5.15 | 4.78 | 4.80 | 4.98 | 5.12 | 5.33 | 5.35 | 4.82 |
| EU % | **57.43** | 56.42 | 52.62 | 52.12 | 53.05 | 53.17 | 54.24 | 54.27 | 54.09 |
| EU $^{-2}$ | **3.96** | 4.11 | 4.73 | 4.32 | 4.64 | 4.95 | 4.51 | 4.50 | 4.62 |
| EU $P\%$ | **5.41** | 5.12 | 4.62 | 4.73 | 4.63 | 4.60 | 4.72 | 4.75 | 4.70 |
| JPY % | **57.95** | 57.30 | 53.22 | 54.29 | 52.48 | 52.33 | 53.46 | 53.47 | 52.10 |
| JPY $^{-2}$ | **4.23** | 4.33 | 5.24 | 5.00 | 4.98 | 5.08 | 4.75 | 4.75 | 5.18 |
| JPY $P\%$ | **5.32** | 5.03 | 4.68 | 4.89 | 4.91 | 4.87 | 4.73 | 4.76 | 4.65 |
| HKD % | **56.37** | 56.31 | 53.50 | 53.95 | 53.88 | 54.02 | 54.21 | 54.22 | 54.13 |
| HKD $^{-2}$ | **4.11** | 4.22 | 4.67 | 4.75 | 4.73 | 4.72 | 4.44 | 4.44 | 4.57 |
| HKD$P\%$ | **5.32** | 5.02 | 4.50 | 4.59 | 4.57 | 4.63 | 4.62 | 4.68 | 4.60 |

As reported before [13,14], the SOMAR model generally outperforms other adaptive methods as also shown in Table 2. The GSOMAR further improves on the SOMAR model in all these performance measures. As can be seen, both GSOMAR and SOMAR consistently outperform other methods by clear margins in the correct prediction percentages and modeling errors. The benefit of using the fuller GSOMAR model is that model variance parameters are readily available to indicate the volatility of the component regressive models and the mixture. Statistical model ARIMA performed the worse on these data sets with the *predicted return* (%) between 50-51% – only slightly better than random guess; while GARCH gave similar performances to SOM+SVM with the *predicted return* (%) between 53-54%.

## 5   Conclusions

A mixture model approach to tackling nonlinear and non-stationary time series has been proposed by using the generalized self-organizing mixture autoregressive (GSOMAR) model. It consists of a number of autoregressive models that are organized and learnt in a self-organized manner by the adaptive LMS algorithm. A correlation-based similarity measure is used for identifying correct AR models, thus making the model more effective and robust compared to the error-based measures. The GSOMAR further generalizes the winner-take-all SOMAR model by learning the mixing weights as well as the model variances. The experiments on various nonlinear, non-stationary time series show that the proposed model can correctly detect and uncover underlying regressive models. The results also show that the proposed method outperforms other methods in terms of modeling errors and prediction performances.

# References

1. Allinson, N.M., Yin, H.: Interactive and semantic data visualization using self-organizing maps. In: Proc. IEE Colloquium on Neural Networks in Interactive Multimedia Systems (1998)
2. Bollerslev, T.: Generalized autoregressive conditional heteroskedasticity. Journal of Econometrics 31, 307–327 (1986)
3. Box, G., Jenkins, G.: Time Series Analysis: Forecasting and Control. Holden-Day, San Francisco (1970)
4. Cao, L.J.: Support vector machines experts for time series forecasting. Neurocomputing 51, 321–339 (2002)
5. Chen, S., Billings, S.A., Cowen, C.F.N., Grant, P.M.: Practical identification of NARMAX models using radial basis functions. Int. Journal of Control 52, 1327–1350 (1990)
6. Dablemont, S., Simon, G., Lendasse, A., Ruttiens, A., Blayo, F., Verleysen, M.: Time series forecasting with SOM and local non-linear models – Application to the DAX30 index prediction. In: Proc. of WSOM 2003, pp. 340–345 (2003)
7. Kohonen, T.: Self-Organizing Maps. Springer, Heidelberg (1997)
8. Enders, W.: Applied Econometric Time Series, 2nd edn. John Wiley & Sons, Chichester (2004)
9. Haykin, S.: Neural Networks – A Comprehensive Foundation, 2nd edn. Prentice-Hall, Englewood Cliffs (1998)
10. Koskela, T.: Time Series Prediction Using Recurrent SOM with Local Linear Models. Helsinki University of Technology (2001)
11. Lampinen, J., Oja, E.: Self-organizing maps for spatial and temporal AR models. In: Proc. of 6$^{th}$ SCIA Scandinavian Conference on Image Analysis, Helsinki, Finland, pp. 120–127 (1989)
12. Martinetz, T., Berkovich, S., Schulten, K.: Neural-gas network for vector quantization and its application to time-series prediction. IEEE Trans. Neural Networks 4, 558–569 (1993)
13. Ni, H., Yin, H.: Time-series prediction using self-organizing mixture autoregressive network. In: Yin, H., Tino, P., Corchado, E., Byrne, W., Yao, X. (eds.) IDEAL 2007. LNCS, vol. 4881, pp. 1000–1009. Springer, Heidelberg (2007)
14. Ni, H., Yin, H.: Self-organizing mixture autoregressive model for non-stationary time series modeling. International Journal of Neural Systems 18, 469–480 (2008)
15. Strickert, M., Hammer, B.: Merge SOM for temporal data. Neurocomputing 64, 39–72 (2005)
16. Voegtlin, T., Dominey, P.F.: Recursive self-organizing maps. Neural Networks 15, 979–991 (2002)
17. Wong, C.S., Li, W.K.: On a mixture autoregressive model. Journal of the Royal Statistical Society, Series B (Statistical Methodology), Part1 62, 95–115 (2000)
18. Yin, H., Allinson, N.M.: Self-organizing mixture networks for probability density estimation. IEEE Trans. on Neural Networks 12, 405–411 (2001)

# An SOM-Hybrid Supervised Model for the Prediction of Underlying Physical Parameters from Near-Infrared Planetary Spectra[*]

Lili Zhang[1], Erzsébet Merényi[2], William M. Grundy[3], and Eliot F. Young[4]

[1] Rice University, Rice Quantum Institute and Department of Electrical & Computer Engineering MS-366, 6100 Main Street, Houston, TX 77005
llzhang@rice.edu
[2] Rice University, Department of Electrical & Computer Engineering MS-380, 6100 Main Street, Houston, TX 77005
erzsebet@rice.edu
[3] Lowell Observatory, 1400 W. Mars Hill Rd., Flagstaff AZ 86001
[4] Space Studies Department, Southwest Research Institute, Boulder, CO 80302

**Abstract.** Near-Infrared reflectance spectra of planets can be used to infer surface parameters, sometimes with relevance to recent geologic history. Accurate prediction of parameters (such as composition, temperature, grain size, crystalline state, and dilution of one species within another) is often difficult because parameters manifest subtle but significant details in noisy spectral observations, because diverse parameters may produce similar spectral signatures, and because of the high dimensionality of the feature vectors (spectra). These challenges are often unmet by traditional inference methods. We retrieve two underlying causes of the spectral shapes, temperature and grain size, with an SOM-hybrid supervised neural prediction model. We achieve 83.0±2.7% and 100.0±0.0% prediction accuracy for temperature and grain size, respectively. The key to these high accuracies is the exploitation of an interesting antagonistic relationship between the nature of the physical parameters, and the learning mode of the SOM in the neural model.

**Keywords:** Self-Organizing Map, parameter prediction, Near-Infrared spectra, New Horizons Space Mission, Pluto-Charon system.

## 1 Machine Learning for Parameter Prediction from Spectra, Motivated by the New Horizons Space Mission

### 1.1 Investigation of Surface Conditions from Near-Infrared Spectra

Near-infrared reflectance spectroscopy offers an extremely powerful remote probe of planetary surfaces. Numerous physical parameters, such as composition,

---

[*] This work was partially supported by grants NNG05GA63G and NNG05GA94G from the Applied Information Systems Research Program, NASA, Science Mission Directorate. Figures are in color, request color copy by email: llzhang@rice.edu, erzsebet@rice.edu

J.C. Príncipe and R. Miikkulainen (Eds.): WSOM 2009, LNCS 5629, pp. 362–371, 2009.
© Springer-Verlag Berlin Heidelberg 2009

texture, and thermal state of surface materials, influence the observable reflectance, from which the parameters can potentially be retrieved. Icy outer Solar System surfaces present a natural application for such retrieval algorithms, since cryogenic ices such as $H_2O$, $N_2$, and $CH_4$ possess distinctive spectral contrasts which change in known ways as a function of temperature [1,2]. NASA's New Horizons spacecraft, which is en route to the Pluto system [3], will map the surfaces of Pluto, Charon, Nix, and Hydra at wavelengths from 1.25 to 2.5 microns with its infrared imaging spectrometer [4] in 2015. By extracting the surface parameters from these spectral maps, it will be possible to determine what processes are at work sculpting the exotic landforms New Horizons will discover.

However, the complexity of the measured spectra, their shapes, the often subtle changes in the spectral curves in response to relevant changes in the underlying causes (the implicit physical parameters of the surface), the interplay between the underlying causes, and the high dimensionality of the feature vectors (spectra), poses significant challenges for accurate retrieval of the parameters. Traditional approaches, based on iteratively inverting spectral mixing models [5] do not give entirely satisfactory results in real world applications [6]. Such techniques work very well for the relatively simple version of the problem posed here, but with the machine learning approach we propose we expect to be able to infer parameters also from complicated noisy real spectra. Specifically, we approach this challenge with a hybrid supervised neural architecture, which has a Self-Organizing Map (SOM) as its hidden layer. In this paper we focus on the inference of two surface parameters, grain size and temperature. We investigate inference capabilities of the neural model for a single material, crystalline water ice, which is common on Solar System surfaces. Results gained will be used in follow-up work to infer parameters from other types of ices ($SO_2$, $CO$, $CO_2$, $NH_3$) possibly occurring in the Pluto system and elsewhere in the Solar System.

## 1.2  Forward Training with Synthetic Spectra and Reverse Engineering from Real Spectra

The training of the neural machine requires a great number of spectra, which should span the meaningful ranges of the physical parameters with appropriate resolutions. However, real spectra collected from icy planetary surfaces are scarce, and not representative of the desired granularity of the prediction. To help this, synthetic or laboratory data have been used for model development in several areas, where real data are hard to obtain [7,8,9]. In our study the SOM-hybrid neural prediction model is trained with synthetic spectra, which are generated through radiative transfer code described in [10] based on the Hapke model, the most common way to represent the interaction of a solid surface with incident sunlight [5,11]. The synthetic spectra are generated on a grid of temperatures and grain sizes where the temperature ranges from 20 to 270 K with 2 K spacing, and the grain size takes 9 values logarithmically spaced from 0.0003 to 3.0 cm. This set of parameters encompasses the range of possible surface conditions of the icy Solar System bodies of our interest, at sufficient resolution for scientific study. The spectral resolution, 230 band passes from 1 to 2.5 $\mu m$, is close to the resolution of the sensor used on the New Horizons spacecraft.

## 1.3   The SOM-Hybrid Supervised Architecture

The supervised neural architecture we use in this study is a fully connected feed-forward network with an SOM as the middle layer and an output layer connected to the SOM by the Widrow-Hoff rule [12]. A 230-band spectrum is taken as an input vector at every learning step. The learning consists of two stages. The first stage is unsupervised, in which the SOM layer captures the structure of the data manifold. The knowledge represented in the SOM is then utilized in the second stage, which is supervised training of the output layer. This construction generally helps achieve good prediction accuracy [13,14]. Its additional merits include ease and economy of training and handling of high-dimensional data, compared to other, more frequently used neural approaches. We use the Conscience variant [15] of the original Kohonen SOM [16]. It introduces a bias to achieve equal winning probabilities across all neural units thus producing more faithful *pdf* matching than the Kohonen SOM. Briefly, the weight vector $\mathbf{w}_i$ of neural unit $i$ in the SOM lattice $A$ of $N$ neural units, is updated iteratively through a two-step procedure. First, a winner (or best matching unit, BMU) $\mathbf{w}_i$ is selected for a given input vector $\mathbf{x}$ such that with the bias $b_j$ for neural unit $j$

$$\| \mathbf{w}_i - \mathbf{x} \|^2 - b_i \leq \| \mathbf{w}_j - \mathbf{x} \|^2 - b_j, \forall j \in A. \tag{1}$$

The bias $b_j$ is computed from the winning frequency $p_j$, of neural unit $j$, as

$$b_j = \gamma(t) \times ((N \times p_j) - 1), \tag{2}$$

where $\gamma$ is a parameter. Second, all weight vectors $\mathbf{w}_j$ are updated:

$$\mathbf{w}_j^{new} = \mathbf{w}_j^{old} + \alpha(t)h_{i,j}(t)(\mathbf{x} - \mathbf{w}_j^{old}). \tag{3}$$

Here, $h_{i,j}(t)$ is a neighborhood function, $\alpha$ is the learning rate. With the Conscience algorithm $h_{i,j}(t)$ can be fixed and of small size (e.g., the immediate neighbors in a diamond or square configuration), instead of a large neighborhood (e.g., Gaussian) that has to decrease with time.

## 2   Relationship of the Underlying Causes of Spectra as Seen from the SOM

The two physical parameters have different influences on the spectral shapes. As shown through two representative examples in Fig. 1, there is a substantial difference in the reflectance values of the spectra between two neighboring grain sizes, deepening of the absorption bands as well as shifting to lower values, in a linear fashion without crossing over. The increase in temperature from 30

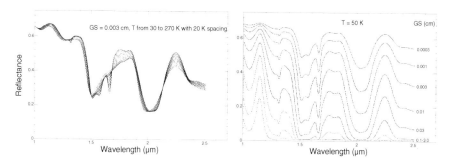

**Fig. 1.** Sample synthetic spectra of crystalline water ice. **Left:** variation in the spectral shape as a function of temperature, for one fixed grain size, 0.003 cm. **Right:** variation of the spectral shape as a function of grain size, at 50 K.

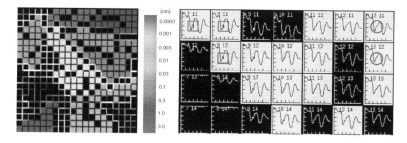

**Fig. 2. Left:** A $20 \times 20$ SOM trained with synthetic spectra of crystalline water ice. The colors represent the known grain size labels as keyed at right. The "fences" drawn between cells have grey scale intensities proportional to the Euclidean distance of the respective weights. White is large distance (dissimilarity). Black cells within double fences separating grain size groups indicate empty weights (no data mapped to them). Other black areas indicate weights representing spectra of other ices ($N_2$, $CH_4$ etc.). **Right:** An example of how spectra are organized within a grain size group, according to temperatures. Shown here are the learned weight vectors in the respective SOM cells, for the 0.003 cm (yellow) group. We can observe a continuous change in the spectral shapes from left to right, caused by increasing temperature. The red boxes and circles exemplify differences in absorption features at low and high temperatures, respectively.

to 270 K causes much smaller changes in reflectance but the direction of the change varies with wavelength. For example, the absorption at 1.65 $\mu m$ gradually weakens with increasing temperature, while in the neighboring window of 1.7 – 1.9 $\mu m$ the reflectances decrease with increasing temperature, resulting in crossovers. Fig. 1 suggests that the (Euclidean) distance between two grain size groups is larger at most wavelengths than the variations caused by temperature within that group. This dominance is reflected in the SOM after unsupervised learning by clearly separated clusters with respect to grain size (Fig. 2, left). Within each grain size cluster, the weight vectors show the continuous change in the spectral shapes caused by temperature, seen in Fig. 1. Fig. 2, right, illustrates this for the 0.003 cm grain size (yellow) group. The weight vectors learned from spectra with low temperatures have a strong absorption at 1.65 $\mu m$ (in red boxes). This feature gradually disappears for high temperatures (in red circles).

# 3   Conjoined Twin Machines

## 3.1   Two Modes of the SOM during Supervised Learning

The SOM can be used in two modes during supervised training. One is the widely used winner-take-all (WTA) mode, in which one SOM neuron fires (has an output signal of 1) in response to a given input vector, the rest send 0 to the weighted sums formed at the output layer. The other possibility is to divide the winning credit among $k$ SOM nodes by assigning each an output value that is inversely proportional to the distance between its associated weight vector and the input vector (such that the credits add up to 1). This can be called "interpolating mode". For practical purposes $k$ can be a number much lower than the number of neurons in the SOM, based on the assumption that the winner weight's Voronoi cell has a relatively low number of neighbor Voronoi cells in data space. For $k > 1$ we will use the term "interpolation on" or "interpolating mode", and use "interpolation off" or "non-interpolating mode" for $k = 1$.

For this data set, the Voronoi cells of all weight vectors of the SOM in Fig. 2, left, except for one, have at most 3 neighbors. We see this from the numbers of connections to Voronoi neighbors (pairs of BMUs and second BMUs formed by weights and their Voronoi neighbors) [17], shown in the order of the most to least connected, in Table 1. The connection strength (the number of data samples selecting a weight and its Voronoi neighbor as a pair of BMU and second BMU) between the one weight that has a fourth neighbor, and that fourth most connected neighbor is 1 (negligible). This justifies $k = 3$ for interpolating mode.

**Table 1.** Number of connections to Voronoi neighbors, from the most connected to the least connected, summed across all SOM weights

| | Most connected | Second most connected | Third most connected | Fourth most connected |
|---|---|---|---|---|
| Number of connections | 280 | 214 | 13 | 1 |

## 3.2   The Effect of SOM Modes on the Prediction Accuracies of Temperature and Grain Size

As observed in Section 2 the grain size dominance on the reflectance spectra causes clustering primarily by grain size in the SOM. Closer inspection reveals that, without exception, all spectra mapped to any weight vector within a grain size cluster have the same grain size label. This provides a good basis for perfect learning of grain size in non-interpolating mode, where only the BMU fires. In interpolating mode, each input spectrum stimulates the second and the third

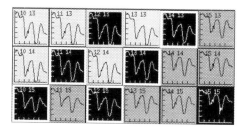

**Fig. 3.** The learned weight vectors shown in the respective SOM cells. The white boxes highlight weight vectors on the boundary between the yellow and orange groups, which represent 0.003 and 0.0001 cm grain sizes, respectively. The light blue box indicates an empty weight vector inside the yellow group.

BMUs, too, which may belong to the boundary area like the empty weight vectors in white boxes in Fig. 3. Their shapes are similar to both neighbors, which makes them candidates to be the second or third BMU for an input vector from either group. This introduces possible confusions in the supervised training. In contrast, the second and third BMUs may help refine the prediction of temperature. Since 126 spectra with different temperature parameters are forced to share approximately 25 – 30 SOM weights in a grain size cluster (Fig. 3, left), each weight forms an average (a mixture) of spectra, and each spectrum is likely to contribute to the mixture in several neighboring weights (smearing across neighbor weights) during training. None of these mixtures will match any specific temperature exactly, but a specific temperature may be reconstructed from several neighboring weight vectors by training the output weights to form their appropriate mixture. This includes empty weight vectors within any grain size group too, such as the one in the light blue box in Fig. 3.

The above discussion suggests conflicting preferences for SOM modes in the prediction of the two parameters. Supervised training results shown as correlations between predicted and true values in Fig. 4 confirm this. Since both physical parameters have large ranges, we quantify the prediction accuracies as

**Table 2.** The prediction accuracies of grain size (GS) and temperature (T) for two separate data sets, containing 9 and 81 grain sizes, respectively, with 20×20 and 40×40 SOMs, each in interpolating and non-interpolating modes. Results for the data with 9 GS are averages of 10 jack-knife runs. Results for the data with 81 GS are from a single run for reasons of time limitations. Further jack-knife runs are in progress.

| | | Data with 9 grain sizes | | Data with 81 grain sizes | |
|---|---|---|---|---|---|
| SOM mode | | Non-interpolating | Interpolating | Non-interpolating | Interpolating |
| 20×20 | GS | **100.0±0.0%** | 76.4±4.4% | 73.0% | **77.9%** |
| SOM | T | 76.2±2.6% | **83.0±2.7%** | 32.9% | **53.7%** |
| 40×40 | GS | - | - | **97.7%** | 54.3% |
| SOM | T | - | - | 61.2% | **77.3%** |

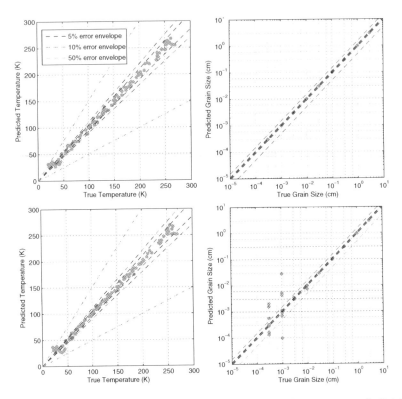

**Fig. 4.** Correlation of predicted (retrieved) and true values of temperature (**left block**) and grain size (**right block**). Data are shown as orange dots. **Top:** results obtained with the SOM in non-interpolating mode. **Bottom:** results obtained with the SOM in interpolating mode. The blue, red and green dashed lines indicate 5%, 10% and 50% error envelopes, respectively. Temperature has a smaller prediction error with interpolating mode. The prediction of grain size is better with non-interpolating mode.

the percentages of test data samples with less than 5% relative error. We achieve $100.0 \pm 0.0\%$ accuracy for grain size in non-interpolating mode, and $83.0 \pm 2.7\%$ for temperature in interpolating mode (Table 2, left block). Contrary to expectation, increasing the grid resolution of the grain size in generating training spectra does not help improve the prediction accuracy in interpolating mode, as shown in Table 2, right block. In fact, with a $20 \times 20$ SOM, the predictions of both temperature and grain size are significantly worse (or at best comparable), in both SOM modes, than results produced with 9 grain sizes. This is likely a result of the softening of boundaries between grain size groups due to an additional eight spectral curves between each two in Fig. 1, right. For this comparison we use an augmented training set generated with 81 grain sizes, equally spaced on a logarithmic scale in the same $0.0003 - 3.0$ cm range as, and including, the training set with 9 grain sizes. Increasing the size of the SOM to $40 \times 40$ still does not improve the prediction of either parameter in interpolating mode. In

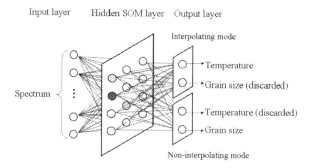

**Fig. 5.** Conceptual diagram of the Conjoined Twin Machines. One head of the Conjoined Twins works in non-interpolating mode, using the output from the BMU (red neuron in the SOM) to predict grain size. The other head, working in interpolating mode, uses, in addition, the second and third BMUs (pink) to predict temperature.

non-interpolating mode grain size prediction recovers to 97.9% because the space in the larger SOM allows the boundaries to become better defined again. However, it is unable to recover to 100% accuracy, because some of the boundary weight vectors still represent spectra with inhomogeneous grain size labels. On average, the 40×40 SOM allocates ∼ 19 weight vectors to each grain size group, which is ∼ 2/3 of the number allocated by the 20×20 SOM for 9 grain size groups. This is likely to be the cause of the unresolved boundaries.

We can conclude from the above that larger SOM size and more grain size samples do not get us closer to better overall prediction with a uniform interpolation scheme. While it is possible that with an even larger SOM we may be able to achieve the same accuracies as in Table 2, left block, the extra resources and time required make that solution undesirable for practical purposes. Instead, we can exploit the knowledge that the two parameters have opposing preferences for interpolation, and encode this duality into the learning machine to combine the advantages of the two SOM modes. This is the concept of the Conjoined Twin Machines (Fig. 5), which includes a shared SOM, containing the learned view of the manifold topology. Two identical "heads" both pull information from this shared SOM but each interprets it somewhat differently. One uses only the output of the BMU, and treats the rest of the SOM outputs as zeros thus not allowing them to influence the learning. This corresponds to non-interpolating mode and therefore will help best predict the grain size. While this grain size specialist head has a second output node identical to its twins', the prediction resulting from that node is discarded. Similarly, the second "head" specializes on temperature, by pulling the outputs of the first three BMUs into the weighted sums for training the output layer. This corresponds to interpolating mode with $k = 3$ and helps predict the temperature accurately while the grain size prediction is discarded. The final output of this machine is the grain size prediction from the first, and the temperature prediction from the second "head".

# 4   Conclusions and Future Work

This paper proposes an effective approach to predict two underlying physical parameters from near-infrared synthetic spectra with high accuracies. The architecture of the learning machine is in the form of Conjoined Twin SOM-hybrid supervised networks. We achieve $100.0\pm0.0\%$ and $83.0\pm2.7\%$ prediction accuracies for grain size and temperature, respectively. This means that for Charon, where temperatures in illuminated regions are likely to range up to 65 K, the neural model should be able predict temperatures with less than $\sim 3$ K error, for $80 - 86\%$ of the measured spectra. This is valuable in resolving diurnal temperature changes on Charon, which provides the boundary condition to discover the processes in Charon's interior and in its atmosphere. To prepare for the real data returned by New Horizons, a noise sensitivity analysis is in progress to gauge the predictive power of learned models (learned with clean as well as noisy data) from spectra obtained in real circumstances.

Because of the observed interplay of the two parameters, in this study we could justify the choices of 1 and 3 for the interpolation granularity $k$ for grain size and temperature, respectively. These choices are obviously data dependent, therefore cannot be automatically applied to other data without prior exploration of the data properties. Future work will include more — potentially interdependent — underlying parameters, in which case the Conjoined Twins can be extended to Conjoined Triplets, Quadruplets, or possibly to other tuplets, each with a different value of $k$. This will in turn motivate looking for automated ways to determine the optimal value of $k$ for each machine, perhaps with meta-learning. Moving from a conceptual prototype to an integrated piece of software in the implementation of the "Conjoined Twins" is a short-term task, that will help with such future extensions.

# References

1. Grundy, W.M., Schmitt, B.: The Temperature-Dependent Near-Infrared Absorption Spectrum of Hexagonal $H_2O$ Ice. J. Geophys. Res. 103, 25809–25822 (1998)
2. Grundy, W.M., Schmitt, B., Quirico, E.: The Temperature Dependent Spectrum of Methane Ice I between 0. 65 and 5 Microns and Opportunities for Near-Infrared Remote Thermometry. Icarus 155, 486–496 (2002)
3. Young, L.A., et al.: New Horizons: Anticipated Scientific Investigations at the Pluto System. Space Science Reviews 140, 93–127 (2008)
4. Reuter, D.C., et al.: Ralph: A Visible/Infrared Imager for the New Horizons Pluto/Kuiper Belt Mission. Space Science Reviews 140, 129–154 (2008)
5. Hapke, B.: Theory of Reflectance and Emittance Spectroscopy. Cambridge University Press, New York (1993)
6. Grundy, W.M., et al.: Near-Infrared Spectra of Icy Outer Solar System Surfaces: Remote Determination of $H_2O$ Ice Temperatures. Icarus 142, 536–549 (1999)
7. Gilmore, M.S., et al.: Effect of Mars Analogue Dust Deposition on The Automated Detection of Calcite in Visible/Near-Infrared Spectra. Icarus 172, 641–646 (2004)
8. English, E.C., Fricke, F.R.: The Interference Index and Its Prediction Using a Neural Network Analysis of Wind-tunnel Data. Journal of Wind Engineering and Industrial Aerodynamics 83, 567–575 (1999)

9. Gilbert, N., Terna, P.: How to Build and Use Agent-Based Models in Social Science. Mind & Society 1(1), 57–72 (2000)
10. Grundy, W.M.: Methane and Nitrogen Ices on Pluto and Triton: a Combined Laboratory and Telescope Investigation. PhD thesis, University of Arizona (1995)
11. Cruikshank, D.P., et al.: Ices on the Satellites of Jupiter, Saturn, and Uranus. In: Solar System Ices, pp. 579–606. Kluwer Academic Publishers, Dordrecht (1998)
12. Widrow, B., Smith, F.W.: Pattern-Recognizing Control Systems. In: Computer and Information Sciences (COINS) Symposium Proceedings, pp. 288–317. Spartan Books, Washington (1964)
13. Howell, E.S., Merényi, E., Lebofsky, L.A.: Classification of Asteroid Spectra Using a Neural Network. Jour. Geophys. Res. 99(E5), 10,847–10,865 (1994)
14. Rudd, L., Merényi, E.: Assessing Debris-Flow Potential by Using AVIRIS Imagery to Map Surface Materials and Stratigraphy in Cataract Canyon, Utah. In: Green, R. (ed.) Proc. 14th AVIRIS Earth Science and Applications Workshop, Pasadena, CA, May 24-27 (2005)
15. DeSieno, D.: Adding a Conscience to Competitive Learning. In: IEEE Intl Conference on Neural Networks, vol. 1, pp. 117–124 (1988)
16. Kohonen, T.: Self-Organizing Maps, 2nd edn. Springer, Heidelberg (1997)
17. Taşdemir, K., Merényi, E.: Exploiting the Data Topology in Visualizing and Clustering of Self-Organizing Maps. IEEE Trans. Neural Networks (in press, 2009)

# Author Index